【博客藏经阁丛书】

ARM Linux入门与实践

——基于TI AM335x处理器

程昌南 沈建华 编著

北京航空航天大学出版社
BEIHANG UNIVERSITY PRESS

内容简介

本书可以算是《ARM Linux入门与实践——一个嵌入式爱好者的自学体验》的姊妹篇，但因将三星ARM9处理器S3C2410改成了TI的Cortex-A8处理器AM335x，所以除保留了前一本书的思路及少部分章节外，大部分的内容都做了修改。

本书分为三篇：ARM硬件、ARM前后台系统、基于Linux系统的应用。第一篇ARM硬件，包括TI官方评估板ARM335x Starter Kit的硬件分析和评估板入门指南；第二篇ARM前后台系统，包括TI官方无操作系统平台下的应用库——StarterWare开发环境搭建及应用分析、启动代码分析、Boot源代码分析、LCD和触摸屏源代码分析、前后台系统应用等；第三篇Linux系统的应用，包括基于PC的操作、应用编程和驱动、嵌入式Linux开发环境创建、嵌入式Linux驱动开发和网络编程等。

本书可作为有志于掌握ARM、Linux等嵌入式技术的高等学校学生和从事相关技术工作的工程师的参考书。

图书在版编目(CIP)数据

ARM Linux入门与实践 ：基于TI AM335x处理器 / 程昌南，沈建华编著. — 北京 ：北京航空航天大学出版社，2018.3

ISBN 978-7-5124-2646-7

Ⅰ. ①A… Ⅱ. ①程… ②沈… Ⅲ. ①Linux操作系统 Ⅳ. ①TP316.85

中国版本图书馆CIP数据核字(2018)第020311号

ARM Linux入门与实践——基于TI AM335x处理器

程昌南　沈建华　编著

责任编辑　杨　昕

*

北京航空航天大学出版社出版发行

北京市海淀区学院路37号(邮编100191)　http://www.buaapress.com.cn

发行部电话：(010)82317024　传真：(010)82328026

读者信箱：emsbook@buaacm.com.cn　邮购电话：(010)82316936

涿州市新华印刷有限公司印装　　各地书店经销

*

开本：710×1 000　1/16　印张：26.25　字数：559千字

2018年3月第1版　2018年3月第1次印刷　印数：3 000册

ISBN 978-7-5124-2646-7　定价：69.00元

前　言

1. ARM 和 Linux 的学习体会

如何才能学好并尽快上手 ARM 和 Linux？这是很多初学者想问的，也是曾经在网上热烈讨论过的。其实学习没有捷径，不管您是否特别聪明，主要还是取决于您的态度和一定的方法，嵌入式 ARM 和 Linux 的学习也一样。在此笔者想根据自己在自学 ARM 与 Linux 过程中的体会并结合网上的讨论做一下总结。

对于 ARM 的学习，如果您已经有了单片机或计算机结构的基本知识，并且也有了 C 语言的基础，那么上手还是比较容易的。首先，应该了解一下 ARM 的体系结构，它有哪些版本，哪些模式，哪些寄存器、异常等，这方面的内容可以参考《ARM 体系结构与编程》一书。这本书介绍的内容还是比较全面的，可以先快速地浏览，特别是汇编指令不需要记住，以后应用时再查。其实很多内容笔者也早已经忘记，但并不妨碍正常理解和应用 ARM。其次，选定一种具体厂家型号的 ARM 处理器，学习该处理器相关开发工具的使用，如开发环境、仿真器及评估板等。然后，可以分析该处理器的外围扩展方法，如 SDRAM、NAND Flash、NOR Flash、SD 卡等，一般官方或第三方都会提供原理图，可以尝试分析原理图。接着，要理解该处理器的启动过程，尝试从一块裸板逐步运行自己的应用程序，也可以参考官方的启动代码及例程。最后，就可以调试、测试该处理器的各种外围部件，如定时器 PWM、串口、LCD 等，特别是中断及执行过程。

对于 Linux 的学习，尽管我们最终应用的是嵌入式 Linux，但还是要先在 PC 下学习，原因是，无论基于 PC 还是嵌入式处理器，其 Linux 开发、编程、调试等都是相近的，而 PC 的性能、资源等都更加丰富，嵌入式 Linux 的开发环境通常也都建立在 PC 的 Linux 环境下，同时有大量 Linux 的相关专著、编程书籍等都是针对 PC 编写的。至于学习的顺序，笔者觉得应该先在 PC 上安装一种常用的 Linux 发行版系统（以前是 RedHat，现在常用 Ubuntu），再买一两本相应的入门书籍，熟悉一下 Linux 环境，学习常用的命令和操作（不一定多，基本、常用的就可以，以后在使用过程中再慢慢积累），理解 Linux 下的目录结构与作用等。其次，学习在 Linux 环境下编程，《GNU/Linux 编程指南》或《UNIX 环境高级编程》都可以，它们都是非常经典的专著，对文件描述符的概念，打开、读、写等操作的系列基础知识都有介绍，如果没有这些基础而直接看《Linux设备驱动程序》会觉得困难。再次，拿本内核的书翻翻，了解一下Linux内核源代码树的目录结构、编译等。最后，学习《Linux设备驱动程序》，理解驱动程序的结构框架等。根据实际需要，读者也可学习网络编程和图形用户界面GUI编程，比较权威的书是《UNIX网络编程》和《C++ GUI QT4编程》；另外由于QT GUI是基于C++的，所以需要学习C++的知识，可参考《C++程序设计教程》一书。如果不想学C++，也可以选择其他如MicroWindows、MiniGUI等。有了PC的Linux基础及编程知识后，就可以直接应用于嵌入式系统了！此时需要一个硬件平台（如开发板或直接的产品等目标系统）和该平台的开发环境。开发环境，一般的处理器原厂或第三方都会提供，所以不需要移植。它包括建立在PC宿主机上的编译等工具和嵌入式Linux内核（包括硬件BSP等驱动）源码。此时只需学习开发环境的建立和使用，并将PC所学的知识应用于具体的嵌入式平台。如果涉及硬件及驱动，那么应该具备一定的硬件调试和解决问题的能力，此时单片机、ARM基础知识及外围设备接口的调试能力将起到很重要的作用。很多朋友（包括非电子、自动化专业的朋友）都希望自己将来能从事Linux驱动方面的工作，认为学习了Linux驱动方面的知识就行了，而往往忽略硬件本身的调试和解决问题的能力，我觉得这是不正确的。因为Linux驱动与前后台控制硬件外设的区别只在于它与应用程序的接口，它要按照一定的结构和规则去驱动，这种规则涉及的也就是通常讲的Linux的驱动知识和技术。它是固定的，容易掌握的，只要去遵循就不会出问题。而硬件及外设是千变万化的，所涉及的技术是多方面的，除处理器本身外，还有各种接口及协议，数字、模拟技术等，在设计、调试时是很容易出现问题的。此时就要求我们有一定的调试、测试手段和方法。所以笔者认为要想成为一名优秀的ARM Linux驱动工程师，首先必须是一名优秀的、具有丰富调试经验的单片机或ARM处理器应用工程师，所以希望那些想从事ARM Linux底层驱动相关工作的朋友在还没有机会接触ARM和Linux时，不要对目前的单片机工作产生抵触情绪，因为如果没有一个扎实的单片机基础和调试能力，即使有一个ARM和Linux的工作机会，您也很难成为这方面最优秀的工程师。

2. 本书的结构与内容

本书共分为三篇，14章。第1章，主要介绍单片机、ARM、Linux等嵌入式概念，TI AM335x处理器和开发资源，以及开发调试方法等。第2章，主要分析AM335x官方开发板的硬件原理及相关的基础知识，如果读者对原理都能认识和理解，那么就有了非常扎实的硬件基础。第3章，主要介绍官方开发板AM335x Starter Kit的入门，包括开发板的使用和仿真调试等。第4章，介绍AM335x在无操作系统平台下的应用库——StarterWare的开发方法，这也是官方针对“裸跑”用户提供的，是非常好用的源代码库。第5章，启动代码分析，这是ARM“裸跑”系统非常重要的一部分代码，有助于读者理解ARM处理器的工作原理、中断向量表等，也是ARM处理器能否稳定运行的关键。第6章，Boot源代码分析，该部分代码基于启动代码之上，是具有比较完整功能的引导代码，初学者往往会混淆Boot源代码与启动代码两者的概念。第7、8章，分别介绍LCD和触摸屏的源代码，LCD和触摸屏也是常见的两种硬件外设。第9章，主要针对BeagleBone Black用户，介绍StarterWare在BeagleBone Black板上运行时容易出现的问题。第10章，基于前后台系统的应用，主要介绍文件系统和图形界面的开发。第11章，介绍PC的Linux学习，PC的Linux的各方面技术都是嵌入式Linux的基础，非常重要，涉及的知识内容也很多。第12章，介绍嵌入式Linux开发环境，也是嵌入式Linux产品开发的工具和基础。第13章，嵌入式Linux驱动开发，介绍驱动设备树和常见的几种外设驱动。第14章，网络编程，介绍常用的TCP客服机/服务器的实现，以及网络的调试方法。

3. 致 谢

感谢德州仪器(TI)大学计划的支持。

感谢华东师范大学沈建华教授的帮助和对本书的审阅。

感谢我亲爱的妻子和可爱的大妞、二妞，没有你们的支持和体谅，就不会有本书的出版，谢谢你们！

由于作者水平有限，书中的错误和不妥之处在所难免，恳请广大读者朋友们批评指正。

程昌南

2017年8月15日

目　录

第一篇　ARM 硬件

第二篇　ARM前后台系统

第三篇　基于 Linux 系统的应用

第一篇

ARM 硬件

第1章

概　述

1.1　嵌入式系统、单片机、ARM及Linux

嵌入式系统的概念、范围很广，大部分相关的书籍都会有介绍，虽然有些不同，但总体的意思是相近的，是指具有计算能力的非PC系统，即具有通常使用的个人电脑的全部或部分特质的专用计算机系统，如生活中的PDA、手机、电视机顶盒、数字电视、数码相机以及工业自动化仪表、医疗仪器等。它分硬件和软件两部分，硬件以嵌入式处理器（相当于PC的CPU）为核心，外扩ROM、RAM（相当于PC的硬盘、内存条）、输入/输出设备（PC的键盘/显示器）、各种通信接口（串口、USB、网络）等。软件由引导程序（PC的BIOS）、嵌入式操作系统（相当于PC的Windows系统）和应用程序（相当于在Windows上运行的程序，如金山词霸、Word等）三部分组成，或只有其中的某部分。

人们通常所指的某种系列单片机和ARM处理器都是嵌入式处理器的一种。通常将主要用于控制功能的8位和16位低端处理器称为单片机，也称微控制器。ARM处理器是32位的，主频、性能相对于单片机更高，不仅能完成单片机的控制功能，而且对运算速度要求较高的场合（如MP3、PDA、GPS、电子书阅读器、智能手机等）也能适用，还能运行如Linux、WinCE、Android等复杂的操作系统。另外，ARM也是一个公司的名字，成立于英国剑桥，主要出售ARM核的芯片设计技术授权。采用ARM技术知识产权（IP）核的微处理器都被称为ARM处理器，如三星公司的S3C44B0X、S3C2410A，TI公司的AM335x，NXP公司的LPC系列，还有Atmel、ST、Freescale、高通等公司的各种处理器。

Linux是一个功能强大、稳定的操作系统，因源代码开放而被广泛移植运行在各种处理器上，不仅可以作为PC的桌面系统，而且也是嵌入式系统中最为常用的操作系统之一。这里指的就是嵌入式Linux，是运行在TI Cortex-A8处理器AM335x上的嵌入式Linux，有别于PC上的桌面Linux系统Ubuntu。

记得最初学习单片机时也没有很特意地找这些名词解释，只知道用单片机可以做些什么，将来能用它开发些什么样的产品等，如课程设计中的温度控制器、各种仪器仪表等，再联想到生活中的遥控器、剃须刀、电动玩具及电冰箱等各种家电。只要能想到的带点自动及智能化的，都可以用单片机去实现。此外，还有比单片机功能更强，能做单片机不能做的，又类似于单片机的处理器，ARM就是其中的一种，使用它可以开发功能更强、更复杂的电子设备，如MP3、PDA、GPS导航和智能手机等，还可以运行Linux等操作系统给开发提供便利，开发出类似于PC那样漂亮、强大的用户界面及系统。

1.2 ARM处理器的选择

初学ARM的朋友往往会犹豫不决，到底选哪个厂家，哪个处理器？ARM7还是ARM9？或是最新的Cortex-M3？Cortex-A8/A9/A15？而不同的ARM处理器厂家及推广商也都说自家的产品好，这更让初学者困惑。对于初学者，我想应遵循以下几个原则：

① 处理器一定要是目前主流的，用的人非常多的。否则，选择一个已经过时或停产的产品，当你出现问题时，很少有人能帮你解决；当你学会后找工作时，也得不到别人的认可。

② 资料一定是很丰富的，这会降低学习难度。当你出现问题、遇到困难时可以很快找到答案，提高学习的积极性。

③ 取决于学习目标。如果你只想将ARM作为更高级的单片机来使用，以后拿它来“裸跑”（前后台系统），或者想运行小型的操作系统如μC/OS等，那么你可以选择某些ARM7或最新的Cortex-M3/M4，因为其本身的定位就是在微控制器领域。如果你想将来能学习Linux、WinCE或Android等功能更强大的操作系统，那么你就应该选择ARM9及以上处理能力、资源更强的处理器。

④ 如果带着最终的产品应用为目的学习，那么就要考虑处理器的性能和外设资源。如处理器的主频，指令/数据Cache的大小，是否集成以太网、SDRAM、DDR/DDR2/DDR3、NAND Flash、LCD控制器，串口通道数，SD卡等，如果是ARM7，那么外部总线是否开放等。

⑤ 学习者还应考虑身边的朋友及同事在应用哪款处理器，这样可以更加方便地请教他们。

本书为什么选用TI公司的AM335x？

由于公司产品研发需要，使我有幸真正认识到了TI，及其高性价比的工业级ARM Cortex-A8处理器AM335x。虽然不是最新的Cortex-A9和Cortex-A15，但作为ARM Linux入门学习之用已经足够，更重要的是AM335x目前具有非常丰富的学习资源，不仅有TI官方发布的大量芯片资料、应用设计、工具软件、培训课程及Linux、Android及WinCE等主流操作系统的SDK，而且有风靡全球的BeagleBone Black等，以及国内外大量的论坛技术支持和维基等帮助。AM335x是一款工业级芯片，这也让它的生命周期比普通的消费类芯片要长许多。另外，长期与TI大学计划合作的华东师范大学沈建华教授及TI大学技术相关负责人也都推荐AM335x，并指出 Cortex-A8 仍是 TI 近几年的主推处理器(特别是在通用市场)。经过多方学习分析，本书最终确定使用 TI 公司的 AM335x。

1.3 AM335x 简介和公版资源

AM335x 是 TI(德州仪器)公司推出的，基于 ARM Cortex-A8 内核的微处理器，属于 TI 的 Sitara 系列，官方链接网址为 http://www.ti.com.cn/lsds/ti_zh/processors/sitara/arm_cortex-a8/am335x/overview.page?paramCriteria=no。

Sitara AM335x ARM Cortex-A8 微处理器有 AM3352、AM3354、AM3356、AM3357、AM3358 和 AM3359 共 6 个型号，具有 Pin-to-Pin(引脚对引脚)兼容和软件兼容等特性，主频有 300 MHz、600 MHz、800 MHz 和 1 GHz 等不同级别，有两种封装，官网给出的最低价格为 5 美元，是目前 ARM Cortex-A8 内核中最具竞争力的价格，相当于 ARM9 内核处理器的价格，广泛应用于家庭自动化、工业自动化、便携式导航设备、智能家电、个人电子产品、流式音频和工业网关等。

AM335x 功能框图如图 1.1 所示。

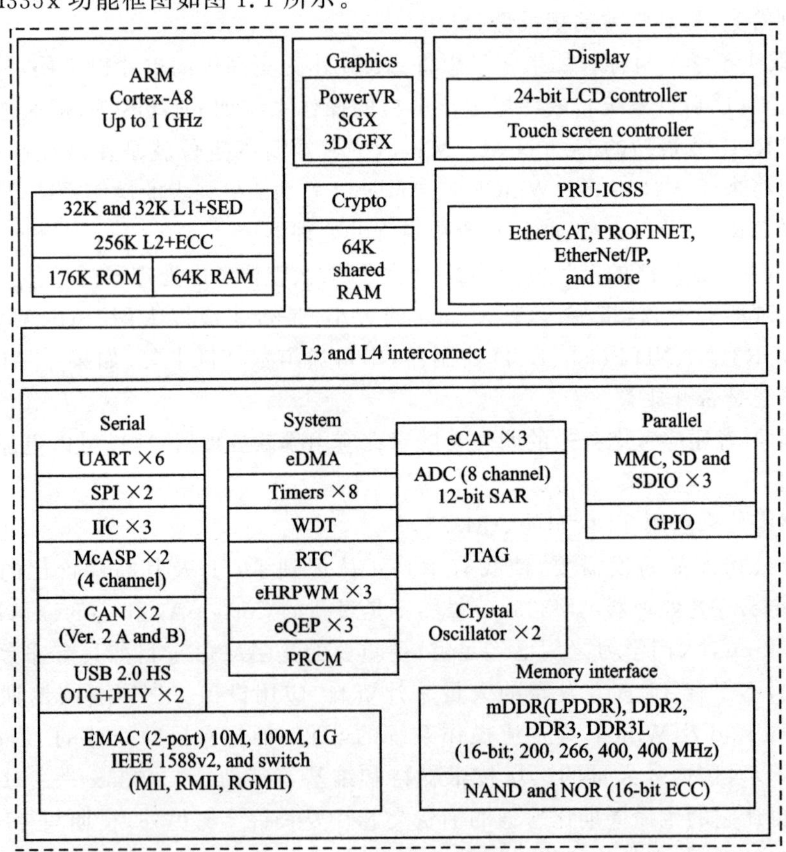

图 1.1　AM335x 功能框图

AM335x包括上限可达1 GHz的ARM Cortex-A8内核，32 KB的L1指令高速缓存、32 KB的数据缓存，以及256 KB的L2指令高速缓存；内置64 KB的RAM和176 KB的ROM，及64 KB共享RAM；具有SGX530 3D图形加速引擎，24位的LCD控制器接口和触摸屏控制器；6个UART，2个SPI，3个IIC，2个4通道的McASP，2个CAN总线和2个USB2.0及OTG接口，2个10M、100M和1G的以太网接口控制器等；可扩展MMC、SD卡，及3个SDIO接口，并具有足够多的GPIO输入/输出引脚；支持mDDR(LPDDR)、DDR2、DDR3或DDR3L，以及NAND和NOR Flash存储器。

AM335x的具体型号及参数如表1.1所列。

表1.1 AM335x型号参数

Model	AM3352	AM3354	AM3356	AM3357	AM3358	AM3359
Applications	Automotive Industrial Personal Electronics	Automotive Industrial Personal Electronics	Automotive Industrial Personal Electronics	Automotive Industrial Personal Electronics	Automotive Industrial Personal Electronics	Automotive Industrial Personal Electron-ics
Operating Systems	Linux Android WinCE	Linux Android WinCE	Linux Android WinCE	Linux Android WinCE	Linux Android WinCE	Linux Android WinCE
ARM CPU	1 ARM Cortex-A8	1 ARM Cortex-A8	1 ARM Cortex-A8	1 ARM Cortex-A8	1 ARM Cortex-A8	1 ARM Cortex-A8
ARM MHz (Max.)	300 600 800 1000	600 800 1000	300 600 800	300 600 800	600 800 1000	800
DRAM	LPDDR DDR2 DDR3 DDR3L	LPDDR DDR2 DDR3 DDR3L	LPDDR DDR2 DDR3 DDR3L	LPDDR DDR2 DDR3 DDR3L	LPDDR DDR2 DDR3 DDR3L	LPDDR DDR2 DDR3 DDR3L
Display Options	LCD	LCD	LCD	LCD	LCD	LCD
Graphics Acceleration		1 3D			1 3D	1 3D
Programmable RealTime Unit Sub-system and Industrial Communication Subsystem			PRU-ICSS	PRU-ICSS+ EtherCAT slave	PRU-ICSS	PRU-ICSS+ EtherCAT slave
EMAC	2-Port 1Gb Switch	2-Port 1Gb Switch	2-Port 1Gb Switch	2-Port 1Gb Switch	2-Port 1Gb Switch	2-Port 1Gb Switch
USB	2	2	2	2	2	2
SPI	2	2	2	2	2	2
IIC	3	3	3	3	3	3
UART (SCI)	6	6	6	6	6	6
CAN(#)	2	2	2	2	2	2

续表 1.1

Model	AM3352	AM3354	AM3356	AM3357	AM3358	AM3359
Operating Temperature Range/℃	0～90 −40～90 −40～105 −40～125	0～90 −40～90 −40～105	0～90 −40～90 −40～105	−40～90 −40～105	0～90 −40～105	−40～105
On-Chip L2 Cache	256 KB (Cortex-A8)	256 KB (Cortex-A8)	256 KB (Cortex-A8)	256 KB (Cortex-A8)	256 KB (Cortex-A8)	256 KB (Cortex-A8)
Other On-Chip Memory	128 KB	128 KB	128 KB	128 KB	128 KB	128 KB
Approx. Price (US$)	6.00｜1ku	7.10｜1ku	8.11｜1ku	11.19｜1ku	9.72｜1ku	14.75｜1ku

这里最让人心动的是PRU－ICSS，它类似于一个独立的FPGA或MCU，可单独运行，通过编程实现串口或工业以太网等各种外设。TI官方提供了实现串口的例程源码，如果实现串口，那么AM335x可以将串口扩展到10个。因为如果用16C554扩展串口，那么一片16C554的价格在40元RMB，且需要占用更多的PCB空间和增加PCB走线。

另外，AM3359价格高是因为集成了EtherCAT slave，而EtherCAT slave本身是要收License费的，但AM335x芯片已经包括该费用了。

如果想更详细地了解AM335x系列各型号器件间的差别，可参见表1.2，或直接参考官方用户手册*Technical Reference Manual.pdf*。

表1.2 AM335x各型号器件间的差异

Subsystem/Co-Processor/Peripheral	AM3352	AM3354	AM3356	AM3357	AM3358	AM3359
ARM MPU Subsystem	Y	Y	Y	Y	Y	Y
Programmable Real-Time Unit and Industrial Communication Subsystem (PRU-ICSS)	N	N	All features excluding EtherCAT	All features including EtherCAT	All features excluding EtherCAT	All features including EtherCAT
Graphics Accelerator (SGX)	N	Y	N	N	Y	Y
Memory Map	Y	Y	Y	Y	Y	Y
Interrupts	Y	Y	Y	Y	Y	Y
Memory Subsystem	Y	Y	Y	Y	Y	Y
Power and Clock Management(PRCM)	Y	Y	Y	Y	Y	Y
Control Module	Y	Y	Y	Y	Y	Y
Interconnects	Y	Y	Y	Y	Y	Y
Enhanced Direct Memory Access(EDMA)	Y	Y	Y	Y	Y	Y
Touchscreeen Controller	Y	Y	Y	Y	Y	Y
LCD Controller	Y	Y	Y	Y	Y	Y

续表 1.2

Subsystem/Co-Processor/Peripheral	AM3352	AM3354	AM3356	AM3357	AM3358	AM3359
Ethernet Subsystem	ZCE: 1 port ZCZ: 2 ports	ZCE: 1 port ZCZ: 2 ports	No ZCE Available ZCZ: 2 ports	No ZCE Available ZCZ: 2 ports	No ZCE Available ZCZ: 2 ports	No ZCE Available ZCZ: 2 ports
Pulse-Width Modulation Subsystem	Y	Y	Y	Y	Y	Y
Universal Serial Bus(USB)	ZCE: 1 port ZCZ: 2 ports	ZCE: 1 port ZCZ: 2 ports	No ZCE Available ZCZ: 2 ports	No ZCE Available ZCZ: 2 ports	No ZCE Available ZCZ: 2 ports	No ZCE Available ZCZ: 2 ports
Interprocessor Communication	Y	Y	Y	Y	Y	Y
Multimedia Card(MMC)	Y	Y	Y	Y	Y	Y
Universal Asynchronous Receiver/Transmitter (UART)	Y	Y	Y	Y	Y	Y
DMTimer	Y	Y	Y	Y	Y	Y
DMTimer 1ms	Y	Y	Y	Y	Y	Y
RTCSS	Y	Y	Y	Y	Y	Y
Watchdog	Y	Y	Y	Y	Y	Y
IIC	Y	Y	Y	Y	Y	Y
Multichannel Audio Serial Port(McASP)	Y	Y	Y	Y	Y	Y
Controller Area Network (CAN)	Y	Y	Y	Y	Y	Y
Multichannel Serial Port Interface(McSPI)	Y	Y	Y	Y	Y	Y
General Purpose Input/Output(GPIO)	Y	Y	Y	Y	Y	Y
Initialization	Y	Y	Y	Y	Y	Y

1.4 AM335x官方开发资源

AM335x不仅是TI史上公开资料最全的一个系列芯片，而且可以说是目前所有ARM处理器当中资料最全的一款。读者可以从官网上下载到几乎所有需要的开发资料，此外还有公开的课程、技术论坛和网络维基等技术支持，以及风靡全球的低成本开源硬件BeagleBone及技术社区。

TI官方关于AM335x系列各处理器的器件功能、性能及各种软硬件资源介绍的网址为http://www.ti.com.cn/lsds/ti_zh/processors/sitara/arm_cortex-a8/am335x/overview.page?paramCriteria=no。

1.4.1 硬件开发板

硬件方面，TI官方目前主要有BeagleBone Black(BBB)、AM335x Starter Kit

(EVM-SK)入门套件和 AM335x Evaluation Module (EVM)评估模块三种,都可以直接在 TI 官网上很方便地使用人民币订购。官方价格分别是¥399.00(RMB)、$206.02(USD)和$995.00(USD),所以前两款的价格都是非常低的。

BeagleBone Black 是一款社区支持的低成本开发平台,面向 ARM Cortex-A8 处理器的开发人员和业余爱好者,官方网站为 http://beagleboard.org/。官方号称不到 10 s 即可启动 Linux,5 min 内即可开始 Sitara AM335x ARM Cortex-A8 处理器的开发。BeagleBone Black 通过板载闪存随 Angstrom Linux 发行版提供,以开始进行评估和开发。BeagleBone Black 还支持许多其他 Linux 发行版和操作系统,包括 Ubuntu、Android 和 Fedora。BeagleBone Black 的功能可以通过一种名为"capes"的插件电路板进行扩展,该插件电路板可以插入到 BeagleBone Black 的两个 46 引脚双排扩展头中。capes 可用于 VGA、LCD、电机控制、原型工具、电池电源,及各种其他功能。

Sitara ARM Processors AM335x Starter Kit 入门套件,通过采用板载加速计支持旋转与倾斜功能的 4.3 in(1 in=2.54 cm) LCD 显示屏,可为智能电器、工业、网络应用以及其他需要触摸屏界面的设备提供低成本平台。该低成本开发平台建立在 Sitara AM3358 ARM Cortex-A8 处理器基础之上,高度集成双千兆以太网、Wi-Fi 以及蓝牙 (Bluetooth)等多个通信选项,适用于创建高度互联的设备。AM335x Starter Kit 入门套件的速度高达 720 MHz,可通过生产就绪型软硬件平台加速设计进程。该入门电路板是一款低成本工具,可快速评估处理器及其配套 TI 组件的特性。无论是经验丰富的设计人员还是设计新手,都能集中精力进行高级图形与连接等产品差异化设计。开发人员在创建互联恒温器、安全面板、无线硬盘驱动器以及便携式热点时,可加速产品上市进程、降低成本。

TI AM335x 入门套件预先通过 FCC 认证,并符合 CE 标准,是具有丰富特性、支持优化材料清单的小型设计工具。

Sitara ARM Processors AM335x Starter Kit 特性如下:

- 基于速度高达 720 MHz 的 AM335x ARM Cortex-A8 处理器,可为图形及工业通信实现引脚对引脚选项的可扩展性;
- 2 Gb DDR3 存储器;
- 通过 AM335x ARM Cortex-A8 处理器上的集成型双端口开关提供 2 Gb 以太网端口;
- 4.3 in LCD 触摸屏(5.1 in×2.6 in)通过板载加速计支持旋转与倾斜功能,使无数应用的实现成为可能;
- 通过集成天线的板载 WL1271 Murata 模块提供具有 Wi-Fi Direct 支持的 Wi-Fi(802.11b/g/n)以及包括蓝牙低能耗在内的蓝牙 v4.0;
- 通过 USB 提供板载 XDS100 JTAG 仿真器功能,可实现无额外成本的调试;
- 提供 USB、USB-UART、导航按钮、用户 LED、音频输出等更多外设。

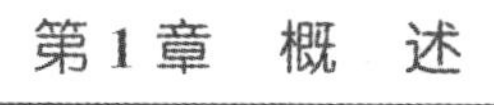

AM335x Starter Kit (EVM-SK)官方主页为 http://www.ti.com.cn/tool/cn/tmdssk3358,目前供货型号为 TMDSSK3358,自带两张分别装有 Linux 和 Android 系统的 μSD 卡。

Sitara ARM Processors AM335x Evaluation Module (EVM)评估模块是 TI 面向高端企业级客户推出的全面型开发板,价格相对比较高,它可使开发人员立即开始评估 AM335x 处理器(AM3352、AM3354、AM3356、AM3358),并开始构建便携式导航、便携式游戏设备和家庭/楼宇自动化等各种应用。

Sitara ARM Processors AM335x Evaluation Module 特性如表 1.3 所列。

表 1.3　AM335x EVM 特性

硬　件	软　件	连　接
AM3358,ARM 微处理器; 512 MB DDR2; TPS65910 电源管理 IC; 7 in 触摸屏 LCD; 通过 WL1271 实现无线(Wi-Fi/BT)连接	Linux EZ SDK; Android	10 M/100 M 以太网(1); UART(4); Wi-Fi/BT(1); SD/MMC(2); USB2.0 OTG/HOST (1/1); 音频输入/输出; JTAG; CAN(2)

AM335x Evaluation Module 目前官网供货型号为 TMDXEVM3358,官方主页为 http://www.ti.com.cn/tool/cn/tmdxevm3358。

1.4.2　软件开发包

软件方面,TI 官方目前主要有前后台系统("裸跑")库 StarterWare、Linux 和 Android 等,另外还有一些第三方的嵌入式操作系统。

AM335x StarterWare 是 TI 官方提供的一个无操作系统平台下的应用库,为 CCS 开发环境下的源代码例程,支持基于 AM335x 处理器的 BeagleBone Black (BBB)、AM335x Starter Kit (EVM-SK)和 AM335x Evaluation Module (EVM)评估板,最新版本为 AM335X_StarterWare_02_00_01_01,官方主页为 http://www.ti.com.cn/tool/cn/starterware-sitara 或 http://processors.wiki.ti.com/index.php/StarterWare;下载地址为 http://software-dl.ti.com/dsps/dsps_public_sw/am_bu/starterware/latest/index_FDS.html。

AM335x 采用的 Sitara Linux SDK(软件开发套件),官方主页为 http://www.ti.com.cn/tool/cn/processor-sdk-am335x;下载地址为 http://software-dl.ti.com/sitara_linux/esd/processor-sdk/PROCESSOR-SDK-LINUX-AM335X/latest/index_FDS.html。

软件开发包主要支持 TI 官方的 AM335x EVM TMDXEVM3358 和 AM335x Starter Kit TMDSSK3358。

Sitara Linux SDK 的亮点如下：

- 长期稳定（LTS)的 Linux 内核支持；
- U-Boot 引导加载程序支持；
- Linaro GNU 编译器集（GCC）工具链；
- 兼容 Yocto Project OE Core 的文件系统。

Linaro 工具链包括强大的商用级工具，这些工具专为 Cortex-A 处理器进行过优化。此工具链得到了 TI 和整个 Linaro 社区的全力支持，包括来自 Linaro 内部工程师、成员公司开发者以及开源社区的其他人员的支持。在最新版的 TI 处理器 SDK 中就提供有 Linaro 工具、软件和测试程序。

Yocto 项目是由 Linux 基金会设立的开源协作项目，旨在简化构建嵌入式 Linux 软件发行版的框架。TI 通过 Arago 发行方式提供。Arago 项目提供经验证和测试的软件包子集，并且是用免费和开放的工具链构建而成的。有关 Yocto 项目和 TI Arago 发行版的其他资源，请访问 http://arago-project.org。

Sitara Linux SDK 特性(其他 Linux 功能)如下：

- 开放的 Linux 支持。
- Linux 内核和引导加载程序。
- 文件系统。
- Qt/Webkit 应用程序框架。
- 3D 图形支持。
- 集成式 WLAN 和蓝牙支持。
- 基于 GUI 的应用程序启动器。
- 示例应用，包括：
  - ARM 基准：Dhrystone、Linpack、Whetstone；
  - Webkit Web 浏览器；
  - 软 Wi-Fi 接入点；
  - 加密：AES、3DES、MD5、SHA；
  - 多媒体：GStreamer/FFMPEG。
- 可编程实时单元（PRU)：
  - 主机工具，包括闪存工具和引脚复用实用程序；
  - 用于 Linux 开发的 Code Composer Studio IDE；
  - 文档。

AM335x 采用的 Sitara Android SDK 开发套件，官方主页为 http://www.ti.com.cn/tool/cn/androidsdk-sitara；下载网址为 http://downloads.ti.com/sitara_android/esd/TI_Android_DevKit/TI_Android_JB_4_2_2_DevKit_4_1_1/index_

FDS.html。

虽然 Android 操作系统起初是专为移动手持终端而设计的，但其仍允许嵌入式应用的设计人员轻松为产品增加高级操作系统。与 Google 联合开发的 Android 是一套可立即实现集成和生产的全面操作系统。

Android 操作系统的亮点如下：

- 完整的开放源码软件解决方案；
- 基于 Linux 内核；
- 针对商业开发的简洁许可条款（Apache）；
- 包含一个完整的应用框架；
- 允许通过 Java 轻松集成定制开发应用；
- 开包即用的多媒体、图形和图形用户界面；
- 大量 Android 和应用开发人员供随时调遣。

TI 的 Android 开发套件是一套完整的软件，Sitara 器件的开发人员可以用其轻松快速地评估基于 Sitara 器件平台的 Android 操作系统。该开发套件提供稳定且经全面测试的软件基础，可广泛用于包括 AM335x EVM 评估模块、AM335x Starter Kit 入门套件和 BeagleBoard Black(beagleboard.org)在内的各种 Sitara 硬件及开发平台。

Sitara Android SDK 开发套件包含：

- Linux 内核；
- U-Boot/x 加载程序；
- 3D 图形 OpenGL(r) 驱动程序和库；
- 用于 Android 的 Adobe Flash 10 库；
- RowboPERF 性能基准应用；
- 3D 图形的示例应用；
- 主机工具；
- 调试选项；
- 完整文档。

其他集成功能，例如对 WLAN、Bluetooth(R)、S 视频、分量视频、快速启动、电源管理、NAND UBIFS、摄像机、HDMI、千兆以太网、音频输入/输出、SATA、PCIe 的支持等可根据使用的硬件平台获得更多功能。有关每个版本具体功能的列表，可在相应版本软件及官方主页中找到相应的发布说明、相关指南和其他文档等。

关于 OpenGL 支持，用户可利用包含在 Android 中的图形功能，获得流畅且强大的用户体验。其还可支持多种 Sitara 器件，开发人员还可利用 ARM(r) Cortex-A8 内核(高达 1 GHz)和多个外设，以使其适用于大多数嵌入式应用。

免费提供 Sitara 器件的 Android 开发套件，不附加任何开发和生产限制。如果希望尝试使用 Sitara 的最新 Android 代码，除 TI 主页外，还可访问 arowboat.org 上的开放源码社区。在该网站上，用户可以下载最新代码，还可与其他开发人员以及Android操作系统的用户进行互动。

1.5 ARM开发工具及调试方法

开发工具就是为用户学习、开发基于某种处理器的应用产品而设计的实验平台及仿真工具，这里主要指硬件仿真器，下载器，集编辑、编译、汇编、链接、调试及工程管理为一体的集成开发环境，评估实验板等。

1.5.1 集成开发环境

以前在ARM7、ARM9等处理器中，常使用STD2.5、ADS1.2等集成开发环境，当时随处可以找到其破解版，后来ARM11、Cortex-A8、Cortex-A9等更高的处理器采用ARM公司的ARM Development Studio 5 (DS-5)，但因软件费用较高，而且国内"裸跑"的用户慢慢变少，所以都直接采用Linux等开发，采用主机+串口终端等调试方法，有的"裸跑"甚至直接采用串口终端。

TI Sitara系列处理器采用TI DSP处理器的Code Composer Studio集成开发环境，通常简写为CCSv5、CCSv6等。

Code Composer Studio是一种集成开发环境 (IDE)，支持TI的微控制器和嵌入式处理器产品系列。Code Composer Studio包含一整套用于开发和调试嵌入式应用的工具。它包含了用于优化的C/C++编译器、源码编辑器、项目构建环境、调试器、描述器以及多种其他功能。直观的IDE提供了单个用户界面，可帮助用户完成应用开发流程的每个步骤。熟悉的工具和界面使用户能够比以前更快地入手。Code Composer Studio将Eclipse软件框架的优点与TI先进的嵌入式调试功能相结合，为嵌入式开发人员提供了一个引人注目、功能丰富的开发环境。

虽然官方也说明CCSv6支持Linux的调试，但我们主要还是用它来调试无操作系统下的StarterWare程序。

另外，还有IAR等集成开发环境也支持基于AM335x的StarterWare等非操作系统下的代码调试。

1.5.2 硬件仿真器

在51系列单片机仿真器中都是使用与目标单片机引脚兼容的仿真头插入目标板，以替代目标单片机，用户程序是在仿真器内部的仿真芯片上运行。ARM核处理器内置ICE(仿真调试模块)，该模块通过标准的JTAG接口引脚与ARM仿真器相连，此时ARM仿真器作为上位调试软件与ARM核芯片之间的协议转换器，用户的目标调试文件被下载到目标板上的存储器(可以是外部的或ARM处理器内部的存储器)中，通过控制目标芯片的仿真模块实现仿真调试，常用的有如下几种：

1. 简易JTAG小板

ARM开发最原始的下载调试工具Wigger、SDT、SEC JTAG等都属于此类，它仅做简单的PC并口与目标板JTAG接口的转换(PC并口直接输入/输出JTAG协议的高低电平信号)，中间用74HC244之类的缓冲芯片对信号进行缓冲。不同类型JTAG小板的区别在于JTAG信号PIN和PC并口PIN的对应关系不同。JTAG小板主要用来在PC上通过JTAG接口烧写目标板上的Flash，常用软件如Flashpgm、SJF2410等，不过SDT JTAG加上JTAG.exe等SERVER软件，也可用SDT2.5软件对S3C44B0X进行源代码级的调试(这也是当初很多朋友说学ARM比学单片机成本更低的理由)，但速度比较低。目前PC的主板基本没有并口了，所以很少人使用，只有小部分还在生产的S3C44B0、S3C2410等老产品中使用，它可通过USB转并口模块实现在无并行接口的PC上的使用。

2. Multi-ICE

Multi-ICE为ARM公司生产的ARM并口仿真器，在当时ARM7、ARM9开发时期使用得非常普遍，支持当时的全系列ARM核处理器，与JTAG小板一样，通过并口与PC连接，不同的是它与PC并口之间不再是直接的串行JTAG协议，而是通过8位数据线与PC实现真正的并口协议通信，再由内部的FPGA实现并口通信到JTAG协议的转换，速度要高很多，下载速度为120 kB/s。当时国内有很多开发板商提供类似的兼容产品，在使用和功能上都是一致的，在稳定性上可能会有所差别，价格相对比较低。由于ARM7、ARM9的应用越来越少，PC基本不再支持打印机并口，所以Multi-ICE也逐渐被带USB接口类的仿真器所取代。

3. RealView ICE

RealView ICE是ARM公司推出的配套RVDS使用的仿真器，支持目前的全系列ARM核处理器，通过网口或USB口与PC连接，内部有MCU和FPGA。由于MCU与PC连接是通过网口或USB连接的，故RealView ICE速度要比Multi-ICE高很多，下载速度可高达600 kB/s，但价格很高。

4. ULINK2

ULINK2是ARM公司推出的配套RealView MDK使用的仿真器，是ULink的升级版本，不仅可以进行在线仿真调试，还可以对Flash进行烧写。

5. TI官方的仿真器

TI官方仿真器主要有XDS100、XDS200和XDS560等，它们的功能都是一样的，主要区别在于价格和仿真调试等下载速度上。它们不仅支持Sitara系列的ARM处理器，同时还支持TI的DSP处理器。

(1) XDS100

XDS100是TI官方仿真器中价格最低的一个系列，目前常用的是XDS100v2，

Blackhawk XDS100v2 货号为 TMDSEMU100V2U-20T,价格仅为＄79.00(USD),官方主页为 http://www.ti.com.cn/tool/cn/TMDSEMU100v2U-20T。

Blackhawk XDS100v2 仿真器实物如图 1.2 所示。

XDS100v2 JTAG Debug Probe (20 pin cTI version)

图 1.2 Blackhawk XDS100v2 仿真器

1) Blackhawk XDS100v2 描述

Blackhawk XDS100v2 是 XDS100 系列 TI 处理器调试探针(仿真器)的第二代产品。XDS100 系列的成本在所有 XDS 调试探针中最低,同时支持传统 JTAG 标准(IEEE1149.1)。此外,对于带有嵌入式缓冲跟踪器(ETB)的所有 ARM 和 DSP 处理器,所有 XDS 调试探针均支持内核和系统跟踪。

注意: 与 cJTAG (IEEE1149.7) 兼容的调试探针标记为 XDS100v3,可从指定的第三方获得。调试 CC26xx、CC2538 或 CC13xx 器件时需要 cJTAG。

Blackhawk XDS100v2 通过 TI 20 引脚连接器连接到目标板,通过 USB2.0 高速(480 Mbps)连接到主机 PC。要在主机 PC 上运行,还需要 Code Composer Studio IDE 许可证。

Blackhawk XDS100v2 封装包含:

➢ XDS100v2 调试板;

➢ USB2.0 电缆;

➢ 快速入门指南。

支持的器件如下:

➢ 基本 JTAG 调试支持大多数 TI 微控制器(MSP430 除外)、所有无线 MCU、大多数 DSP(C54x、C62x、C67x 除外)以及所有 ARM 处理器;

➢ 可以通过 ETB 在指定的 ARM 和 DSP 处理器中进行内核和系统跟踪。

有关特定器件跟踪功能的更多信息,请参阅其技术参考手册(TRM)。

您将需要:

➢ 免费的 Code Composer Studio IDE 许可证(也可以使用付费许可证);

➢ 达到 Code Composer Studio IDE 最低要求的主机 PC；
➢ 带有兼容的 JTAG 接头的目标板。

可能需要：

➢ 用于连接目标板，带有不同 JTAG 接头的适配器。

发货信息：

➢ 产品包装盒尺寸：160 mm×160 mm×32 mm(6.0 in×6.0 in×1.3 in)；
➢ 包装产品重量：230 g(8.0 oz，1 oz=28.349 5 g)；
➢ ECCN：3A992A；
➢ HTS/时间表 B：8542.31.0000。

2）XDS100v2 特性

XDS100 产品系列采用了可实现全面 JTAG 连接的低成本设计，非常适合 TI 微控制器、处理器和无线设备的入门级调试。

在 XDS100 系列中，XDS100v2 是第一款支持所有带 JTAG 调试端口的 TI 器件的产品，取代了此前的 XDS100v1 技术。此外，它还可以对支持嵌入式缓冲跟踪器（ETB）的所有 ARM 和 DSP 器件进行内核和系统跟踪。

鉴于市场上有大量 JTAG 接头，XDS100 系列的各种型号分别带有 TI 14 引脚、TI 20 引脚、ARM 10 引脚和 ARM 20 引脚连接器。某些型号带有两种连接器，灵活性更高。

XDS100 系列支持传统的 IEEE1149.1（JTAG）和 IEEE1149.7（cJTAG），运行时的接口电平为 +1.8 V 和 3.3 V。

与传统 JTAG 相比，cJTAG 有巨大的进步。因为它仅需使用两个引脚即可支持所有功能，可用于某些指定的 TI 无线连接微控制器中。仅 XDS100v3 支持 cJTAG，这是 XDS100 系列低成本 JTAG 调试探针（仿真器）中的第三种。XDS100v3 调试探针可从指定的 TI 第三方获得。

所有型号的 XDS100 均支持与主机的 USB2.0 全速（11 Mbps）或高速（480 Mbps）连接。

XDS100 系列与 TI 的 Code Composer Studio IDE 完全兼容。这一组合提供了完整的硬件开发环境，包含集成调试环境、编译器以及针对 TI 微控制器、处理器和无线连接微控制器的完整硬件调试和跟踪功能（指定器件上）。

（2）XDS560

XDS560 是 TI 官方仿真器中速度最高、最高端的一个系列，目前常用的是 XDS560v2，Blackhawk XDS560v2 货号为 TMDSEMU560V2STM-U，价格为＄955.00（USD），官方主页为 http://www.ti.com.cn/tool/cn/TMDSEMU560V2STM-U#0。

Blackhawk XDS560 仿真器实物如图 1.3 所示。

1）Blackhawk XDS560v2 描述

XDS560v2 System Trace 是 XDS560v2 系列高性能 TI 处理器调试探针（仿真

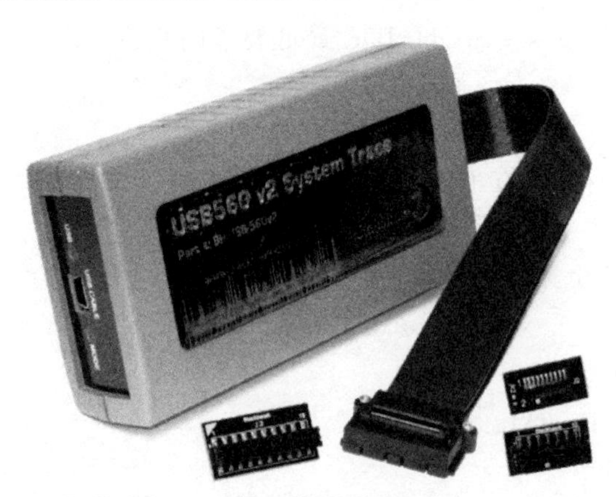

图 1.3 Blackhawk XDS560v2 仿真器

器)的第一种型号。XDS560v2 是 XDS 系列调试探针中性能最高的一款,同时支持传统 JTAG 标准 (IEEE1149.1) 和 cJTAG (IEEE1149.7)。

XDS560v2 System Trace 在其巨大的外部存储器缓冲区中加入了系统引脚跟踪。这种外部存储器缓冲区适用于指定的 TI 器件,通过捕获相关器件信息,获得准确的总线性能和吞吐量,并对内核和外设进行电源管理。此外,对于带有嵌入式缓冲跟踪器 (ETB) 的所有 ARM 和 DSP 处理器,所有 XDS 调试探针均支持内核和系统跟踪。

Blackhawk XDS560v2 System Trace 通过 MIPI HSPT 60 引脚连接器(带有适合 TI 14 引脚、TI 20 引脚和 ARM 20 引脚的多个适配器)连接到目标板,通过 USB2.0 高速(480 Mbps)连接到主机 PC。要在主机 PC 上运行,还需要 Code Composer Studio IDE 许可证。

Blackhawk XDS560v2 System Trace 的封装包含:

- XDS560v2 System Trace 调试器;
- MIPI HSPT 60 引脚至 TI 14 的引脚转换器适配器;
- MIPI HSPT 60 引脚至 TI 20 的引脚转换器适配器;
- MIPI HSPT 60 引脚至 ARM 20 的引脚转换器适配器;
- USB2.0 电缆;
- 含有设备驱动程序的 CDROM;
- 快速入门指南;
- 保修和产品注册信息。

支持的器件如下:

- 基本 JTAG 调试支持大多数 TI 微控制器(MSP430 除外)、所有无线 MCU、大多数 DSP(C54x、C62x、C67x 除外)以及所有 ARM 处理器;

- cJTAG 调试支持所有 CC26xx、CC2538 和 CC13xx 器件；
- 系统引脚跟踪可用于启用系统跟踪的器件，例如 AM335x、AM437x、66AK2x、C66x、DM81x、AM38x；
- 可以通过 ETB 在指定的 ARM 和 DSP 处理器中进行内核和系统跟踪。

有关特定器件跟踪功能的更多信息，请参阅其技术参考手册（TRM）。

您将需要：

- 带有完整许可证或 90 天评估许可证的 Code Composer Studio IDE（不能使用免费有限许可证）；
- 达到 Code Composer Studio IDE 最低要求的主机 PC；
- 带有一种兼容型 JTAG 接头的目标板。

可能需要：

- 用于连接目标板，带有不同 JTAG 接头的适配器。

发货信息：

- 产品包装盒尺寸：280 mm×210 mm×51 mm (11 in×8.0 in×2.0 in)；
- 包装产品重量：450 g (1.01 b，1 b=453.6 g)；
- ECCN：3A991.A.2；
- HTS/时间表 B：8542.31.0000。

2) XDS560v2 特性

XDS560v2 是 XDS560 系列高性能 TI 处理器调试探针（仿真器）的最新型号。XDS560v2 具有整个系列中最高的速度和最多的功能，对于 TI 微控制器、处理器和无线连接微控制器的调试来说，它是最全面的解决方案。

XDS560v2 是 XDS560 调试探针系列中最先提供系统跟踪（STM）功能的一款，这种类型的跟踪可以通过捕获系统事件（例如处理内核的状态、内部总线和外设）来监控整个设备。大多数 XDS560v2 模型还提供系统引脚跟踪模式，在这种模式中，系统跟踪数据被送到 XDS560v2 内的外部存储器缓冲区（128 MB），因此能够捕获大量系统事件。系统引脚跟踪数据连接需要通过额外的接线连接 JTAG 连接器。

在 XDS560 调试探针系列中，XDS560v2 PRO Trace 是提供内核引脚跟踪功能（指令和数据）的第二代产品，这种跟踪可以捕获内核执行的所有指令并将其发送到 XDS560v2 PRO Trace 内的外部存储器缓冲区（1 GB）。内核引脚跟踪并不干扰系统的实时行为，而且可以捕获更多的指令。内核引脚跟踪数据连接需要通过额外的接线连接 JTAG 连接器。

为了支持所有类型的引脚跟踪（指令和系统），XDS560v2 的所有型号都提供标准的 60 引脚 MIPI HSPT 连接器作为与目标之间的主要 JTAG 连接。此外，所有型号都提供针对 TI 和 ARM 标准 JTAG 连接器的模块化目标适配器（提供的适配器因型号而异）。

XDS560v2 支持传统的 IEEE1149.1（JTAG）仿真和 IEEE1149.7（cJTAG），运

行时的 JTAG 接口电平为 1.2～4.1 V。

与传统 JTAG 相比，紧凑型 JTAG（cJTAG）有巨大的进步。因为它仅需使用两个引脚即可支持所有功能，可用于某些指定的 TI 无线连接微控制器中。

所有 XDS560v2 型号均支持 USB2.0 高速连接（480 Mbps）或以太网 10/100 Mbps，某些型号还同时支持二者。另外，某些型号支持 PoE（以太网供电），增加了灵活性。

XDS560v2 System Trace 单元与 TI 的 Code Composer Studio IDE 完全兼容。这一组合提供了完整的硬件开发环境，包含集成调试环境、编译器以及针对 TI 微控制器、处理器和无线连接微控制器的完整硬件调试和跟踪功能（指定器件上）。

另外，还有配合 IAR EWARM 的 J-Link 仿真器，以及功能强大的 TRACE 32 仿真器等。

1.5.3 前后台系统的调试方法

1. 利用硬件仿真器调试

如果读者有仿真器，或目标板上自带仿真器，那么它的调试和 8 位单片机的调试方法基本相同。以 CCSv6 集成开发环境与 XDS100v2 仿真器为例，首先将仿真器通过 JTAG 接口与目标板（如 BeagleBone Black）相连，通过 USB 接口与 PC 相连，然后给目标板上电。其次，在 PC 中启动 CCSv6 集成开发环境，打开需要在线调试的 StarterWare 源码例程项目，选择 New→Target Configuration File，在对话框中输入配置文件名，并勾选 Use shared location。选择仿真器 XDS100v2 和调试板或芯片，然后单击右边的 Save Configuration 的 Save 按钮。完成以上操作以后，可以单击 Test Connection 按钮进行连接测试，测试 CCS 是否能够与仿真器相连接。如果连接正常，则会弹出一个对话框，接着就可以开始在线仿真调试了。

2. 无硬件仿真器的调试

在开发 51 系列单片机时，如果没有硬件仿真器，一样可以利用 Keil 强大的软件仿真功能来仿真单片机的内部外设（如定时器、I/O 口、中断等）顺利调试应用程序，待软件仿真达到要求后，再对目标单片机编程验证结果，这对中、小项目而言是经济、有效的。在 ARM 的开发工具中，RealView MDK 源于 Keil，所以读者也可以用 51 的开发方式去完成一些小的项目。当项目较大或使用不具备软件仿真 ARM 处理器内部外设的开发环境（如 ADS1.2）时，可以采用下述介绍的方法。

首先，需要一个监控程序（如 S3C44B0X 的 44Bmon，S3C2410A 的 u241mon），它通过 Console（控制台，如具有 USB/串口下载功能的终端软件 DNW 或 PC 自带的超级终端、串口调试软件等）与我们交互，可以接收我们发送的应用程序，去引导、执行该应用程序，而我们则通过输入命令或选项控制它。这个监控程序可大可小，用户可以将其设计成只具有串口下载和引导执行的基本功能，也可以像引导 Linux、WinCE

操作系统的 Bootloader 一样，完全取决于实践需求。

其次，将该监控程序烧写到目标板的 Flash 中，目标板通过串口与 PC 相连（如果需要通过 USB 或以太网下载程序，则还需连接 USB 电缆或网线），启动 PC 的终端软件，然后给目标板上电运行监控程序。

接着，我们就可以在 ADS1.2 等开发环境中编写程序，编译生成目标文件了。通过在终端软件中输入命令，下载预调试的目标文件。成功下载完毕后，可以输入命令使其运行验证结果。那么如何确切地像有仿真器那样在仿真窗口观察程序的执行状况呢？我们通常是在要调试的代码中增加往终端打印信息的代码（一般是往串口发送数据，在终端软件中显示），来确定程序的执行状况。

最后，读者可能会问为什么要加监控程序，而不直接烧写用户程序呢？原因是应用程序通常比较大，这样用 JTAG 烧写 Flash 需要很长的时间，也容易出错。而且应用程序可能需要反复地被修改、调试，而如果每修改一次就烧写一次 Flash 显然不好。

在 TI Sitara AM335x 处理器中，可以直接用 CCSv6 编辑、编译源代码，然后链接生成目标文件。此时需要两个目标文件，一个是启动引导代码，被命名为 MLO，TI 的很多例程中都提供。另一个是用户程序的目标文件，编译生成后需要修改为 app 文件名。此时只需将这两个文件复制到 μSD 卡插入到目标板上，再让目标板从 μSD 卡上引导启动即可运行。我们可以在源代码中加入串口打印信息来调试各种代码。

1.5.4 嵌入式 Linux 的开发调试方法

在嵌入式 Linux 下的调试有点类似于无硬件仿真器情况下的前后台调试，只不过它更加复杂，判断程序的运行情况依然是在调试代码中加入往终端打印信息的方法。

首先，在 PC（又称宿主机）上安装一个桌面的 Linux 操作系统（如 Ubuntu 14.04 LTS），这样就有了 Linux 环境及常用工具，有些没有涉及 ARM 处理器硬件的程序，也可以先在 PC 上调试，成功后再下载到目标板的 Linux 上运行。在 PC 的 Linux 上建立嵌入式 Linux 的交叉编译工具，用于编译嵌入式 Linux 的内核、驱动及应用程序等，此时我们也称该 PC 为宿主机。

其次，在目标板上烧写 Bootloader、内核、根文件系统（具体的方法参考后面章节）。用串口和以太网将宿主机和目标板连接起来，在宿主机上启动 Minicom（Linux 下的超级终端）程序，再启动目标板 Linux 系统，此时 Minicom 将作为目标板 Linux 系统的控制台，与之交互。

接着，当我们需要调试一个在目标板上运行的程序时，只需在宿主机上交叉编译生成目标平台的可执行文件，然后通过网络下载到目标板的可读目录下执行即可。有时为了方便，也可以建立 NFS（网络文件系统），将目标板的目录挂载到宿主机的目录下，将编译生成的可执行文件直接放入该目录而免去下载的操作（相当于在网络上共享文件）。

最后，我们可以将最终的驱动编译进内核，最终的应用程序放入文件系统，再更新目标板的内核和文件系统。

上述方法也经常被称作宿主机＋目标机的交叉调试方式。

1.6 "实践再实践"在ARM学习中的意义

目前TI专门为入门学习者及工程师推出的BeagleBone Black及其中国版的BeagleBone Black或Starter Kit都很廉价，学习者只须亲自去分析原理图，读懂原理图的方方面面，学会从零调试一块电路板等各项工作，这就是所谓的实践、实践，再实践！它是嵌入式学习的必经过程。

在嵌入式技术方面，如何判断一个工程师技术水平的高低呢？一位国内单片机应用先驱的答案是：解决问题的能力。确实，在工程师日渐增多，网络资源如此便利、丰富的今天，不在乎您掌握多少技术，也不在乎您会ARM还是Linux，只要能够解决别人都不能解决的问题，您才能真正体现出价值。而解决问题的能力又如何培养？这不仅需要理论，更需要实践（或经验）的积累，需要亲自去设计，去制作、调试，去不断地发现问题，在解决问题的过程中不断积累经验。

在任21IC BBS ARM版版主的近三年时间里，我面对过很多朋友提的问题，以及身边同事们的问题，这些问题都不是很难，主要是没有理解ARM处理器底层的知识和与硬件相关的知识，没有亲自去经历在一块还不能正常运行的裸板上，让ARM处理器顺利工作，让Flash、SDRAM DDR读/写正常，让Bootloader、Linux操作系统正常启动，以及让Linux正常驱动外设等的整个调试过程。没有经历过当出现问题后如何去思考、判断、排除的整个实践过程。所以阿南（作者网名）在此真诚地希望买开发板学习的朋友，都能够亲自去学习、去分析这些更底层方面的知识，试着将开发板的整个系统破坏（不要舍不得这个板子，没有什么比知识更加值钱的），把它变成没有任何程序的一块裸板，然后亲自去体验这些过程。只有这样你才算是真正地学习ARM，掌握ARM处理器的应用，今后才能顺利（无论是官方或开发板商的资源都有存在BUG的可能）地开发出基于ARM处理器的应用产品；否则，你只能算是一个C语言的编程者，一个不在乎是ARM处理器，还是单片机微控制器或PC的编程者。

提高解决问题的能力，最有效的方法是亲自设计、制作自己理想中的ARM学习平台（学习板），或者刨根问底地分析现成的开发板。在亲自分析原理图中的每一个部分，CPU的每一个引脚定义（特别是有些关键引脚），存储器、外设的扩展方法，考虑电源、功率等之后，也就相当于掌握了该ARM处理器的硬件技术。从绘制原理图，绘制PCB，再到焊接、调试的整个过程，你的硬件整体水平（包括解决问题的能力）也都会进一步地提高。即使我们不是亲自制作，也可以去了解原理图的绘制，

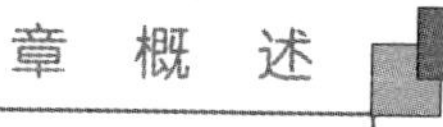

PCB图工具的使用，学会检查PCB图，焊接等基本功在电路调试过程中也是需要的。而在器件（即使是常用的电容、二极管）的选择、采购上也都有很大的学问，如果不亲自去实践，真的很难想象其中的奥妙。最后当你看着这个自己费尽心血完成的ARM实验系统，或“吃透”一块开发板后，心里会无比的自豪和有成就感，充满自信。

那么，亲自制作或刨根问底地分析现成的开发板，对于我们来讲难吗？首先从技术难度来看，这和每个朋友的基础有关，有些朋友可能本身就是做硬件的，那么原理图、PCB、焊接等自然都不是问题，而有些朋友可能只是做底层软件的没有做过硬件，还有些学生朋友可能在很多方面都是新手等，可能就会觉得比较困难。其实，阿南从入行以来，主要的工作职责也是嵌入式软件部分，硬件等其他方面的经验也都很欠缺，是新手，但是阿南有很强的学习欲望，不耻下问，希望自己在各方面都能不断地进步。所以当年在规划《ARM Linux入门与实践——一个嵌入式爱好者的自学体验》（简称《ARM Linux自学体验》）一书的时候，就决定亲自为其设计实验平台，而在规划本书时就打算刨根问底地分析所用的开发板，这样自己就可以学习更多新的知识，还可以将这个过程中的问题、心得记录下来与读者共同分享、交流和提高。阿南可以做到，读者自然也都可以做到，因为现在的网络、通信是如此的方便。总之，只要我们勇于去实践，困难总是可以战胜的。

第2章

AM335x Starter Kit 实验平台硬件分析

前面,我们已经提到 TI Sitara AM335x 处理器的官方开发板主要有 BeagleBone Black、Starter Kit 和 EVM。那么,针对学习,我们应该如何从这 3 块开发板中选择最适合我们自己的呢? EVM 由于价格高,对于一般学习者和普通企业开发人员来讲都无法接受,所以率先排除。相比于 EVM,Starter Kit 的价格要低很多,也带有触摸屏和 LCD 屏,只不过与 EVM 相比小了点儿,为 4.3 in,但其基本的功能外设在使用上与 EVM 相差不是很大,足够学习评估 AM335x 的大部分资源,因此对于有触摸屏和 LCD 需求的学习者来讲是最佳选择。BeagleBone Black 不仅有全球版,还有中国版,而且价格也是最低的,并且开发资源也非常丰富,非常适合入门者学习使用,国外有很多玩家将它当作嵌入式电脑,接上鼠标、键盘和带有 HDMI 接口的显示器,玩得不亦乐乎,他们更偏向于将其当作廉价的开源硬件,重点用于嵌入式软件的学习。随着国外玩家的普及,国内几个开发板商和网络论坛的推广,以及中国版 BeagleBone Black 的推出,国内也有更多的工程师、学生、嵌入式爱好者选择 BeagleBone Black。但阿南在网络论坛和博客中收集 BeagleBone Black 相关信息及资料时发现,很多国内玩家并没有像国外玩家那样玩得深入,很多都是论坛组织推广活动时低价买入的,但拿到手后只是跑了一下,好点的跑了一下其他人测试通过的例程,反正一块板也就是请朋友出去吃顿饭的钱。只有少部分工程师,认认真真地深入研究,并做出各种扩展的"cape"展示在论坛上。综上考虑,如果不考虑触摸屏和 LCD 屏的需求,最低价的 BeagleBone Black 是个不错的学习选择,但如果要自己再扩展个触摸屏和 LCD 屏就有点勉强了,所以本书选择 AM335x Starter Kit (EVM-SK)入门套件作为实验平台。

2.1 AM335x Starter Kit 实验平台概述及功能组成

在前面章节,我们大致地介绍了 AM335x Starter Kit (TMDSSK3358)的一些情况及特性,它是 TI 推出的基于 AM335x 处理器的低成本的入门级的评估、开发平台套件,但能基本满足 AM335x 的所有硬件资源的评估。

TMDSSK3358 由主板和 LCD 显示板组成，主板包括 AM3358 处理器、外设接口及芯片、电源及电源管理芯片等。LCD 显示板贴有 4.7 in 的 LCD 和电阻式触摸屏。这里的 LCD PCB 板并没有走线，而只是起固定作用，LCD 和触摸屏通过软排线连接到主板的 FPC 座上。LCD PCB 板通过铜柱与主板实现固定。如图 2.1 所示为 TMDSSK3358 学习板正面图。

图 2.1 AM335x Starter Kit 正面(TMDSSK3358)

TMDSSK3358 主板底层图如图 2.2 所示。

图 2.2 AM335x Starter Kit 主板底层图(TMDSSK3358)

去掉 LCD 板后，TMDSSK3358 主板的顶层图如图 2.3 所示。

在功能上，TMDSSK3358 以 AM3358 为核心，带有 2 Gb DDR3 内存、TF 接口座，并带有 Wi-Fi/BT 模块、音频解码器、3 轴数字加速度计。在 USB HUB 及 USB 接口中，扩展有两个以太网及接口、24 bit 真彩 LCD 和触摸屏，以及时钟、复位及电源、电源管理 IC，还有可以用户定义的按钮输入及 LED 输出等。如图 2.4 所示为 TMDSSK3358 功能方框图。

图 2.3　AM335x Starter Kit 主板顶层图(TMDSSK3358)

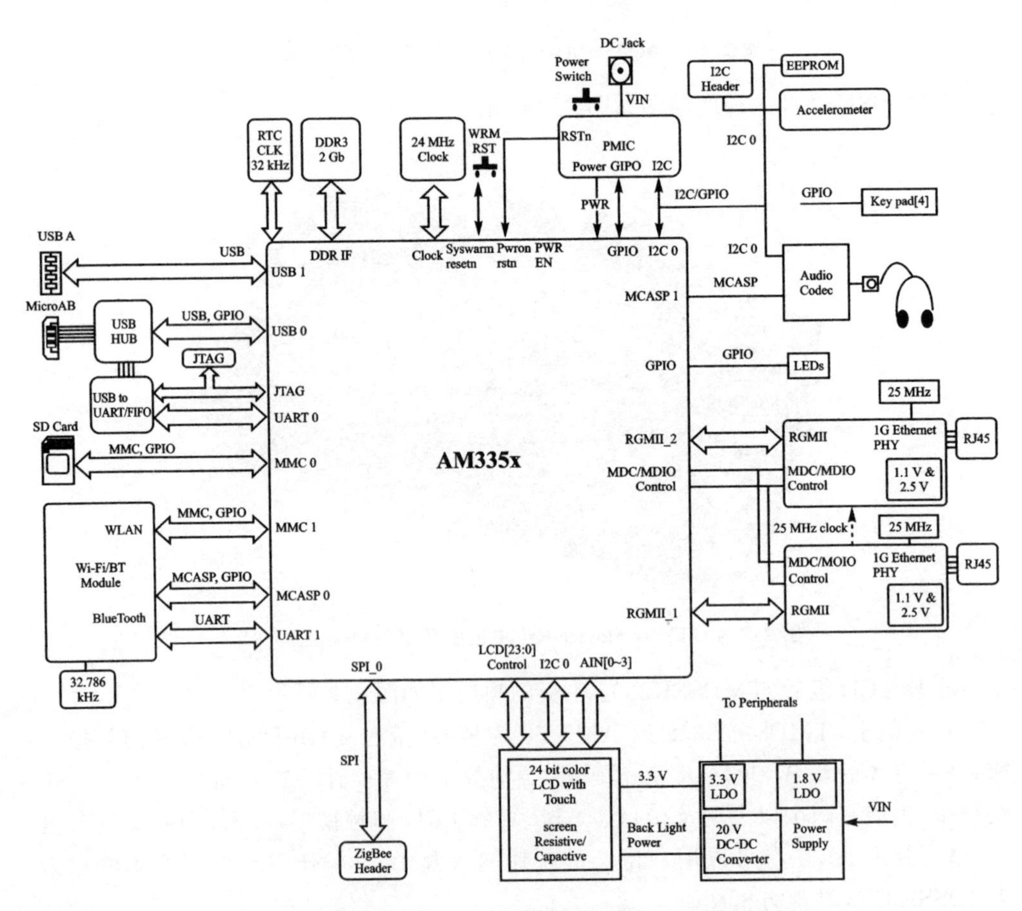

图 2.4　AM335x Starter Kit 功能方框图(TMDSSK3358)

2.2　地址空间分配

2.2.1　AM335x 处理器内存映射

AM335x 处理器内存映射部分读者可以参考 TI 官方的 AM335x 用户手册 *Technical Reference Manual.pdf* 的第 177 页“Memory Map”一章，了解更多的相关知识，因为它很重要。

如表 2.1 所列为 L3 内存映射(Memory Map)。

表 2.1　AM335x L3 内存映射

块名称	起始地址	结束地址	大　小	描　述
GPMC (External Memory)	0x0000_0000	0x1FFF_FFFF	512 MB	8/16 bit External Memory (Ex/R/W)
Reserved	0x2000_0000	0x3FFF_FFFF	512 MB	Reserved
Boot ROM	0x4000_0000	0x4001_FFFF	128 KB	
	0x4002_0000	0x4002_BFFF	48 KB	32 bit Ex/R-Public
Reserved	0x4002_C000	0x400F_FFFF	848 KB	Reserved
Reserved	0x4010_0000	0x401F_FFFF	1 MB	Reserved
Reserved	0x4020_0000	0x402E_FFFF	960 KB	Reserved
Reserved	0x402F_0000	0x402F_03FF	64 KB	Reserved
SRAM internal	0x402F_0400	0x402F_FFFF		32 bit Ex/R/W
L3 OCMC0	0x4030_0000	0x4030_FFFF	64 KB	32 bit Ex/R/W OCMC SRAM
Reserved	0x4031_0000	0x403F_FFFF	960 KB	Reserved
Reserved	0x4040_0000	0x4041_FFFF	128 KB	Reserved
Reserved	0x4042_0000	0x404F_FFFF	896 KB	Reserved
Reserved	0x4050_0000	0x405F_FFFF	1 MB	Reserved
Reserved	0x4060_0000	0x407F_FFFF	2 MB	Reserved
Reserved	0x4080_0000	0x4083_FFFF	256 KB	Reserved
Reserved	0x4084_0000	0x40DF_FFFF	5 888 KB	Reserved
Reserved	0x40E0_0000	0x40E0_7FFF	32 KB	Reserved
Reserved	0x40E0_8000	0x40EF_FFFF	992 KB	Reserved
Reserved	0x40F0_0000	0x40F0_7FFF	32 KB	Reserved
Reserved	0x40F0_8000	0x40FF_FFFF	992 KB	Reserved
Reserved	0x4100_0000	0x41FF_FFFF	16 MB	Reserved
Reserved	0x4200_0000	0x43FF_FFFF	32 MB	Reserved

续表 2.1

块名称	起始地址	结束地址	大 小	描 述
L3F CFG Regs	0x4400_0000	0x443F_FFFF	4 MB	L3Fast configuration registers
Reserved	0x4440_0000	0x447F_FFFF	4 MB	Reserved
L3S CFG Regs	0x4480_0000	0x44BF_FFFF	4 MB	L3Slow configuration registers
L4_WKUP	0x44C0_0000	0x44FF_FFFF	4 MB	L4_WKUP
Reserved	0x4500_0000	0x45FF_FFFF	16 MB	Reserved
McASP0 Data	0x4600_0000	0x463F_FFFF	4 MB	McASP0 Data Registers
McASP1 Data	0x4640_0000	0x467F_FFFF	4 MB	McASP1 Data Registers
Reserved	0x4680_0000	0x46FF_FFFF	8 MB	Reserved
Reserved	0x4700_0000	0x473F_FFFF	4 MB	Reserved
USBSS	0x4740_0000	0x4740_0FFF	32 KB	USB Subsystem Registers
USB0	0x4740_1000	0x4740_12FF		USB0 Controller Registers
USB0_PHY	0x4740_1300	0x4740_13FF		USB0 PHY Registers
USB0 Core	0x4740_1400	0x4740_17FF		USB0 Core Registers
USB1	0x4740_1800	0x4740_1AFF		USB1 Controller Registers
USB1_PHY	0x4740_1B00	0x4740_1BFF		USB1 PHY Registers
USB1 Core	0x4740_1C00	0x4740_1FFF		USB1 Core Registers
USB CPPI DMA Controller	0x4740_2000	0x4740_2FFF		USB CPPI DMA Controller Registers
USB CPPI DMA Scheduler	0x4740_3000	0x4740_3FFF		USB CPPI DMA Scheduler Registers
USB Queue Manager	0x4740_4000	0x4740_7FFF		USB Queue Manager Registers
Reserved	0x4740_8000	0x477F_FFFF	4 MB~32 KB	Reserved
Reserved	0x4780_0000	0x4780_FFFF	64 KB	Reserved
MMCHS2	0x4781_0000	0x4781_FFFF	64 KB	MMCHS2
Reserved	0x4782_0000	0x47BF_FFFF	4 MB~128 KB	Reserved
Reserved	0x47C0_0000	0x47FF_FFFF	4 MB	Reserved
L4_PER	0x4800_0000	0x48FF_FFFF	16 MB	L4 Peripheral(see L4_PER table)
TPCC(EDMA3CC)	0x4900_0000	0x490F_FFFF	1 MB	EDMA3 Channel Controller Registers
Reserved	0x4910_0000	0x497F_FFFF	7 MB	Reserved

续表 2.1

块名称	起始地址	结束地址	大　小	描　述
TPTC0(EDMA3TC0)	0x4980_0000	0x498F_FFFF	1 MB	EDMA3 Transfer Controller 0 Registers
TPTC1(EDMA3TC1)	0x4990_0000	0x499F_FFFF	1 MB	EDMA3 Transfer Controller 1 Registers
TPTC2(EDMA3TC2)	0x49A0_0000	0x49AF_FFFF	1 MB	EDMA3 Transfer Controller 2 Registers
Reserved	0x49B0_0000	0x49BF_FFFF	1 MB	Reserved
Reserved	0x49C0_0000	0x49FF_FFFF	4 MB	Reserved
L4_FAST	0x4A00_0000	0x4AFF_FFFF	16 MB	L4_FAST
Reserved	0x4B00_0000	0x4B13_FFFF	1 280 KB	Reserved
Reserved	0x4B14_0000	0x4B15_FFFF	128 KB	Reserved
DebugSS_DRM	0x4B16_0000	0x4B16_0FFF	4 KB	Debug Subsystem: Debug Resource Manager
DebugSS_ETB	0x4B16_2000	0x4B16_2FFF	4 KB	Debug Subsystem: Embedded Trace Buffer
Reserved	0x4B16_3000	0x4BFF_FFFF	15 MB~396 KB	Reserved
EMIF0	0x4C00_0000	0x4CFF_FFFF	16 MB	EMIF0 Configuration registers
Reserved	0x4D00_0000	0x4DFF_FFFF	16 MB	Reserved
Reserved	0x4E00_0000	0x4FFF_FFFF	32 MB	Reserved
GPMC	0x5000_0000	0x50FF_FFFF	16 MB	GPMC Configuration registers
Reserved	0x5100_0000	0x52FF_FFFF	32 MB	Reserved
Reserved	0x5300_0000	0x530F_FFFF	1 MB	Reserved
	0x5310_0000	0x531F_FFFF	1 MB	Reserved
Reserved	0x5320_0000	0x533F_FFFF	2 MB	Reserved
Reserved	0x5340_0000	0x534F_FFFF	1 MB	Reserved
	0x5350_0000	0x535F_FFFF	1 MB	Reserved
Reserved	0x5360_0000	0x54BF_FFFF	22 MB	Reserved
ADC_TSC DMA	0x54C0_0000	0x54FF_FFFF	4 MB	ADC_TSC DMA Port
Reserved	0x5500_0000	0x55FF_FFFF	16 MB	Reserved
SGX530	0x5600_0000	0x56FF_FFFF	16 MB	SGX530 Slave Port
Reserved	0x5700_0000	0x57FF_FFFF	16 MB	Reserved
Reserved	0x5800_0000	0x58FF_FFFF	16 MB	Reserved

续表 2.1

块名称	起始地址	结束地址	大 小	描 述
Reserved	0x5900_0000	0x59FF_FFFF	16 MB	Reserved
Reserved	0x5A00_0000	0x5AFF_FFFF	16 MB	Reserved
Reserved	0x5B00_0000	0x5BFF_FFFF	16 MB	Reserved
Reserved	0x5C00_0000	0x5DFF_FFFF	32 MB	Reserved
Reserved	0x5E00_0000	0x5FFF_FFFF	32 MB	Reserved
Reserved	0x6000_0000	0x7FFF_FFFF	512 MB	Reserved
EMIF0 SDRAM	0x8000_0000	0xBFFF_FFFF	1 GB	8/16 bit External Memory (Ex/R/W)
Reserved	0xC000_0000	0xFFFF_FFFF	1 GB	Reserved

另外，L4_WKUP Peripheral Memory Map、L4_PER Peripheral Memory Map、L4 Fast Peripheral Memory Map 等 L4 外设内存映射图，请参考 *Technical Reference Manual.pdf* 中的相关内容。

GPMC 为通用内存控制器，在 AM335x 中非常重要，不仅可用于外扩 NOR Flash、NAND Flash 等存储器，还可以用于外扩其他具有地址线、数据线及控制线的外部接口设备(如以太网接口芯片、CPLD 或 FPGA 等)，起始地址为 0x00000000，结束地址为 0x1FFFFFFF，共 512 MB 存储空间。我们一般会通过 CSn 片选信号将程序代码保存在该地址空间的存储器中，或在该地址空间定义外部接口设备，用于读/写访问控制等。AM335x 的 GPMC 控制器共有 7 个 CSn，CSn0 的地址空间为 0x0000_0000～0x0400_0000，CSn1 的地址空间为 0x0600_0000～0x0A00_0000，……，以此类推。如图 2.5 所示为 GPMC 接口信号图。

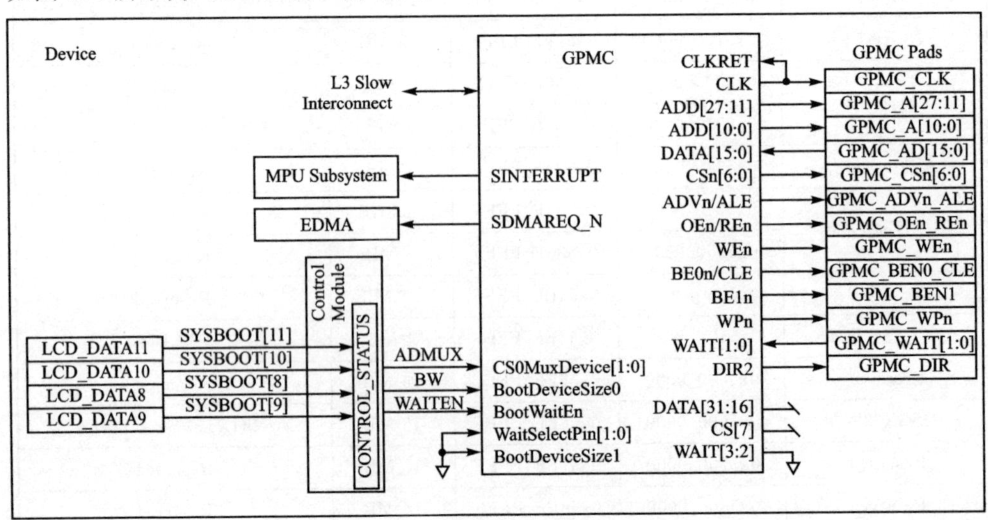

图 2.5 AM335x GPMC 接口信号图

如图 2.6 所示为通过 GPMC 扩展 NAND 示例图。

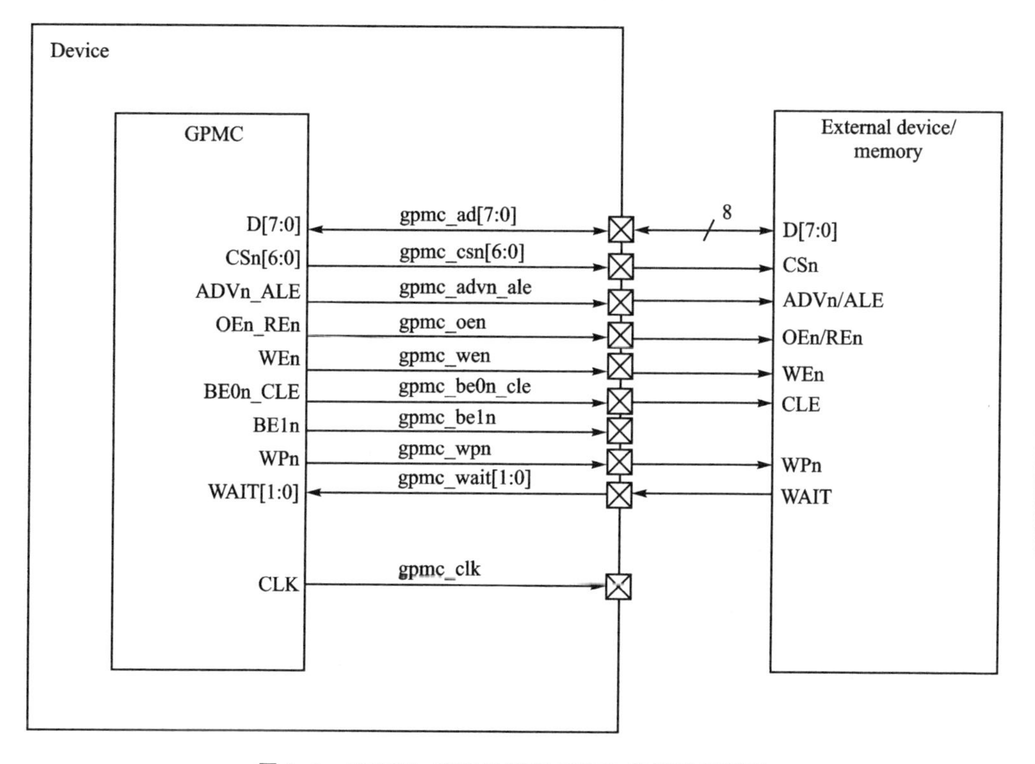

图 2.6 AM335x GPMC 扩展 NAND 存储器示例图

如图 2.7 所示为通过 GPMC 扩展 NOR Flash 示例图。

GPMC 总线上挂的设备(如 NOR Flash,NAND Flash)共享了数据线、地址线和一些控制线(如 CLE、ALE、WP 和 CLK 等),然后由片选信号控制、使能对应的设备。这里需要提到一点,不同片选的地址空间配置、时序配置,都是分开的,可参考 *Technical Reference Manual.pdf* 中关于 GPMC_CONFIG1_i～GPMC_CONFIG7_i 的配置说明,i 的取值决定了写入的地址区间不同,其对应的就是不同的片选。

Boot ROM 为处理器内部的 ROM,用于存放处理器自身的引导代码(出厂后已固化,例如,在 StarterWare 文档里称它为 RBL,即 Read Only Memory BootLoader),也是处理器上电及复位后最先执行的一段引导代码,ROM 内存映射图和异常向量表如图 2.8 所示。

SRAM Internal 和 L3 OCMC0 是处理器内部的 RAM,RBL 会将例如 Sarter-Ware 中的 MLO 从存储介质中装载到该区域执行,Public RAM 内存映射图和异常向量表如图 2.9 所示。

0x402F0400～0x4030B800 共 109 KB,RBL 将 MLO 装载到此区域,所以 MLO 最大不能超过 109 KB。

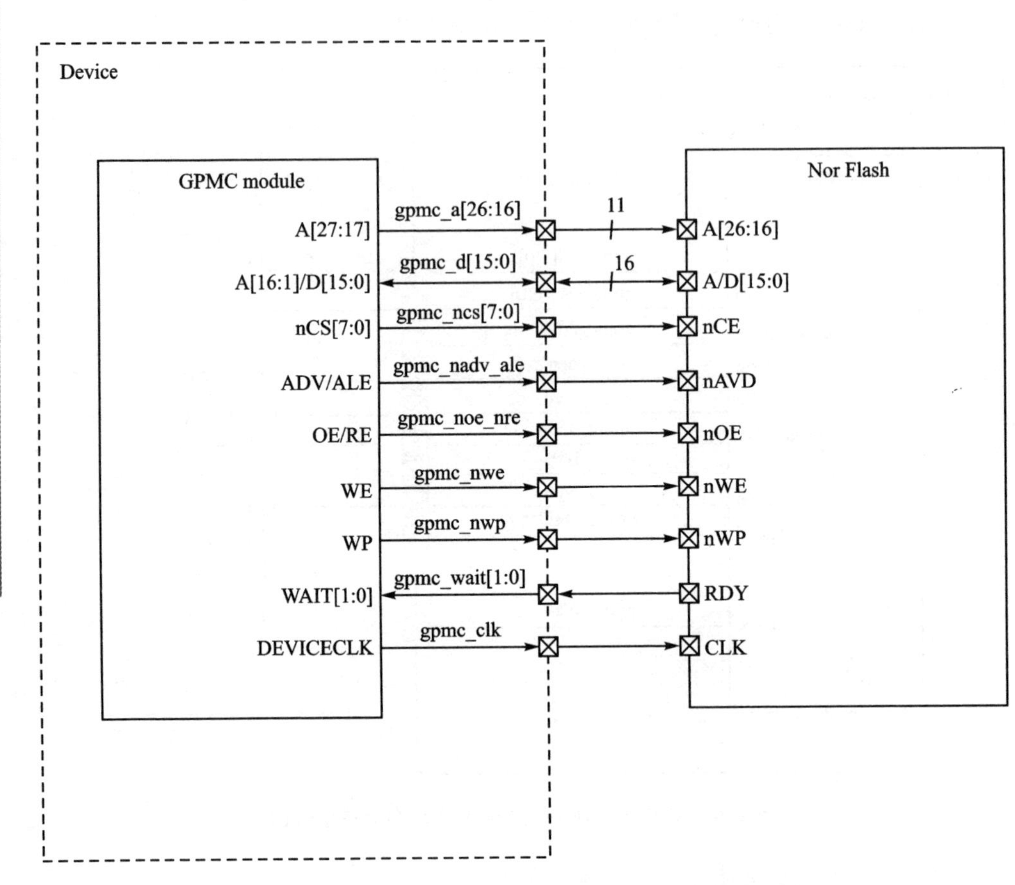

图 2.7　AM335x GPMC 扩展外部 NOR Flash 存储器示例图

ROM Memory Map

0x2BFFF	ROM Version
0x2BFFC	
	Code
0x20100	Dead loops
0x20080	Public ROM CRC
0x20020	ROM Exc. Vectors
0x20000	

ROM Exception Vectors

Address	Exception	Content
20000h	Reset	Branch to the Public ROM Code startup
20004h	Undefined	PC=4030CE04h
20008h	SWI	PC=4030CE08h
2000Ch	Pre-fetch abort	PC=4030CE0Ch
20010h	Data abort	PC=4030CE10h
20014h	Unused	PC=4030CE14h
20018h	IRQ	PC=4030CE18h
2001Ch	FIQ	PC=4030CE1Ch

图 2.8　Boot ROM 内存映射图和 ROM 异常向量表

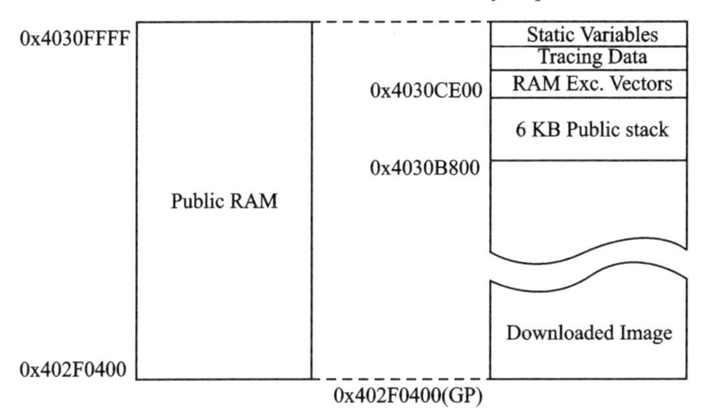

RAM Exception Vectors

Address	Exception	Content
4030CE00h	Reserved	Reserved
4030CE04h	Undefined	PC=[4030CE24h]
4030CE08h	SWI	PC=[4030CE28h]
4030CE0Ch	Pre-fetch abort	PC=[4030CE2Ch]
4030CE10h	Data abort	PC=[4030CE30h]
4030CE14h	Unused	PC=[4030CE34h]
4030CE18h	IRQ	PC=[4030CE38h]
4030CE1Ch	FIQ	PC=[4030CE3Ch]

图 2.9　Public RAM 内存映射图和 RAM 异常向量表

EMIF0 SDRAM 外部存储器接口 SDRAM 控制器，用于扩展如 DDR2 或 DDR3 SDRAM 存储器，起始地址为 0x80000000，如 StarterWare 是由 MLO 将 app 装载到此开始区域执行的。

与 S3C2410 相比，AM335x 的内存映射要难理解一些。S3C2410 通过 nGCS0～nGCS7 均匀地将地址空间分成 8 个存储体 Memory Banks，每个存储体的大小都固定为 128 MB，且复位后直接从 0x0 地址开始执行 Bootloader，详细内容请参考《ARM Linux 自学体验》一书，或介绍 S3C2410 的相关书籍。

2.2.2　AM335x Starter Kit 平台地址空间分配

AM335x Starter Kit 可能是为了降低板上的 BOM 物料成本，在板上并没有 NOR Flash 和 NAND Flash，而只有 μSD 卡接口，所以通常将程序代码保存在 μSD（即 TF）卡中，通过设置启动模式使处理器上电时从 μSD 卡中装载启动用户程序。如图 2.10 所示为 Starter Kit 的 SYSBOOT[15:0]引脚定义，其启动引导时检查设备的顺序为 MMC0→SPI0→UART0→USB0。

在电源上电复位完成时，LCD_DATA[15:0]引脚分别作为 SYSBOOT[15:0]功

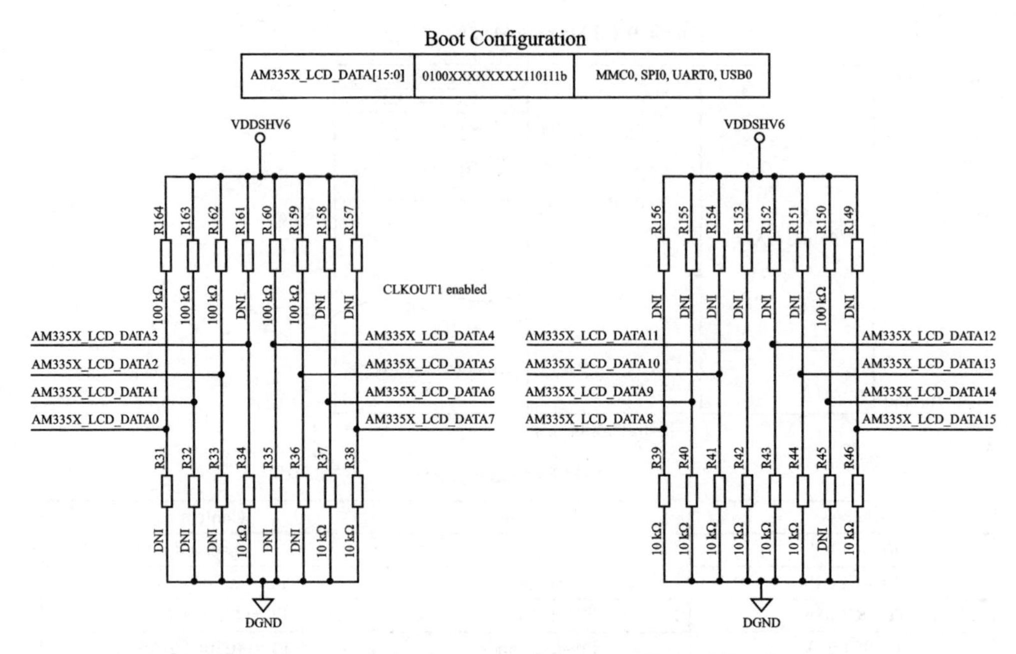

图 2.10　AM335X Starter Kit 的 SYSBOOT[15:0]引脚定义

能被锁定为 Boot 引导配置定义。这里的 MMC0(Multimedia Card 0)接口即为 μSD 接口，如图 2.11 所示。

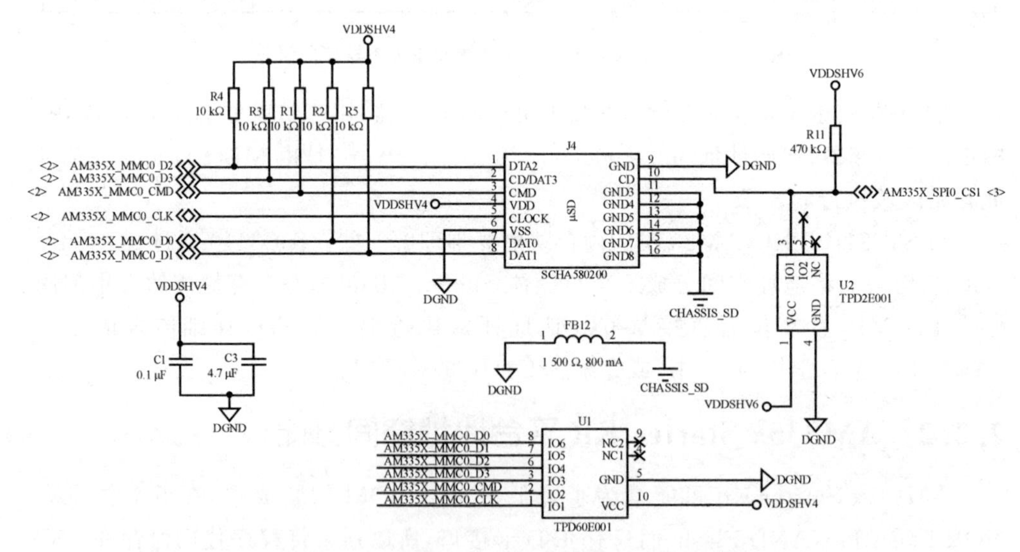

图 2.11　AM335X Starter Kit 的 μSD 卡接口图

除代码存储器外，AM335x Starter Kit 还带有 2 Gb 的 DDR3 SDRAM 数据存储器 MT41J128M16，如图 2.12 所示。

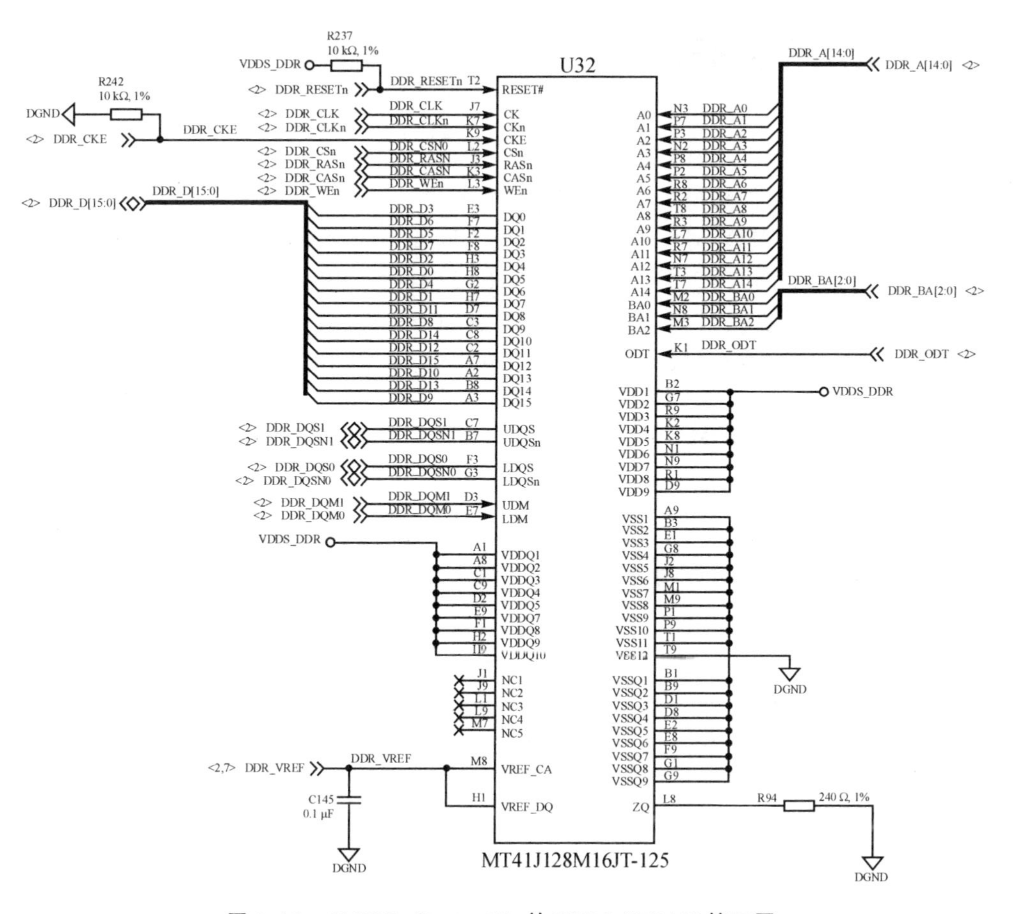

图 2.12　AM335x Starter Kit 的 DDR3 SDRAM 接口图

AM335x 是通过 EMIF0 SDRAM 控制器扩展 DDR3 SDRAM 的，所以它的映射地址为 0x80000000。

总结，AM335x Starter Kit 平台地址空间分配如下：

Boot ROM　　　　　0x4000_0000～0x4001_FFFF 共 128 KB；

SRAM internal　　　0x402F_0400～0x402F_FFFF 共 63 KB；

L3 OCMC0　　　　　0x4030_0000～0x4030_FFFF 共 64 KB；

EMIF0 SDRAM　　　0x8000_0000～0x8FFF_FFFF 共 256 KB。

注：OCMC 为 On-Chip Memory Controller，详情请参考 *Technical Reference Manual.pdf* 的第 820 页“7.2 OCMC-RAM”；内存映射相关内容除参考 *Technical Reference Manual.pdf* 的第 177 页“2.1 ARM Cortex-A8 Memory Map”外，还可参考第 4 096 页“26.1.3 Memory Map”。

2.3 常用元件概述

阿南在亲自设计原理图及选择器件时，发现自己对很多基础元件的了解还不够，虽然以前也经常见，但都没考虑得太多，这次要自己针对电路参数进行选择时，才发现有很多细节和知识是必须清楚的，于是就上网找，去元器件生产商网站查看等，花费了不少时间，现在也将它记录下来，希望对没有做过硬件的朋友有所帮助，如果有错误的地方也请读者指出并来信告诉阿南，共同学习提高。当然，即使读者们不亲自设计原理图和选择器件，仅仅阅读、分析官方原理，也是很有帮助的。

2.3.1 电阻标称值

在设计电路时可能需要自己确定某个电阻的阻值(如使用可调节电源 IC 时)，除了 22 Ω、10 kΩ 等常用阻值是我们知道的标称阻值外，其他一些阻值可能无法确定其是否为标称值，是否可以购买得到，因此列出表 2.2 和表 2.3 的电阻国家标称值，以供参考。

表 2.2 精度为 5%的碳膜电阻标称值

1.0 Ω	1.1 Ω	1.2 Ω	1.3 Ω	1.5 Ω	1.6 Ω	1.8 Ω	2.0 Ω	2.2 Ω	2.4 Ω
2.7 Ω	3.0 Ω	3.3 Ω	3.6 Ω	3.9 Ω	4.3 Ω	4.7 Ω	5.1 Ω	5.6 Ω	6.2 Ω
6.8 Ω	7.5 Ω	8.2 Ω	9.1 Ω	10 Ω	11 Ω	12 Ω	13 Ω	15 Ω	16 Ω
18 Ω	20 Ω	22 Ω	24 Ω	27 Ω	30 Ω	33 Ω	36 Ω	39 Ω	43 Ω
47 Ω	51 Ω	56 Ω	62 Ω	68 Ω	75 Ω	82 Ω	91 Ω	100 Ω	110 Ω
120 Ω	130 Ω	150 Ω	160 Ω	180 Ω	200 Ω	220 Ω	240 Ω	270 Ω	300 Ω
330 Ω	360 Ω	390 Ω	430 Ω	470 Ω	510 Ω	560 Ω	620 Ω	680 Ω	750 Ω
820 Ω	910 Ω	1 kΩ	1.1 kΩ	1.2 kΩ	1.3 kΩ	1.5 kΩ	1.6 kΩ	1.8 kΩ	2 kΩ
2.2 kΩ	2.4 kΩ	2.7 kΩ	3 kΩ	3.2 kΩ	3.3 kΩ	3.6 kΩ	3.9 kΩ	4.3 kΩ	4.7 kΩ
5.1 kΩ	5.6 kΩ	6.2 kΩ	6.6 kΩ	7.5 kΩ	8.2 kΩ	9.1 kΩ	10 kΩ	11 kΩ	12 kΩ
13 kΩ	15 kΩ	16 kΩ	18 kΩ	20 kΩ	22 kΩ	24 kΩ	27 kΩ	30 kΩ	33 kΩ
36 kΩ	39 kΩ	43 kΩ	47 kΩ	51 kΩ	56 kΩ	62 kΩ	68 kΩ	75 kΩ	82 kΩ
91 kΩ	100 kΩ	110 kΩ	120 kΩ	130 kΩ	150 kΩ	160 kΩ	180 kΩ	200 kΩ	220 kΩ
240 kΩ	270 kΩ	300 kΩ	330 kΩ	360 kΩ	390 kΩ	430 kΩ	470 kΩ	510 kΩ	560 kΩ
620 kΩ	680 kΩ	750 kΩ	820 kΩ	910 kΩ	1 MΩ	1.1 MΩ	1.2 MΩ	1.3 MΩ	1.5 MΩ
1.6 MΩ	1.8 MΩ	2 MΩ	2.2 MΩ	2.4 MΩ	2.7 MΩ	3 MΩ	3.3 MΩ	3.6 MΩ	3.9 MΩ
4.3 MΩ	4.7 MΩ	5.1 MΩ	5.6 MΩ	6.2 MΩ	6.8 MΩ	7.5 MΩ	8.2 MΩ	9.1 MΩ	10 MΩ
15 MΩ	22 MΩ								

表 2.2 精度为 1%的金属膜电阻标称值

10 Ω	10.2 Ω	10.5 Ω	10.7 Ω	11 Ω	11.3 Ω	11.5 Ω	11.8 Ω	12 Ω	12.1 Ω
12.4 Ω	12.7 Ω	13 Ω	13.3 Ω	13.7 Ω	14 Ω	14.3 Ω	14.7 Ω	15 Ω	15.4 Ω
15.8 Ω	16 Ω	16.2 Ω	16.5 Ω	16.9 Ω	17.4 Ω	17.8 Ω	18 Ω	18.2 Ω	18.7 Ω
19.1 Ω	19.6 Ω	20 Ω	20.5 Ω	21 Ω	21.5 Ω	22 Ω	22.1 Ω	22.6 Ω	23.2 Ω
23.7 Ω	24 Ω	24.3 Ω	24.7 Ω	24.9 Ω	25.5 Ω	26.1 Ω	26.7 Ω	27 Ω	27.4 Ω
28 Ω	28.7 Ω	29.4 Ω	30 Ω	30.1 Ω	30.9 Ω	31.6 Ω	32.4 Ω	33 Ω	33.2 Ω
34 Ω	34.8 Ω	35.7 Ω	36 Ω	36.5 Ω	37.4 Ω	38.3 Ω	39 Ω	39.2 Ω	40.2 Ω
41.2 Ω	42.2 Ω	43 Ω	43.2 Ω	44.2 Ω	45.3 Ω	46.4 Ω	47 Ω	47.5 Ω	48.7 Ω
49.9 Ω	51 Ω	51.1 Ω	52.3 Ω	53.6 Ω	54.9 Ω	56 Ω	56.2 Ω	57.6 Ω	59 Ω
60.4 Ω	61.9 Ω	62 Ω	63.4 Ω	64.9 Ω	66.5 Ω	68 Ω	68.1 Ω	69.8 Ω	71.5 Ω
73.2 Ω	75 Ω	75.5 Ω	76.8 Ω	78.7 Ω	80.6 Ω	82 Ω	82.5 Ω	84.5 Ω	86.6 Ω
88.7 Ω	90.9 Ω	91 Ω	93.1 Ω	95.3 Ω	97.6 Ω	100 Ω	102 Ω	105 Ω	107 Ω
110 Ω	113 Ω	115 Ω	118 Ω	120 Ω	121 Ω	124 Ω	127 Ω	130 Ω	133 Ω
137 Ω	140 Ω	143 Ω	147 Ω	150 Ω	154 Ω	158 Ω	160 Ω	162 Ω	165 Ω
169 Ω	174 Ω	178 Ω	180 Ω	182 Ω	187 Ω	191 Ω	196 Ω	200 Ω	205 Ω
210 Ω	215 Ω	220 Ω	221 Ω	226 Ω	232 Ω	237 Ω	240 Ω	243 Ω	249 Ω
255 Ω	261 Ω	267 Ω	270 Ω	274 Ω	280 Ω	287 Ω	294 Ω	300 Ω	301 Ω
309 Ω	316 Ω	324 Ω	330 Ω	332 Ω	340 Ω	348 Ω	350 Ω	357 Ω	360 Ω
365 Ω	374 Ω	383 Ω	390 Ω	392 Ω	402 Ω	412 Ω	422 Ω	430 Ω	432 Ω
442 Ω	453 Ω	464 Ω	470 Ω	475 Ω	487 Ω	499 Ω	510 Ω	511 Ω	523 Ω
536 Ω	549 Ω	560 Ω	562 Ω	565 Ω	578 Ω	590 Ω	604 Ω	619 Ω	620 Ω
634 Ω	649 Ω	665 Ω	680 Ω	681 Ω	698 Ω	715 Ω	732 Ω	750 Ω	768 Ω
787 Ω	806 Ω	820 Ω	825 Ω	845 Ω	866 Ω	887 Ω	909 Ω	910 Ω	931 Ω
953 Ω	976 Ω	1 kΩ	1.02 kΩ	1.05 kΩ	1.07 kΩ	1.1 kΩ	1.13 kΩ	1.15 kΩ	1.18 kΩ
1.2 kΩ	1.21 kΩ	1.24 kΩ	1.27 kΩ	1.3 kΩ	1.33 kΩ	1.37 kΩ	1.4 kΩ	1.43 kΩ	1.47 kΩ
1.5 kΩ	1.54 kΩ	1.58 kΩ	1.6 kΩ	1.62 kΩ	1.65 kΩ	1.69 kΩ	1.74 kΩ	1.78 kΩ	1.8 kΩ
1.82 kΩ	1.87 kΩ	1.91 kΩ	1.96 kΩ	2 kΩ	2.05 kΩ	2.1 kΩ	2.15 kΩ	2.2 kΩ	2.21 kΩ
2.26 kΩ	2.32 kΩ	2.37 kΩ	2.4 kΩ	2.43 kΩ	2.49 kΩ	2.55 kΩ	2.61 kΩ	2.67 kΩ	2.7 kΩ
2.74 kΩ	2.8 kΩ	2.87 kΩ	2.94 kΩ	3.0 kΩ	3.01 kΩ	3.09 kΩ	3.16 kΩ	3.24 kΩ	3.3 kΩ
3.32 kΩ	3.4 kΩ	3.48 kΩ	3.57 kΩ	3.6 kΩ	3.65 kΩ	3.74 kΩ	3.83 kΩ	3.9 kΩ	3.92 kΩ
4.02 kΩ	4.12 kΩ	4.22 kΩ	4.32 kΩ	4.42 kΩ	4.53 kΩ	4.64 kΩ	4.7 kΩ	4.75 kΩ	4.87 kΩ
4.99 kΩ	5.1 kΩ	5.11 kΩ	5.23 kΩ	5.36 kΩ	5.49 kΩ	5.6 kΩ	5.62 kΩ	5.76 kΩ	5.9 kΩ
6.04 kΩ	6.19 kΩ	6.2 kΩ	6.34 kΩ	6.49 kΩ	6.65 kΩ	6.8 kΩ	6.81 kΩ	6.98 kΩ	7.15 kΩ

续表 2.2

7.32 kΩ	7.5 kΩ	7.68 kΩ	7.87 kΩ	8.06 kΩ	8.2 kΩ	8.25 kΩ	8.45 kΩ	8.66 kΩ	8.8 kΩ
8.87 kΩ	9.09 kΩ	9.1 kΩ	9.31 kΩ	9.53 kΩ	9.76 kΩ	10 kΩ	10.2 kΩ	10.5 kΩ	10.7 kΩ
11 kΩ	11.3 kΩ	11.5 kΩ	11.8 kΩ	12 kΩ	12.1 kΩ	12.4 kΩ	12.7 kΩ	13 kΩ	13.3 kΩ
13.7 kΩ	14 kΩ	14.3 kΩ	14.7 kΩ	15 kΩ	15.4 kΩ	15.8 kΩ	16 kΩ	16.2 kΩ	16.5 kΩ
16.9 kΩ	17.4 kΩ	17.8 kΩ	18 kΩ	18.2 kΩ	18.7 kΩ	19.1 kΩ	19.6 kΩ	20 kΩ	20.5 kΩ
21 kΩ	21.5 kΩ	22 kΩ	22.1 kΩ	22.6 kΩ	23.2 kΩ	23.7 kΩ	24 kΩ	24.3 kΩ	24.9 kΩ
25.5 kΩ	26.1 kΩ	26.7 kΩ	27 kΩ	27.4 kΩ	28 kΩ	28.7 kΩ	29.4 kΩ	30 kΩ	30.1 kΩ
31.9 kΩ	31.6 kΩ	32.4 kΩ	33 kΩ	33.2 kΩ	33.6 kΩ	34 kΩ	34.8 kΩ	35.7 kΩ	36 kΩ
36.5 kΩ	37.4 kΩ	38.3 kΩ	39 kΩ	39.2 kΩ	40.2 kΩ	41.2 kΩ	42.2 kΩ	43 kΩ	43.2 kΩ
44.2 kΩ	45.3 kΩ	46.4 kΩ	47 kΩ	47.5 kΩ	48.7 kΩ	49.9 kΩ	51 kΩ	51.1 kΩ	52.3 kΩ
53.6 kΩ	54.9 kΩ	56 kΩ	56.2 kΩ	57.6 kΩ	59 kΩ	60.4 kΩ	61.9 kΩ	62 kΩ	63.4 kΩ
64.9 kΩ	66.5 kΩ	68 kΩ	68.1 kΩ	69.8 kΩ	71.5 kΩ	73.2 kΩ	75 kΩ	76.8 kΩ	78.7 kΩ
80.6 kΩ	82 kΩ	82.5 kΩ	84.5 kΩ	86.6 kΩ	88.7 kΩ	90.9 kΩ	91 kΩ	93.1 kΩ	95.3 kΩ
97.6 kΩ	100 kΩ	102 kΩ	105 kΩ	107 kΩ	110 kΩ	113 kΩ	115 kΩ	118 kΩ	120 kΩ
121 kΩ	124 kΩ	127 kΩ	130 kΩ	133 kΩ	137 kΩ	140 kΩ	143 kΩ	147 kΩ	150 kΩ
154 kΩ	158 kΩ	160 kΩ	162 kΩ	165 kΩ	169 kΩ	174 kΩ	178 kΩ	180 kΩ	182 kΩ
187 kΩ	191 kΩ	196 kΩ	200 kΩ	205 kΩ	210 kΩ	215 kΩ	220 kΩ	221 kΩ	226 kΩ
232 kΩ	237 kΩ	240 kΩ	243 kΩ	249 kΩ	255 kΩ	261 kΩ	267 kΩ	270 kΩ	274 kΩ
280 kΩ	287 kΩ	294 kΩ	300 kΩ	301 kΩ	309 kΩ	316 kΩ	324 kΩ	330 kΩ	332 kΩ
340 kΩ	348 kΩ	357 kΩ	360 kΩ	365 kΩ	374 kΩ	383 kΩ	390 kΩ	392 kΩ	402 kΩ
412 kΩ	422 kΩ	430 kΩ	432 kΩ	442 kΩ	453 kΩ	464 kΩ	470 kΩ	475 kΩ	487 kΩ
499 kΩ	511 kΩ	523 kΩ	536 kΩ	549 kΩ	560 kΩ	562 kΩ	576 kΩ	590 kΩ	604 kΩ
619 kΩ	620 kΩ	634 kΩ	649 kΩ	665 kΩ	680 kΩ	681 kΩ	698 kΩ	715 kΩ	732 kΩ
750 kΩ	768 kΩ	787 kΩ	806 kΩ	820 kΩ	825 kΩ	845 kΩ	866 kΩ	887 kΩ	909 kΩ
910 kΩ	931 Ω	953 Ω	976 kΩ	1.0 MΩ	1.5 MΩ	2.2 MΩ			

2.3.2 肖特基二极管

嵌入式系统中经常会采用集成开关电源(DC/DC)转换器作为系统的主电源,因为它比线性电源有更高的效率,而在开关电源的输出端 SW 都会有一个整流(有时也称"续流")二极管。为了最大限度地减少由于二极管的正向电压和反向恢复时间等带来的损耗,往往都采用肖特基二极管(Schottky Diode)。肖特基二极管与普通整流二极管相比具有正向压降小(一般为 0.4 V 和 0.5 V)和反向恢复时间可以忽略不计的特点,以及不存在开关过程中电荷储存问题等,特别适用于 5 V 左右的低电压输出的 DC/DC 电路。

比较常用的肖特基二极管有1N5817、1N5818、1N5819系列，它们的不同之处是可持续承受的反向峰值电压（Peak Repetitive Reverse Voltage，V_{RRM}）不同，分别为20 V、30 V和40 V，正向平均整流电流（Average Rectified Forward Current，I_F）都为1 A。还有ON Semiconductor的MBRS130、MBRS140、MBRS330、MBRS340等系列，MBRS130、MBRS140的I_F都为1 A，V_{RRM}分别为30 V和40 V，而MBRS330、MBRS340的I_F都为3 A，V_{RRM}分别为30 V和40 V。另外，同等参数（3 A，30 V和40 V）的还有Diodes Inc的B330和B440。在实际的DC/DC应用中，一般要选择I_F大于负载电流，而V_{RRM}要大于DC/DC芯片的输入电压，如AN2410MB选择3 A的MP1423，外部电源输入电压为12 V选择MBRS330。

2.3.3 功率电感

在DC/DC转换电路中，还有一个必需的器件——功率电感（有时也称储能电感，我们将其他电路上使用的普通电感称为信号电感）。它不仅扼制突变电流，更为重要的作用是像电容一样作为储能元件，为输出端提供连续的能量供应。它和电容、肖特基二极管一起将被开关信号斩成方波的输入电源整流、滤波成平滑稳定的电源输出。

功率电感的取值一般取决于电路对纹波的要求和PCB板的空间等。电感值越大，电源输出纹波越小，但其电感尺寸也相应越大，等效的直流阻抗也越大，一般的DC/DC转换器数据手册会给出参考值。知名的功率电感厂家有Sumida、Toko等，他们在产品数据手册中一般都能给出形状、具体的尺寸、电感值和额定直流电流等具体参数。

2.3.4 铁氧体磁珠

铁氧体磁珠（Ferrite Bead）是一种阻抗随频率变化的电阻器，广泛应用的EMI（电磁干扰）静噪元件，与电感有相同的电路原理和功能，但是频率特性是不同的。铁氧体磁珠在高频时呈现电阻性，相当于品质因数很低的电感器，能在相当宽的频率范围内保持较高的阻抗，比普通电感具有更好的高频滤波特性。电感是储能元件，多用于电源滤波回路，侧重于抑止传导性干扰。而铁氧体磁珠是能量转换（消耗）器件，多用于信号回路，主要用于EMI方面，用来吸收高频干扰信号，如RF电路、PLL、振荡电路、高速存储器电路（DDR、SDRAM等）的电源输入部分。另外，在DC/DC模块的电源输入端，地与地之间也能看到铁氧体磁珠，因为它在衰减较高频干扰信号的同时，却让较低频信号或电源几乎无阻碍地通过，即提高高频滤波特性的同时不会在系统中产生新的零点，不会破坏系统原有的稳定性。

铁氧体磁珠的单位是欧姆，因为它的参数规格是按照某一具体频率（通常是100 MHz，在厂家的数据手册上一般会提供频率和阻抗的特性曲线图）产生的阻抗来标称的，而阻抗的单位是欧姆。在电路原理图的表示上，铁氧体磁珠直接使用电感的电路符号，但通常会在型号上标出使用的是磁珠，如在 AN2410 中标明“FB/200/3A”时，表明它为 200 Ω、3 A 的铁氧体磁珠。目前知名的磁珠厂家有 muRata、Toko 等，他们通常有详细的选型指南和数据手册，可以根据参数直接查找到对应的型号。如果购买国产或不知名厂家的磁珠，需有阻抗和额定电流等，如铁氧体磁珠 200 Ω，3 A 等。

2.3.5　自恢复保险丝 PPTC

自恢复保险丝 PPTC (Polymeric Positive Temperature Coefficient，也称 PolySwitch PPTC)是一种正温度系数聚合物热敏电阻，作过流保护用，用于代替电流保险丝。电路正常工作时它的阻值很小(压降很小)，当电路出现过流使它温度升高时，阻值急剧增大几个数量级，使电路中的电流减小到安全值以下，从而使后面的电路得到保护；过流消失后自动恢复为低阻值，免除电流保险丝经常更换的麻烦。

自恢复保险丝的主要特性参数如下：

- 保持电流(Hold Current)I_H：不会使电阻值突变的最大电流。
- 触发电流(Trip Current)I_T：能使电阻值突然变大的最小电流，一般为保持电流的 2 倍。
- 额定电压(Rated Voltage)V_{MAX}：在额定电流下能承受的不会损害 PPTC 器件本身的最大电压。
- 最大电流(Maximum Current)I_{MAX}：在额定电压下能承受的不会损害 PPTC 器件本身的最大电流。
- 动作功率(Typical Power)P_d：动作状态下消耗的功率。
- 最大动作时间(Max Time to Trip)T_{trip}：规定电流下的最大动作时间。
- 静态电阻(Resistance Tolerance)R_{MIN}、R_{MAX}：在不加电情况下的静态电阻最小值 R_{min} 和最大值 R_{max}，器件的实际值在该范围之内，即 $R_{min} \leqslant R \leqslant R_{max}$。

在使用中，要先根据线路最大工作电压 V、正常工作电流 I、故障电流 I_f 及最高使用环境温度 T_M 选择合适的型号。在最高使用环境温度 T_M 时，应满足：

$$I_H(T_M) \approx I$$

$$I_T(T_M) \leqslant I_f \leqslant I_{max}$$

$$V_{MAX} \geqslant V$$

这里的线路电压、电流是指直流或交流的有效值。

比较知名的 PolySwitch PPTC 厂商有 Raychem(瑞侃)，还有 Fuzetec、muRata 等。选择(或采购)PPTC 时，一般至少要确定封装和保持电流，或根据参数直接查找某个厂商的具体型号。

2.3.6　有源和无源蜂鸣器

在使用上经常将蜂鸣器分为有源蜂鸣器和无源蜂鸣器，它们的根本区别是驱动的输入信号要求不一样。有源蜂鸣器工作的理想信号是直流电，在通常的 MCU 控制时，输出一个高电平(或加三极管驱动)就可以使它发声，控制方便。因为有源蜂鸣器内部有一个振荡电路，能将恒定的直流电转化成一定频率的脉冲信号，从而产生磁场交变，带动钼片振动发声。无源蜂鸣器工作的理想信号是方波，因为它内部没有驱动电路，如果直接给直流信号，内部磁路是恒定的，钼片不会振动发声。在 MCU 控制中，一般输出 PWM 信号去控制无源蜂鸣器，相对复杂一些，但是由于内部少了驱动电路，所以成本相对较低，而且可以通过调节节奏产生悦耳的音效。

2.4 AM335x Starter Kit(TMDSSK3358)原理图分析

2.4.1 MPU AM3358

MPU AM3358 部分需要读者参考 AM335x 用户手册的“4.3 Signal Descriptions”一节，理解 AM335x 处理器各个引脚信号的定义。AM3358 共有多达 324 个引脚，所以在绘制原理图时，通常按功能将其分成多个部分。

1. MPU(DDR/GPMC/MMC/JTAG/Clock/RESET/EXTINT)

如图 2.13 所示为 AM3358 处理器的第 1 部分原理图。

各信号引脚的功能定义和说明如下(从芯片图左侧开始从上往下)：

XTALIN[OSC0_IN]：处理器高频主时钟输入引脚。

XTALOUT[OSC0_OUT]：处理器高频主时钟输出引脚。

VSS_OSC：主时钟 OSC0 提供的地连接。

RTC_XTALIN[OSC1_IN]：处理器低频(32.768 kHz)RTC 实时时钟输入引脚。

RTC_XTALOUT[OSC1_OUT]：处理器低频(32.768 kHz)RTC 实时时钟输出引脚。

VSS_RTC：RTC 时钟 OSC1 提供的地连接。

DDR_A[15:0]：DDR SDRAM 行/列地址输出总线。

DDR_BA[2:0]：DDR SDRAM 块地址输出。

DDR_CLK：DDR SDRAM 时钟输出，差分线对正端。

DDR_CLKn：DDR SDRAM 时钟输出，差分线对负端。

DDR_CKE：DDR SDRAM 时钟使能输出。

DDR_CSn0：DDR SDRAM 片选 0 输出(低电平有效)。

DDR_CASn：DDR SDRAM 列地址选通输出(低电平有效)。

DDR_RASn：DDR SDRAM 行地址选通输出(低电平有效)。

DDR_WEn：DDR SDRAM 写使能输出(低电平有效)。

DDR_DQM0：DDR SDRAM 写使能/DDR_D[7:0]的数据屏蔽。

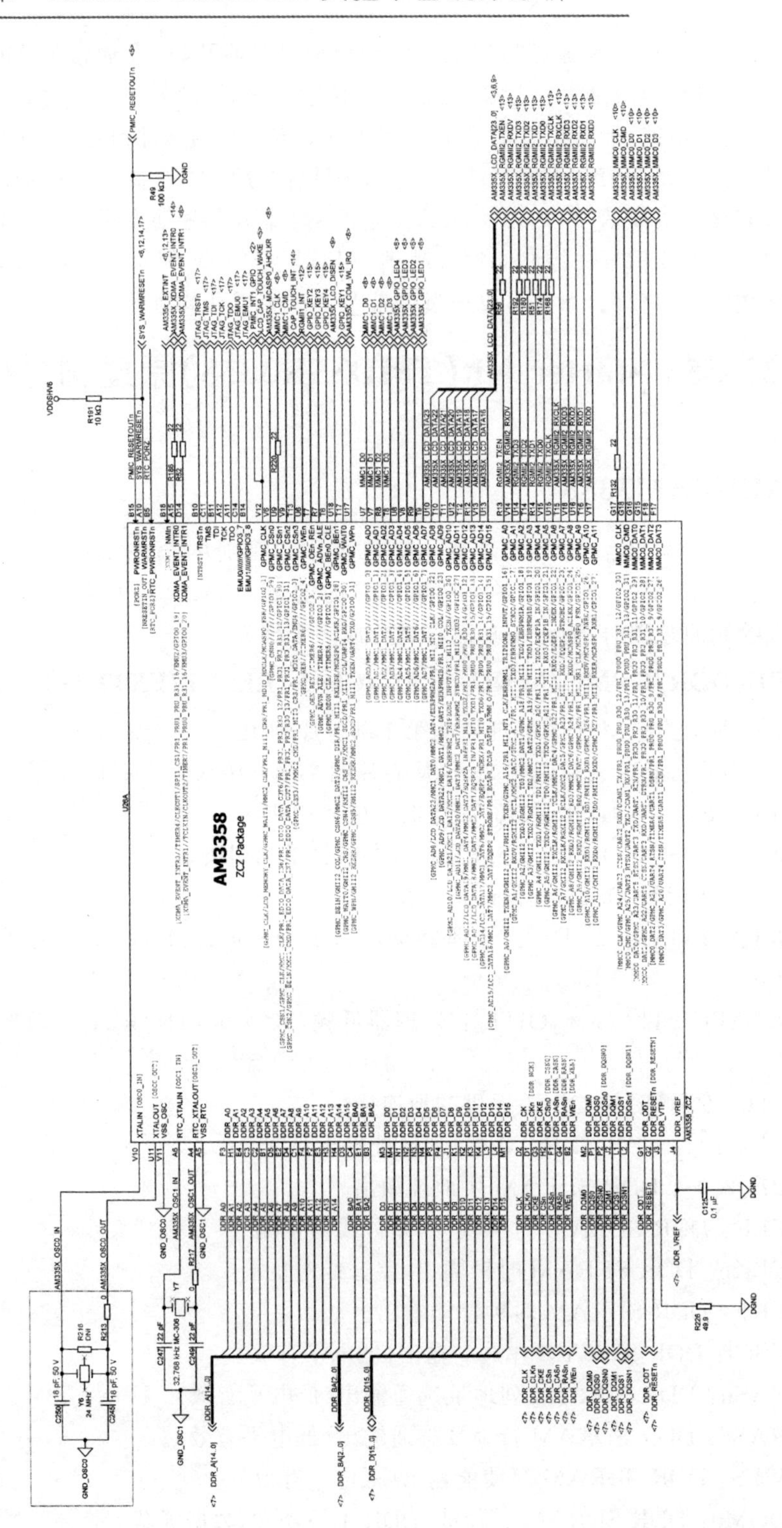

图2.13 MPU(DDR/GPMC/MMC/JTAG/Clock/RESET/EXTINT)

DDR_DQM1：DDR SDRAM 写使能/DDR_D[15:8]的数据屏蔽。

DDR_DQS0：DDR SDRAM 数据 DDR_D[7:0]选通输出，差分线对正端。

DDR_DQSn0：DDR SDRAM 数据 DDR_D[7:0]选通输出，差分线对负端。

DDR_DQS1：DDR SDRAM 数据 DDR_D[15:8]选通输出，差分线对正端。

DDR_DQSn1：DDR SDRAM 数据 DDR_D[15:8]选通输出，差分线对负端。

DDR_ODT：DDR SDRAM ODT(On-Die Termination)终端电阻输出。

我们知道 DDR SDRAM 的主板上面为了防止数据线终端反射信号，需要大量的终端电阻，大大增加了主板的制造成本。实际上，不同的内存模块对终端电路的要求是不一样的，终端电阻的大小决定了数据线的信号比和反射率：终端电阻小，则数据线信号反射率低，但是信噪比也较低；终端电阻大，则数据线的信噪比高，但是信号反射率也会增高。因此主板上的终端电阻并不能非常好地匹配内存模块，还会在一定程度上影响信号品质。DDR2 可以根据自己的特点内建合适的终端电阻，这样可以保证最佳的信号波形。使用 DDR2 不但可以降低主板成本，还得到了最佳的信号品质，这是 DDR 不能比拟的。

DDR_RESETn：DDR3/DDR3L 复位输出(低电平有效)。

DDR_VTP：VTP(Voltage,Temperature,and Process)补偿电阻连接端。

DDR 内存控制器是能够控制 I/O 口输出阻抗的，这个特性就允许 DDR 内存控制器去调节 I/O 口的输出阻抗去和 PCB 板做阻抗匹配。I/O 口的输出阻抗控制是非常重要的特性，因为阻抗匹配能够减少信号的反射，创造一个纯净的板级设计。同时，I/O 口输出阻抗的测算校准补偿也将减少 DDR 内存控制器的功率消耗。而该补偿是通过配置电压、温度和数字电路的运算处理过程(VTP)的关系实现的。VTP 信息参数的获取来自 I/O 的输出阻抗控制。

关于 VTP 的更详细信息，读者可以参考 TI 官方的 *TMS320DM644x DMSoC DDR2 Memory Controller User's Guide* 中第 29 页的“2.11 VTP IO Buffer Calibration”的相关内容。

DDR_VREF：DDR SDRAM 参考电压输入。

MMC0_CLK：MMC/SD/SDIO 时钟。

MMC0_CMD：MMC/SD/SDIO 命令。

MMC0_DAT[7:0]：MMC/SD/SDIO 数据总线。

MMC0_POW：MMC/SD 电源切换控制。

MMC0_SDCD：SD 卡检测输入端。

MMC0_SDWP：SD 卡写保护输入端。

GPMC_A[27:0]：GPMC 控制器地址。

GPMC_AD[15:0]：GPMC 控制器地址和数据复合引脚。

GPMC_WPn：GPMC 控制器写保护，低电平有效。

GPMC_WAIT[1:0]：GPMC 控制器信号等待。

GPMC_BEn1：GPMC 控制器字节使能信号 1。

GPMC_BEn0_CLE：GPMC 控制器字节使能信号 0/命令锁存使能信号。

GPMC_ADVn_ALE：GPMC 控制器地址有效/地址锁存使能信号。

GPMC_OEn_REn：GPMC 控制器输出/读使能信号。

GPMC_WEn：GPMC 控制器写使能信号，低电平有效。

GPMC_CSn[6:0]：GPMC 控制器片选信号，低电平有效。

GPMC_CLK：GPMC 控制器时钟信号。

注：在 PCB 中，上述各总线如果走线太长，可以串联 22 Ω 电阻（或排阻），且靠近 CPU（驱动端），以延长信号的上升/下降时间，使过冲信号变得较为平滑，从而减小输出波形的高频谐波幅度，有效抑制 EMI。更详细的原因请读者参考电磁兼容（EMC）、电磁干扰（EMI）及微波理论等关于阻抗匹配、信号反射等内容。

EMU[1:0]：用于 JTAG 仿真器引脚。

TDO：JTAG 测试数据输出信号。

TCK：JTAG 测试时钟信号。

TDI：JTAG 测试数据输入信号。

TMS：JTAG 测试模式选择输入信号。

TRSTn：JTAG 测试复位输入信号，低电平有效。

XDMA_EVENT_INTR[2:0]：外部 DMA 事件或中断输入信号。

NMIn：去往 ARM Cortex-A8 内核的外部中断输入信号。

[NRESETIN_OUT]WARMRSTn：处理器（热启动）复位输出。注：该引脚为双向的（热启动）复位信号。当作为输入时，它连接外部复位芯片（或复位电路），当外部复位信号低电平有效时，复位处理器；当作为输出时，该引脚能输出一个低电平的复位信号，来复位外部的设备，也可以作为复位电路是否正常工作的检测指示。

关于热启动和冷启动的区别，请参考官方 *Technical Reference Manual* 的第 1 151 页“8.1.7.4.1 External Warm Reset”的相关内容。

[PORZ]PWRONRSTn：处理器上电复位输入端，低电平有效。

[RTC_PORZ]RTC_PWRONRSTn：RTC 复位输入，外部电路如图 2.14 所示，SN74AUP2G08 为两输入正“与”门，只有两输入都为高电平时输出才为高电平。

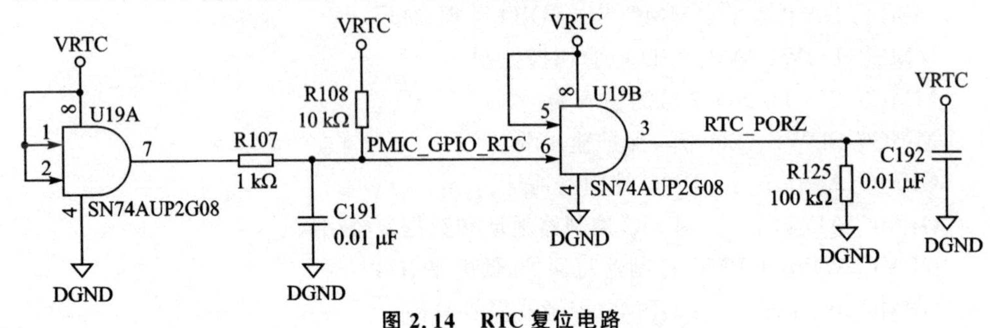

图 2.14　RTC 复位电路

2. MPU(AIN(TP)/UART/LCD/ IIC/MCASP/SPI/USB/MII)

如图 2.15 所示为 AM335x 处理器的第 2 部分原理图。

各信号引脚的功能定义和说明如下：

PMIC_POWER_EN：PMIC 电源管理芯片的电源使能信号输出，用于使能外部电源管理芯片。

EXT_WAKEUP：外部唤醒输入。

AIN[7:0]：A/D 模拟信号输入，其中 AIN[3:0]也分别为触摸屏的左、右、上和下的模拟信号输入端。

VREFP：A/D 模拟信号的参考电压，正端。

VREFN：A/D 模拟信号的参考电压，负端。

SPI0_SCLK：SPI 接口 0 时钟。

SPI0_D[1:0]：SPI 接口 0 数据。

SPI0_CS[1:0]：SPI 接口 0 片选信号。

UART0 TXD：串口 UART0 的数据发送端。

UART0_RXD：串口 UART0 的数据接收端。

UART0_CTSn：串口 UART0 在硬件流控制时的清除发送信号。

UART0_RTSn：串口 UART0 在硬件流控制时的请求发送信号。

UART1_TXD：串口 UART1 的数据发送端。

UART1_RXD：串口 UART1 的数据接收端。

UART1_CTSn：串口 UART1 在硬件流控制时的清除发送信号。

UART1_RTSn：串口 UART1 在硬件流控制时的请求发送信号。

I2C0_SCL：IIC 接口 0 的时钟信号。

I2C0_SDA：IIC 接口 0 的数据信号。

USB0_DP：USB 总线接口 0 的数据线 DATA+。

USB0_DM：USB 总线接口 0 的数据线 DATA－。

USB0_CE：USB 总线接口 0 的充电使能输出，高电平有效。

读者可参考 *Technical Reference Manual.pdf* 的第 1 360 页“Figure 9－1 USB Charger Detection”。

USB0_ID：USB 总线接口 0 的 USB OTG 迷你插座的 ID 标识。

USB0_DRVVBUS[GPIO0_18]：USB 总线接口 0 的 USB OTG 充电使能，当 OTG 作为 HOST 时，外部电源驱动控制信号。

USB0_VBUS：USB 总线接口 0 的 USB OTG 迷你插座的 VBUS 电源。

USB1_DP：USB 总线接口 1 的数据线 DATA+。

USB1_DM：USB 总线接口 1 的数据线 DATA－。

USB1_CE：USB 总线接口 1 的充电使能输出，高电平有效。

USB1_ID：USB 总线接口 1 的 USB OTG 迷你插座的 ID 标识。

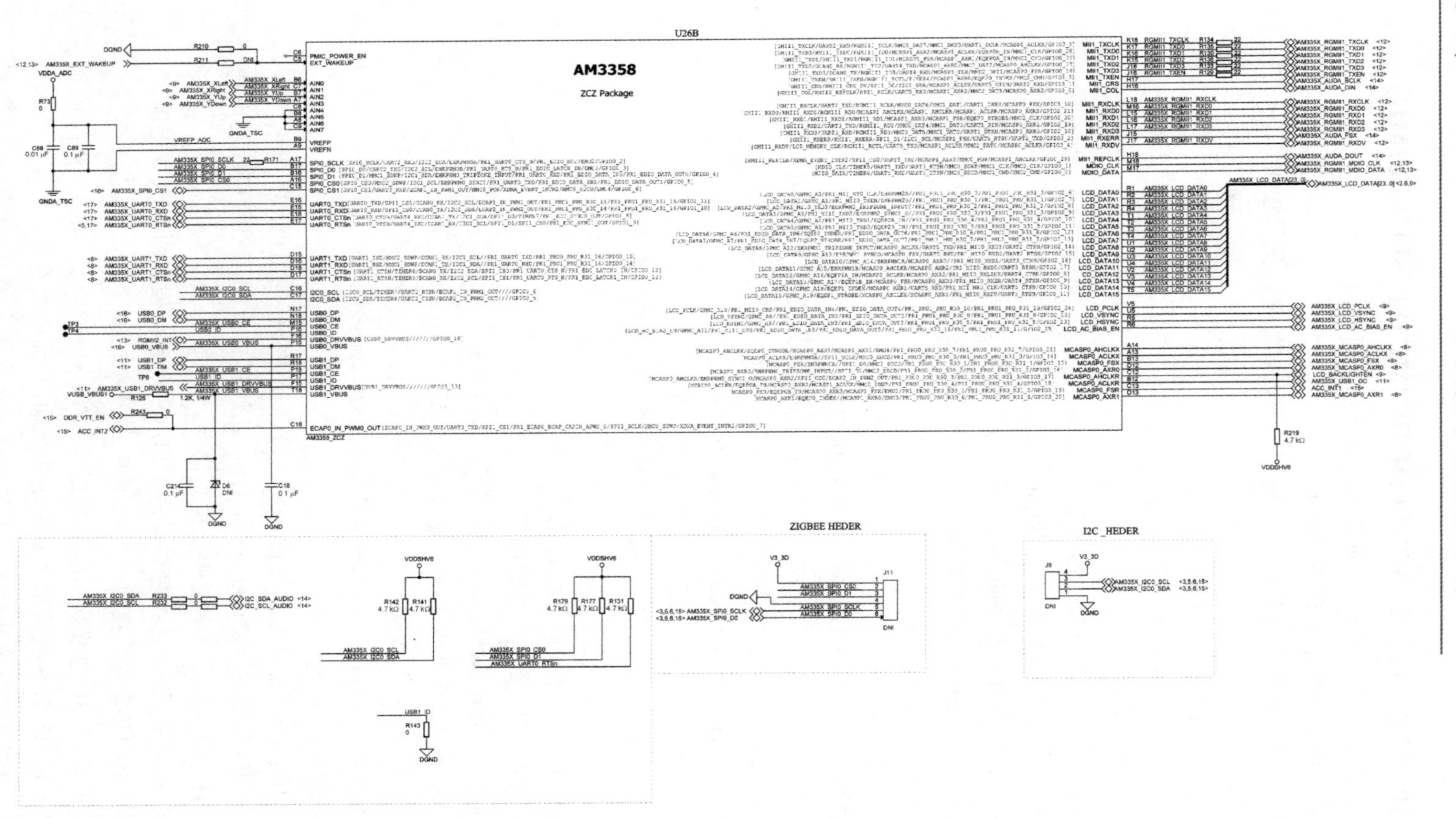

图2.15　MPU(AIN(TP)/UART/LCD/IIC/MCASP/SPI/USB/MII)

USB1_DRVVBUS[GPIO3_13]：USB 总线接口 1 的 USB OTG 充电使能，当 OTG 作为 HOST 时，外部电源驱动控制信号。

USB1_VBUS：USB 总线接口 1 的 USB OTG 迷你插座的 VBUS 电源。

ECAP0_IN_PWM0_OUT：外部捕获 0 输入使能信号，或脉宽调制 PWM0 输出。

MII1_TXCLK：MII1 数据发送时钟。

MII1_TXD[3:0]：MII1 发送数据位。

MII1_TXEN：MII1 发送使能输出。

MII1_CRS：MII1 载波监听信号输入。

MII1_COL：MII1 冲突检测信号输入。

MII1_RXCLK：MII1 数据接收时钟

MII1_RXD[3:0]：MII1 接收数据位。

MII1_RXERR：MII1 接收数据错误输入。

MII1_RXDV：MII1 接收数据有效输入。

MII1_REFCLK：MII1 参考时钟。

MDIO_CLK：Management data clock，MDIO 串行接口时钟。

MDIO_DATA：MDIO 串行接口数据。

LCD_DATA[15:0]：LCD 的 16 位数据总线，R(红)、G(绿)、B(蓝)可分别占用 5 位、6 位和 5 位，组成一般 TFT LCD 屏常用的 565 格式(R 为 5 位数据，G 为 6 位数据，B 为 5 位数据)。

LCD_PCLK：LCD 像素时钟信号。像素是 LCD 屏的最小单位，通常讲的 LCD 屏分辨率为 480×272，指的就是一整屏共有 480×272 个像素点，而每个像素点又由 RGB 颜色数据(VD[23:0])构成，而 LCD 屏的驱动 IC 就是通过采集 LCD_PCLK 信号来接收 RGB 数据的。

LCD_ HSYNC：LCD 水平同步信号(Horizontal synchronous signal)，当 AM335x 的 LCD 控制器发送完一行的数据后，输出该信号告知 LCD 屏驱动器一行的结束。

LCD_VSYNC：LCD 垂直同步信号(Vertical synchronous signal)，当 AM335x 的 LCD 控制器发送完一帧的数据后，输出该信号告知 LCD 屏驱动器一帧的结束。一帧也就是一个屏幕，写满整个屏幕的数据也就称为一个"帧"的数据。

LCD_AC_BIAS_EN：AC 偏置使能输出。

MCASP0_AHCLKX：McASP0 发送主时钟。

MCASP0_ACLKX：McASP0 发送位时钟。

MCASP0_FSX：McASP0 发送帧同步信号。

MCASP0_AXR0：McASP0 串行数据位 0。

MCASP0_AHCLKR：McASP0 接收主时钟。

MCASP0_ACLKR：McASP0 接收位时钟。

MCASP0_FSR：McASP0 接收帧同步信号。

MCASP0_AXR1：McASP0 串行数据位 1。

MCASP：Multichannel Audio Serial Port，多通道音频串行接口。

3. MCU(POWER/EINT)

如图 2.16 所示为 AM335x 处理器的第 3 部分原理图。

各信号引脚的功能定义和说明如下：

VDD_CORE：提供给内核逻辑的电源。

VDD_MPU：提供给 MPU 内核逻辑的电源。

VDD_MPU_MON：当需要监控 MPU 电源时，输出控制电源管理 IC。详情请参考数据手册第 105 页“6.1.3 VDD_MPU_MON Connections”。

CAP_VDD_SRAM_CORE：AM335x 内部集成低压差电源调节器(LDO)，为了稳定其输出，所以需要外接电容器。该引脚连接为内核静态存储器提供电源(CORE SRAM LDO)的输出电容器。PCB 布板时需要将电容器尽量靠近该引脚。详情请参考数据手册第 96 页“Figure 5-1 shows an example of the external capacitors”。

VDDS_PLL_MPU：提供 MPU 锁相环 DPLL 所需的电源。

VDDS_SRAM_MPU_BB：为 MPU 内部 SRAM 所提供的 LDOs 的电源。

CAP_VDD_SRAM_MPU：该引脚连接为 MPU 静态存储器提供电源(MPU SRAM LDO)的输出电容器。

CAP_VBB_MPU：该引脚连接 Back Bias LDO 电源输出的电容器。

VDDS_SRAM_CORE_BG：内核静态存储器电源(CORE SRAM LDO)输出。

VDDA3P3V_USB0：为 USB0 物理层(USBPHY)接口 3.3 V 提供电源。

VDDA1P8V_USB0：为 USB0(物理层)接口和外部设备锁相环(USBPHY and PER DPLL)的 1.8 V 提供电源。

VDDA3P3V_USB1：为 USB1 物理层(USBPHY)接口 3.3 V 提供电源。

VDDA1P8V_USB1：为 USB1(物理层)接口和锁相环(USBPHY and PER DPLL)的 1.8 V 提供电源。

VSSA_USB：USB 物理层接口电源的接地端。

VDDS_PLL_DDR：DDR 锁相环供电电源，为 1.8 V。

VDDS_DDR：为 DDR 模块接口(DDR I/O)提供电源，典型值为 2.1 V。

VSS：处理器数字地。

RESERVED：保留，用于测试输出。

VDDS_OSC：为系统晶体振荡器提供电源。

RTC_KALDO_ENn：RTC 内部 LDO 电源使能引脚(低电平有效)。当该引脚接低电平时，处理器内部 LDO 给实时时钟 RTC 核提供电源；当该引脚接高电平时，需要外部电源给 RTC 核供电。

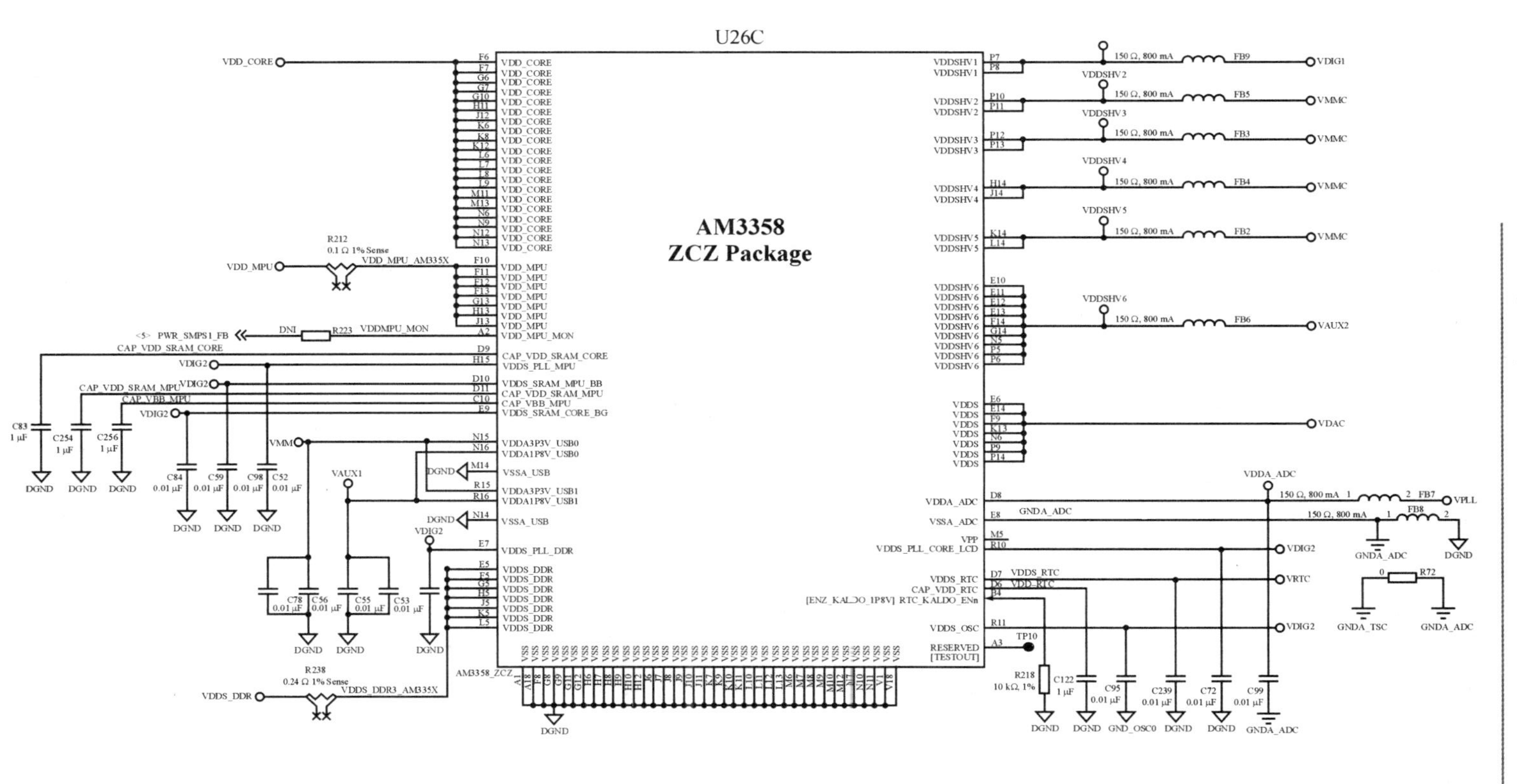

图2.16　MCU(POWER)

CAP_VDD_RTC：连接 RTC 提供电源的输出电容器。

VDDS_RTC：为 RTC(实时时钟)提供电源。

VDDS_PLL_CORE_LCD：为内核锁相环(CORE PLL)、LCD 锁相环(LCD PLL)提供电源。

VPP：为只读存储器熔丝(FUSE ROM)提供电源。

VDDA_ADC：为模/数转换器 ADC 提供电源。

VSSA_ADC：为模/数转换器 ADC 电源的接地端。

VDDS：为所有双电压输入/输出接口(all dual-voltage I/O domains)外设提供电源，推荐值为 1.8 V。

VDDSHV[1:6]：为双电压输入/输出接口外设提供电源，共 6 路。推荐值为 1.8 V、3.3 V。

2.4.2 时钟电路

AM335x 器件有两个时钟源输入，包括主时钟和 RTC 实时时钟。它们可使用外部晶振(无源晶振)或外部时钟源(有源晶振)，主时钟的时钟频率可以选择 19.2 MHz、24 MHz、25 MH 或 26 MHz，RTC 实时时钟通常选择 32.768 kHz。处理器可通过内部 PLL 电路倍频，使系统运行速度更高，CPU 最高操作时钟频率可达通常所说的 600 MHz、800 MHz 或 1 GHz。

如图 2.17 所示为 AM335x 处理器 24 MHz 主时钟和 32.768 kHz 的 RTC 实时时钟的外围电路，以及与处理器引脚的原理图。

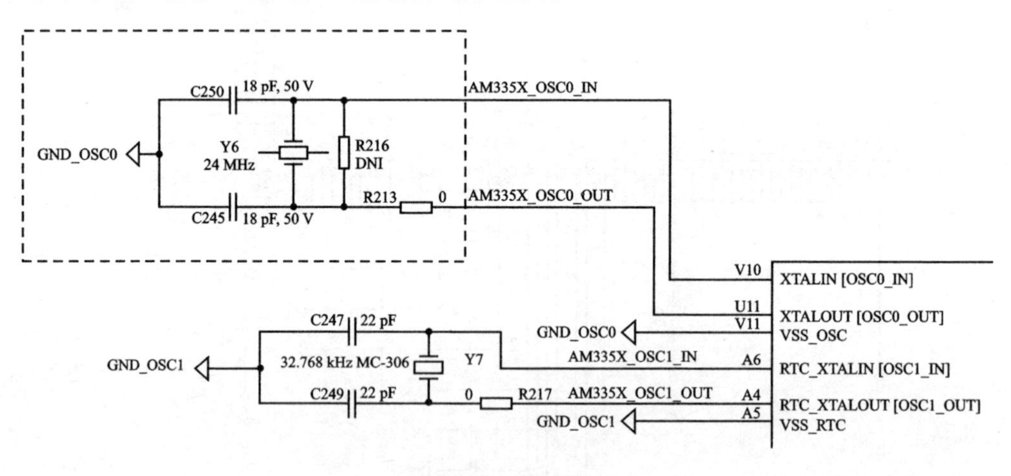

图 2.17 时 钟

图 2.17 中 Y6 为 24 MHz 的外部晶振，电容 C250 和 C245 为匹配电容，取值范围通常在 15～22 pF，具体取值要根据晶振厂家给出的负载电容值通过公式计算得到，严格讲是需要咨询晶振厂家的，但通常我们对它的要求不是很高，这里取 18 pF。

在晶振两端的并联电阻 R216(通常可取 1 MΩ、5.1 MΩ 或 10 MΩ)可以保证晶振更容易起振或避免起振延迟。串联的 R213 电阻,常用来预防晶振被过分驱动,晶振过分驱动将逐渐损耗减少晶振的接触电镀而引起频率的上升,并导致晶振的早期失效。这里 R213 取 0 Ω,可以根据实际情况焊接适量阻值的电阻。详细情况也可以参考数据手册的第 107 页“6.2.2 Input Clock Requirements”。

RTC 实时时钟部分原理图 Y7、C247、C249 和 R217 的原理、作用和主时钟部分是一样的,这里不再重复叙述,取值范围及要求等详细情况请参考数据手册的第 107 页“6.2.2 Input Clock Requirements”。

2.4.3 上电引导模式配置

AM335x 的 Boot 引导配置引脚和 LCD 数据总线共用,也被称作 SYSBOOT 引脚。在电源上电复位完成时,LCD_DATA[15:0]引脚分别作为 SYSBOOT[15:0]功能被锁定为配置定义。如图 2.18 所示为启动配置定义,各引脚通过 100 kΩ 电阻上拉到电源,通过 10 kΩ 电阻下拉到地。图 2.18 中 DNI 为空,因此 Boot Configuration 的值为 0100 0000 0011 0111b,该值被定义为复位启动时检查启动设备的顺序,MMC0→SPI0→UART0→USB0,也就是说当 AM335x 处理器复位后进入引导模式时,先检查 MMC0(μSD)接口是否可以正常启动,如果正常则从 MMC0 中启动,反之则继续检查 SPI0 接口。如果此时有 SPI Flash,则从 SPI Flash 中引导启动,否则再检查 UART0,一直检测到 USB0 接口设备。

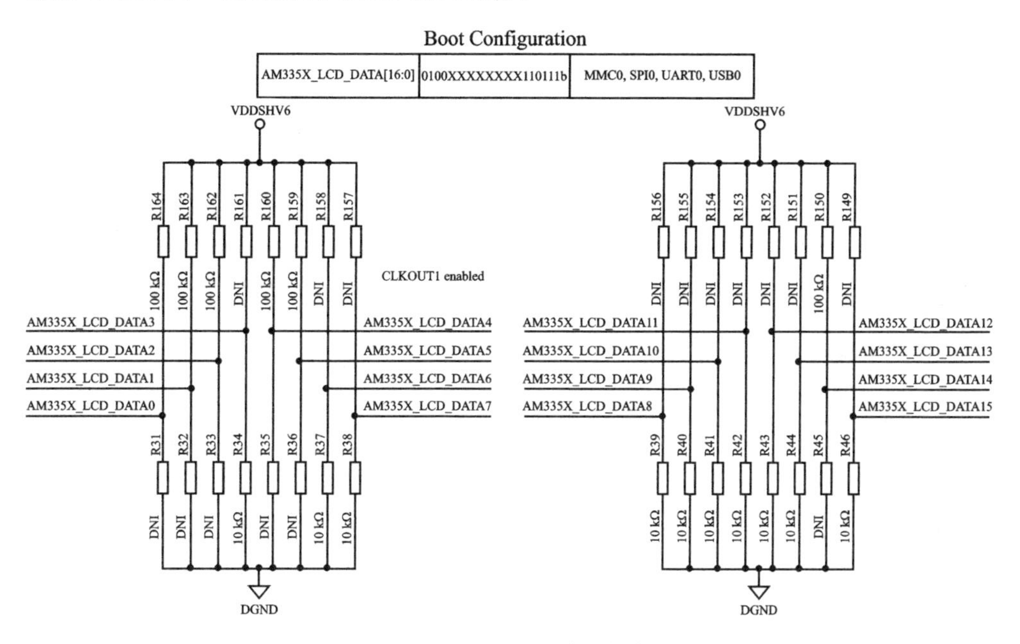

图 2.18 AM335x 启动配置定义

AM335x 详细的启动引导过程，请参考后续的相关章节。

2.4.4 JTAG 接口电路

任何一款 ARM 处理器都会内置有标准 JTAG 接口的 Embedded ICE 跟踪调试模块，AM335x 也是。用户可以使用 XDS100v2 等硬件仿真器通过 JTAG 接口对处理器实时在线仿真调试，也可使用仿真器或其他烧写工具通过 JTAG 接口对其外扩的 Flash 存储器进行在线编程等。

而 S3C2410A 处理器及原理图中的 JTAG 接口为标准的 ARM 公司定义的 20 针 JTAG 接口，可以被任何 ARM 公司及第三方开发工具支持。但 TI 有自己定义的 JTAG 接口，被 TI 自己的 DSP/ARM 开发工具所支持，与 ARM 公司的 JTAG 接口在信号功能上是一样的，只是引脚数和信号排列不一样而已。当然，目前 TI 公司出厂的仿真器(XDS100v2、XDS560v2 等)也都有可选的支持接口，如 TMDSEMU100v2U-ARM 为支持 ARM 公司的 JTAG 接口的 XDS100v2 仿真器，TMDSEMU100v2U-20T 为支持 TI 公司的 20 针 JTAG 接口的 XDS100v2 仿真器。ARM 使用标准的 20 针 JTAG 接口，读者可参考《ARM Linux 自学体验》，也可以上网或通过其他资料查找。如图 2.19 所示为 TMDSSK3358 原理图中的 JTAG 接口图。

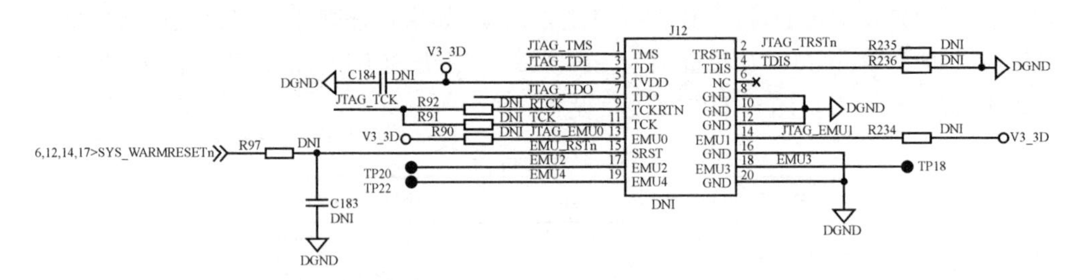

图 2.19 Pin JTAG 接口定义

注：在标准的 ARM JTAG 接口电路中，TMS、TDI、TCK 等都需要接 10 kΩ 的上拉电阻，但 TI 的接口电路中并没有接。

2.4.5 μSD 卡接口

TMDSSK3358 可能是为了降低成本而没有在板上设计 NAND Flash，而直接利用 AM335x 处理器的 MMC0 接口扩展出一个 μSD 卡接口座，让用户插入 μSD 卡，将程序烧写保存到 μSD 卡中。当处理器复位后，从 μSD 卡中引导启动系统。ARM9、ARM11 的处理器往往都是将程序保存到 NAND 中；但 Cortex-A8、Cortex-A9 的启动方式更多且灵活，往往直接将其程序保存到 SD 卡等存储介质中引导启动。

目前 μSD(即 TF)卡广泛应用于数码相机、GPS 导航等设备,身边的一些朋友也经常将 SD 卡适配器作为大容量的 U 盘使用。SD 卡有 SD bus 和 SPI bus 两种通信模式(前者有 4 条数据线 DATA0～DATA3,而后者只有 1 条数据线 DATA0),就是说处理器可以通过 SD bus 或 SPI bus 对 SD 存储卡进行读/写。SPI bus 在没有集成 SD 卡控制器的处理器(如 8 位 MCU)上使用得较多,而 AM335x 等高端处理器通常集成了 SD 卡控制器的 MMC 接口,它通过 4 位数据线的 SD bus 与 SD 卡接口,如图 2.20 所示为它们的接口电路。J4 为插入 μSD 存储卡的卡座(在画 PCB 时要注意,SD 卡座有很多种,一定要确认使用的卡座封装),AM335X_MMC0_CLK 为 SD 卡控制器提供的时钟信号,AM335X_MMC0_CMD 为双向的命令/应答信号,AM335X_MMC0_D[3:0]为 4 位数据信号,SD 卡可以只需 1 位数据 DTA0 实现通信,上电复位后的默认方式就是 1 位,控制器在初始化时可以改变该总线宽度。卡座的 CD 引脚(AM335X_SPI0_CS1)为卡插入检测信号,当卡座上有卡插入时,它将与接地的引脚短接使之为低电平,处理器可以通过外部中断来判断 SD 卡的插入。这里没有将 WP 信号引出(WP 为写保护检测信号,当 SD 卡写保护时,它也将与接地的引脚短接)。另外,SD 卡的每个信号都接 10 kΩ 的上拉电阻,而 TPD6E001 为 ESD 保护器件。

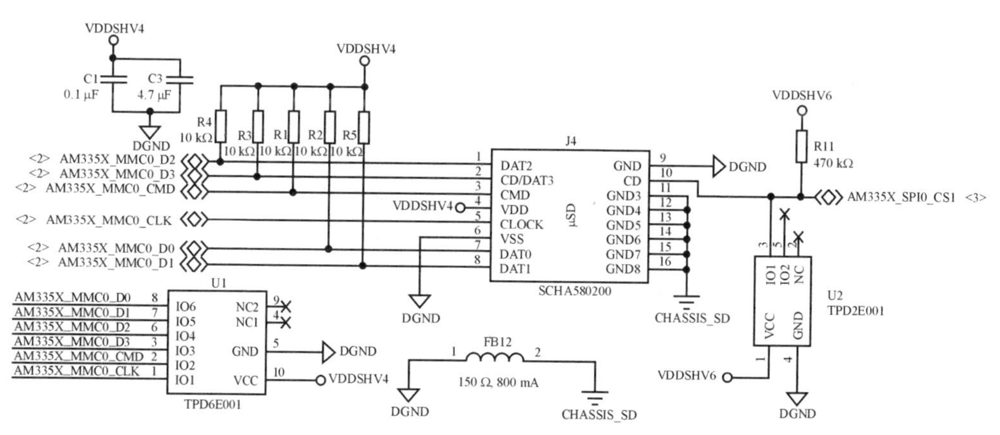

图 2.20 SD 卡接口

2.4.6 DDR3 SDRAM 存储器

AM335x 集成了 mDDR(LPDDR)、DDR2、DDR3、DDR3L SDRAM 内存控制器及接口,所以用户不需要额外地增加 DDR3 SDRAM 控制器,就可以和 DDR3 SDRAM 存储器相连接。如图 2.21 所示为 TMDSSK3358 扩展 DDR3 SDRAM 的接口原理图,如图 2.22 所示为 DDR3 SDRAM 存储器信号的终端电阻及参考电压等电路。MT41J128M16JT - 125 为 Micron(美国镁光)半导体有限公司的 DDR3 SDRAM 存储产品,单片容量为 2 Gb(256 MB),数据长度为 16 位,即 128 Mb×16。

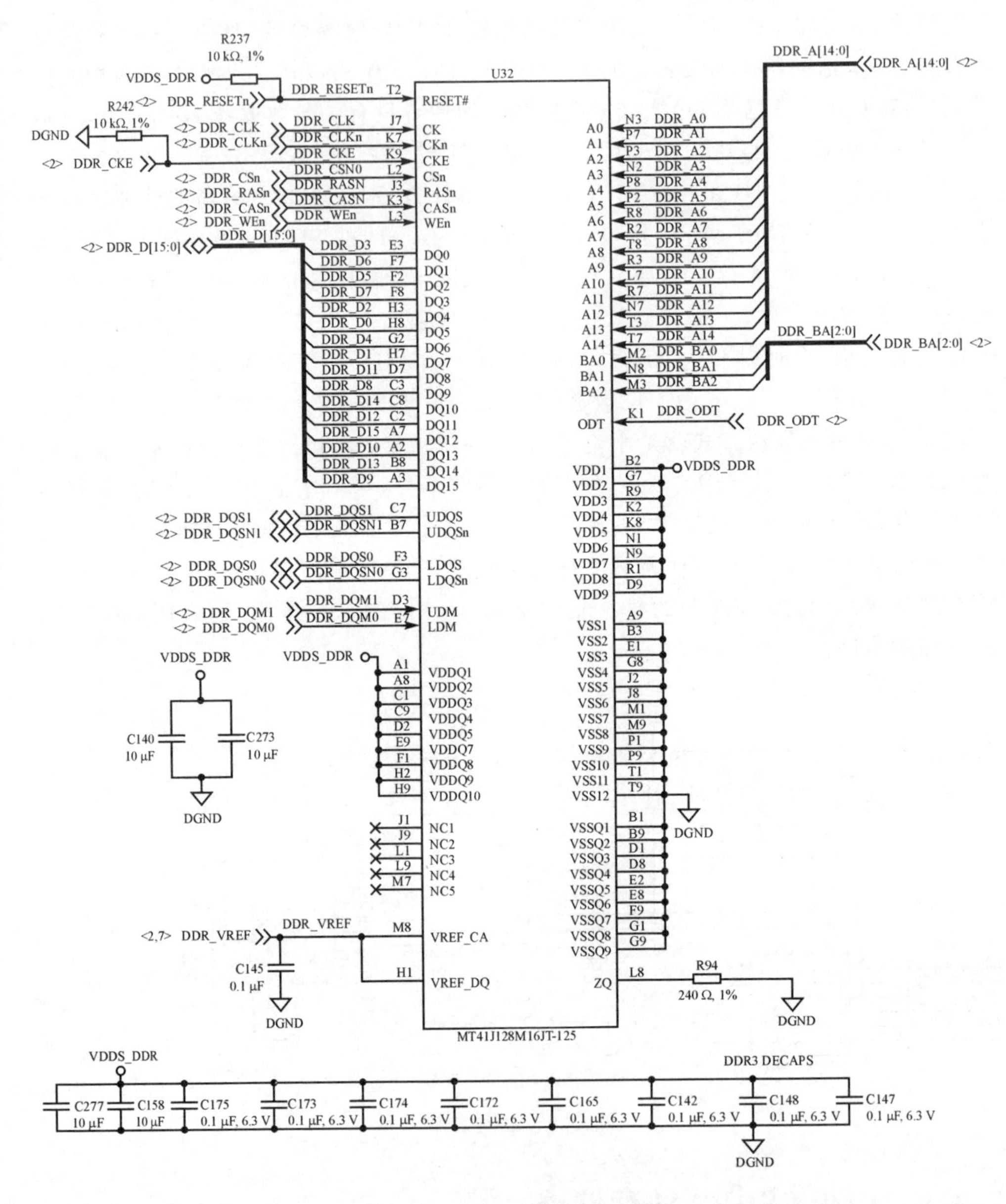

图 2.21 DDR3 SDRAM 存储器

AM335x 的内存起始地址被固定为 0x80000000。

这里的 DDR3 SDRAM 为 16 位数据长度，常见的还有 32 位，或者 2 片 16 位组成 32 位，或者 4 片 8 位组成 32 位。

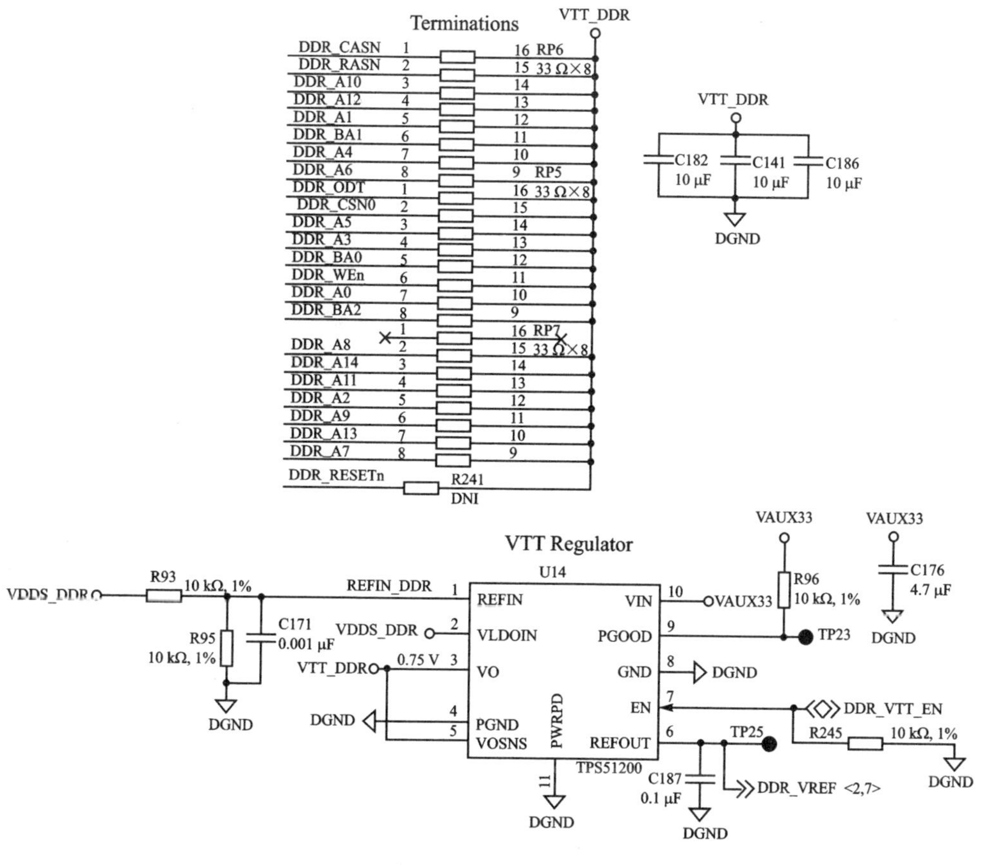

图 2.22　DDR3 SDRAM 存储器补充电路

2.4.7　调试串口

AM335x 集成了 6 个 UART (Universal Asynchronous Receiver and Transmitter)，在实际应用中，通常将 UART0 作为调试串口。目前由于 PC 主板的串口使用得越来越少，所以新配的计算机基本不带 RS232 串口。为使开发者无需额外的 USB 转串口调试器，所以 TMDSSK3358 板上带有 USB 转串口的芯片 FT2232HL，如图 2.23 所示。

电路中将 UART0 的 RTSn 和 CTSn 信号引出以支持自动流控制(AFC)的应用，在 AFC 中，RTSn 由接收器的接收情况来控制，CTSn 则控制了发送器的工作。

FT2232 是一款 USB 到 UART/FIFO 的转换电路，FT_DM、FT_DP 分别为 USB 的数据线 DATA－、DATA＋，连接到 USB 的 HUB 芯片 USB2412，最后才连接到 USB 的接口座，如图 2.24 所示。

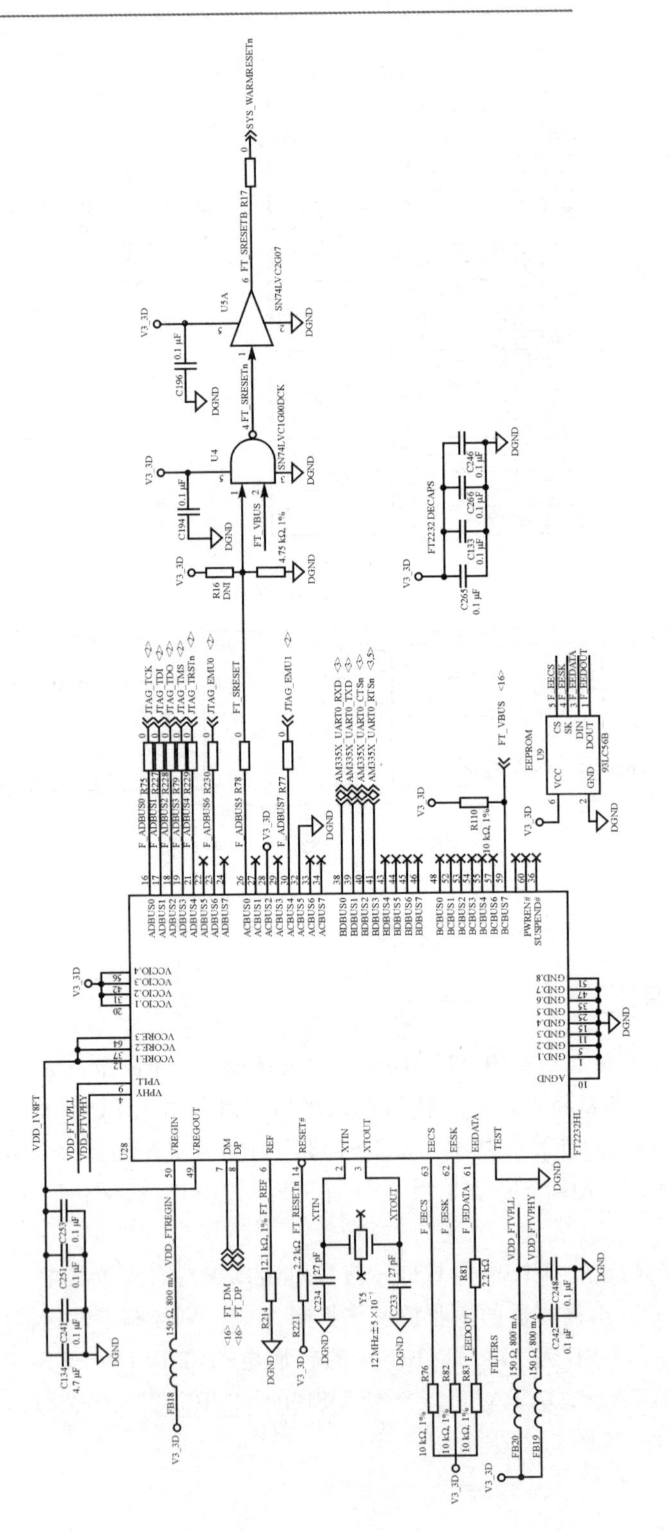

图2.23　调试串口UART0

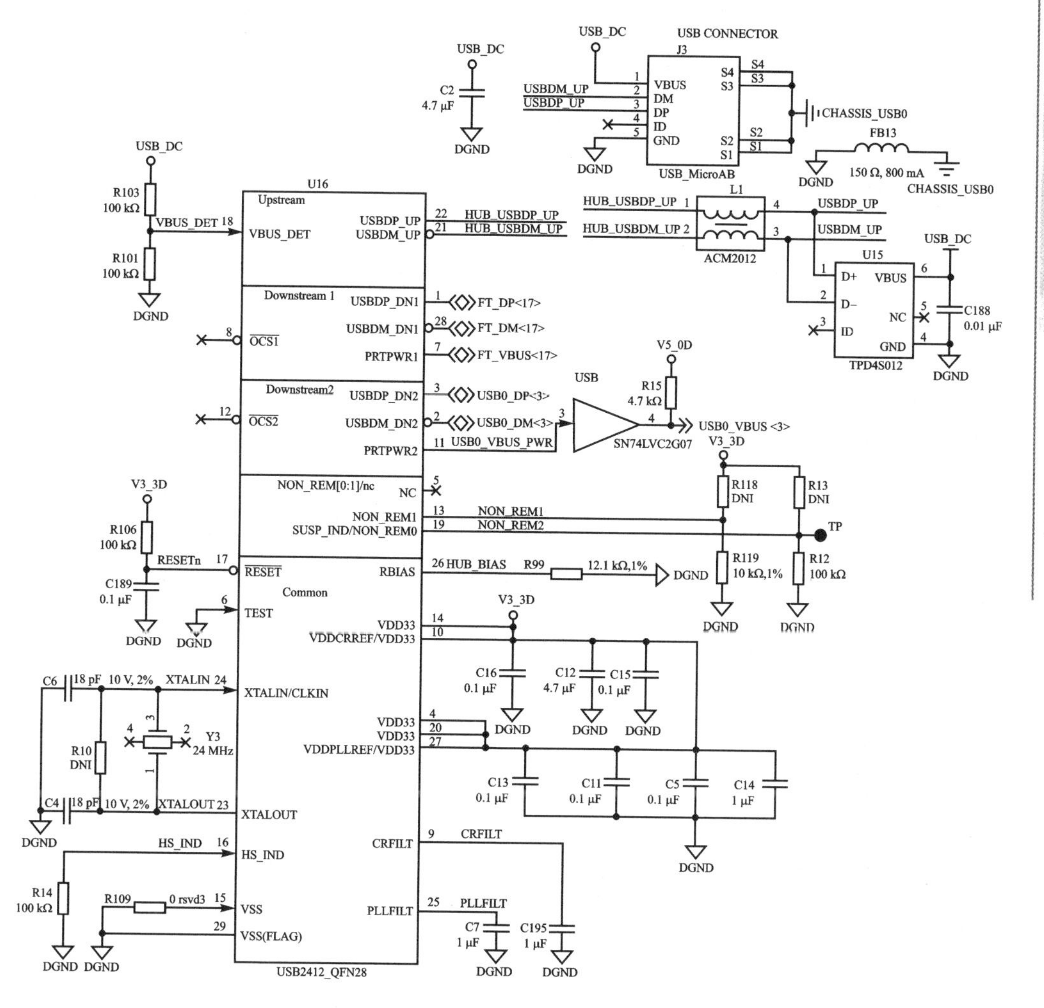

图 2.24 USB HUB 接口电路

2.4.8 以太网接口

在 Linux 的开发及应用中网络是非常重要的，Linux 与前后台系统相比，很重要的优势就在于对网络的支持，同时网络在 Linux 的开发调试过程中也起到了举足轻重的作用（如可以在宿主机与目标板间下载文件系统等几兆甚至几十兆的大文件，建立 NFS 网络文件系统，使用 ztelnet 登录传送文件等，给交叉调试带来极大的便利）。而且，目前双网口在嵌入式系统中也越来越普遍，可以有效提升产品的网络性能。而 AM335x 就集成了两个 10M/100M/1 000M 的 EMAC 以太网控制器，支持 GMII、RGMII 和 RMII 媒体接口。TMDSSK3358 板上支持两个以太网接口，如图 2.25 所示。

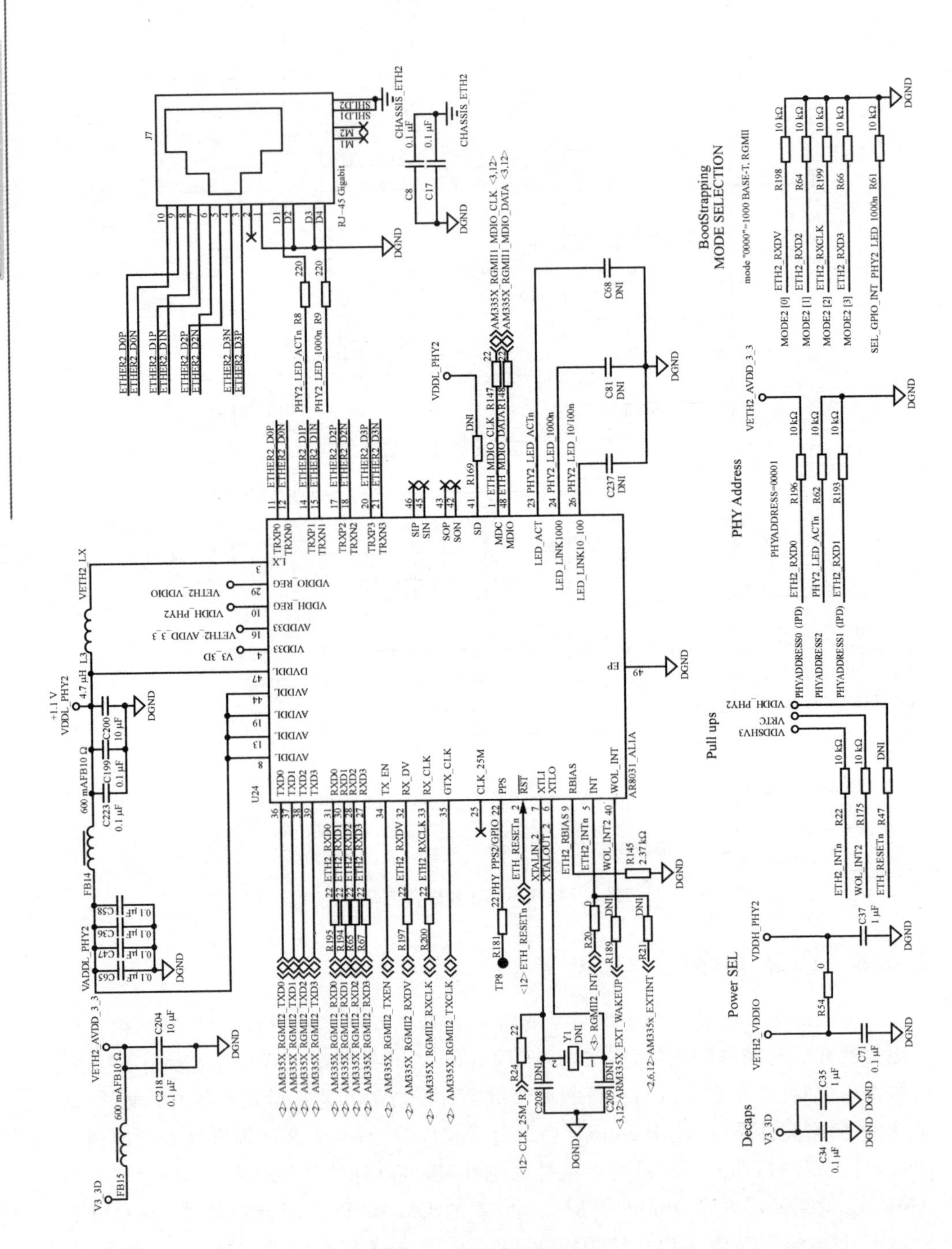

图 2.25　以太网接口电路

AR8031 为千兆网络的接口芯片，虽然 AM335x 已经集成网络控制器，但还是需要在物理层上有网络接口芯片。TMDSSK3358 的 Linux SDK(Software Development Kit)中已经包含了网络及 AR8031 的双网口驱动程序，所以我们不需要研究它的驱动及过多的网络知识就可以使用和学习基于网络应用上的编程。

AM335X_RGMII1_TXD[0:3]：为 AM335x 处理器以太网端口 1 的发送。

AM335X_RGMII1_RXD[0:3]：为 AM335x 处理器以太网端口 1 的接收。

ETHER_D[0:3]P：为网口 1 连接到网络变压器端的发送/接收信号的正端，与 ETHER_D[0:3]N 一起构成差分对信号。

ETHER_D[0:3]N：为网口 1 连接到网络变压器端的发送/接收信号的负端，与 ETHER_D[0:3]P 一起构成差分对信号。

AM335X_RGMII2_TXD[0:3]：为 AM335x 处理器以太网端口 2 的发送。

AM335X_RGMII2_RXD[0:3]：为 AM335x 处理器以太网端口 2 的接收。

ETHER2_D[0:3]P：为网口 2 连接到网络变压器端的发送/接收信号的正端，与 ETHER_D[0:3]N 一起构成差分对信号。

ETHER2_D[0:3]N：为网口 2 连接到网络变压器端的发送/接收信号的负端，与 ETHER_D[0:3]P 一起构成差分对信号。

另外，还有其他信号引脚请读者直接参考原理图，AM335x Starter Kit 支持双网口。另一块网络接口芯片电路图与此基本相同，也请读者自行参考相关原理图。

2.4.9 按键 GPIO

如图 2.26 所示为按键 GPIO 电路。按键 GPIO_KEY[4:1]连接 AM335x 处理

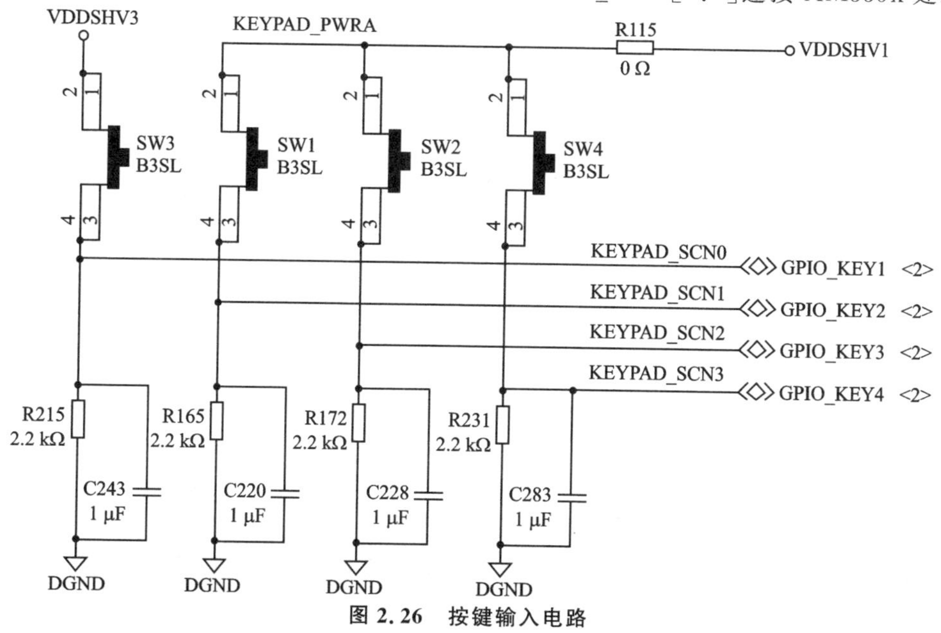

图 2.26　按键输入电路

器的普通 GPIO 端口,我们可用这 4 个按键完成普通 GPIO 口的输入实验和外部中断实验,或者其他应用中的按键触发。当没有按键按下时,GPIO 端口线为高电平,按下后为低电平。

2.4.10 LED 显示

如图 2.27 所示为 GPIO 驱动的 LED 显示电路。AM335X_GPIO_LED[1:4]分别驱动 4 个场效应晶体管 BSS138,当然这里也可以用普通的三极管替代。

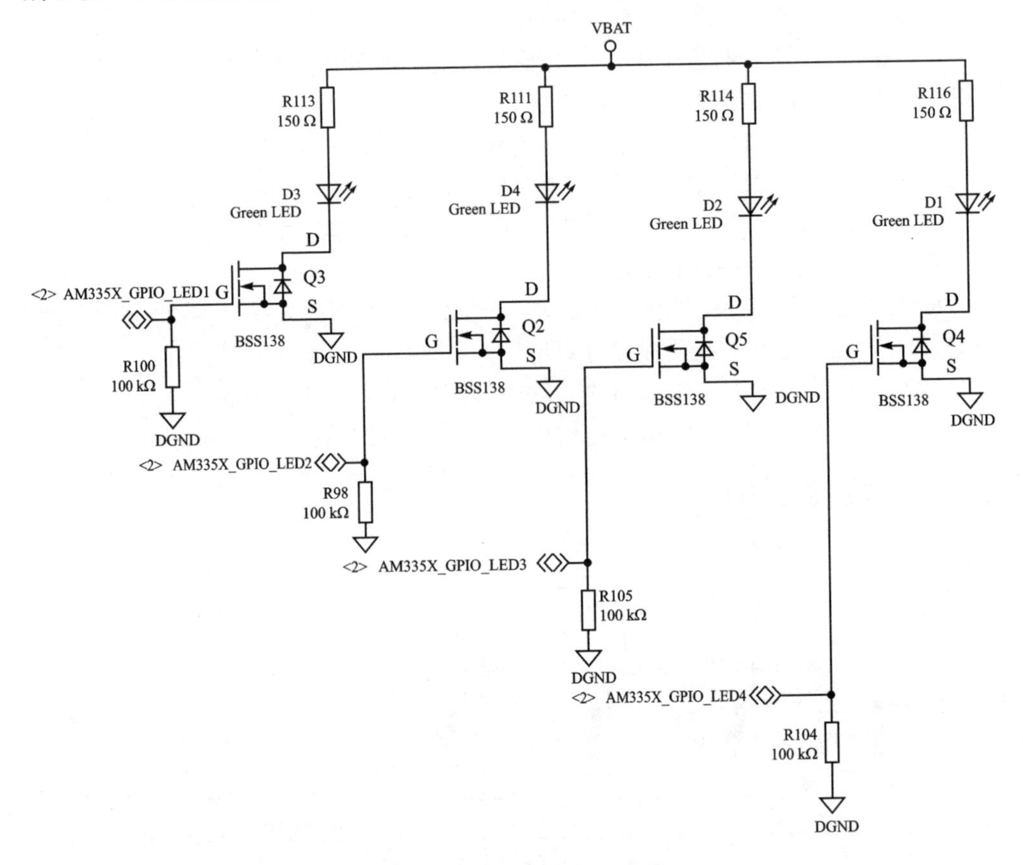

图 2.27 用户 LED 显示电路

2.4.11 IIC 总线的 EEROM 存储器

如图 2.28 所示为 AM335x 通过 IIC 总线 AM335X_I2C0_SCL、AM335X_I2C0_SDA 信号扩展的 EEROM 存储器 CAT24C256W 和加速度计 LIS331DLH。CAT24C256W 容量为 256 Kb,即 32 KB。这里 TMDSSK3358 板用 EEROM 存储器保存一些如板子版本号、设备 ID 等信息。LIS331DLH 为 3 轴加速度计,为 MEMS 数字输出运动传感器。

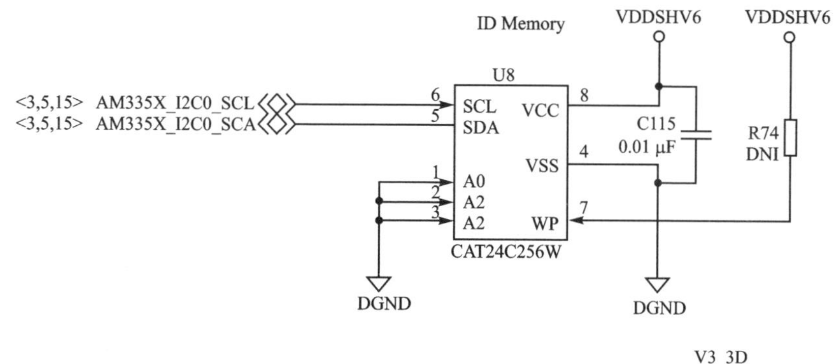

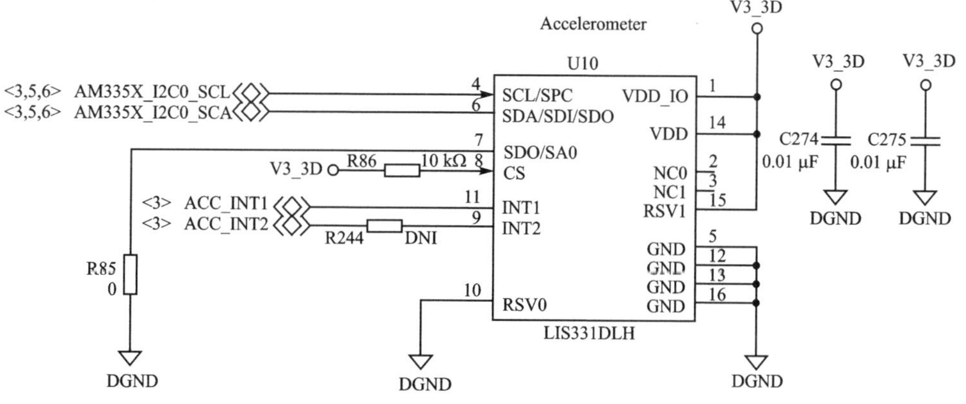

图 2.28　IIC 接口扩展 EEROM 和加速度计电路

2.4.12　复位电路

如图 2.29 所示，为 AM335x 处理器的复位信号输入端，AM335x 的 WARMRSTn 输入端口处通过一个 10 kΩ 的上拉电阻连接到 VDDSHV6 电源。

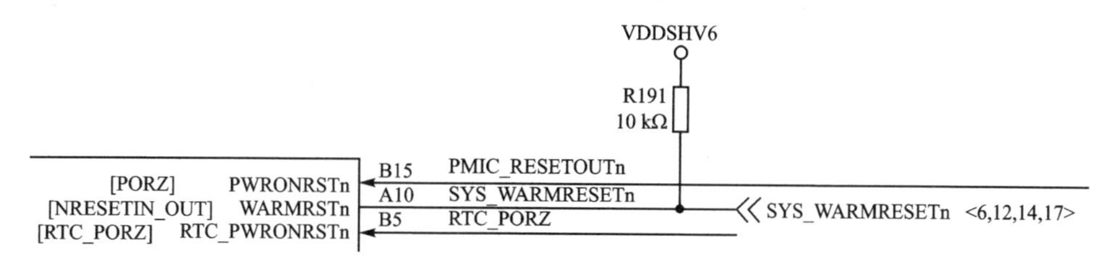

图 2.29　ARM335x 处理器复位信号输入端

如图 2.30 所示，SW7 为复位按键，它连接到图 2.29 中 AM335x 的复位输入端。C279 可与图 2.29 中 R191 组成阻容方式的复位电路，当系统刚上电时，通过电容的充电延时完成处理器的复位。平时，当复位按键 SW7 按下时，一低电平有效的复位信号会使处理器复位。

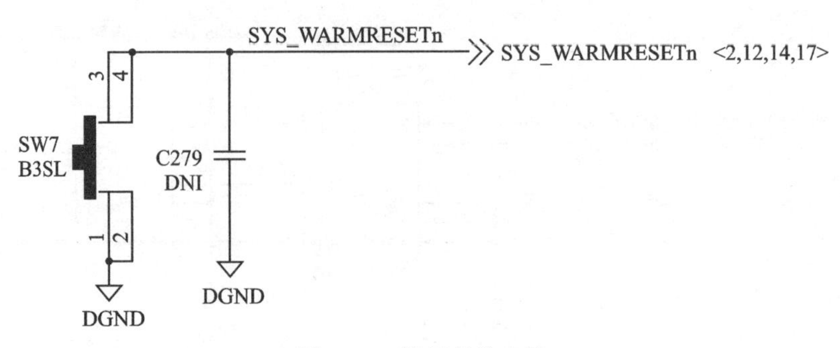

图 2.30　按键复位电路

2.4.13　按键中断输入

如图 2.31 所示，SW6 按键可以控制 AM335x_EXTINT 信号的高电平或低电平，以完成处理器的外部中断。

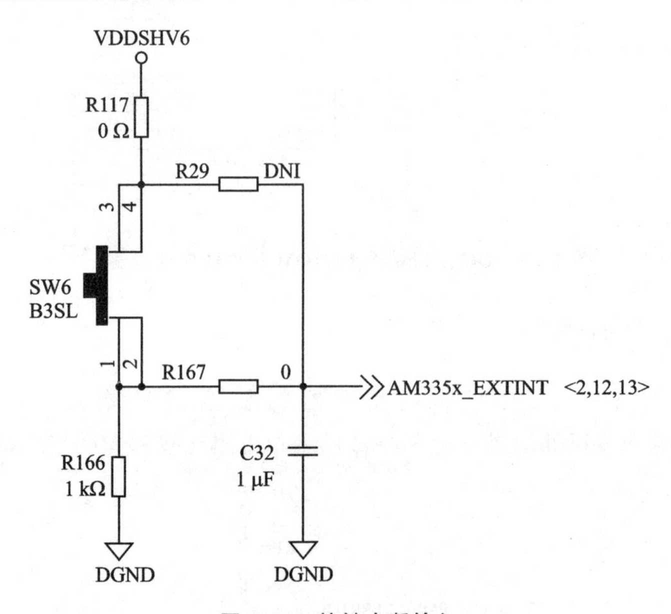

图 2.31　按键中断输入

2.4.14　电源输入及 PMIC 电源管理芯片

参考图 2.16，AM335x 处理器电源部分的原理图主要的电源有：VDD_CORE、VDD_MPU、VDDS_PLL_MPU（VDIG2）、VDDS_SRAM_MPU_BB（VDIG2）、VDDS_SRAM_CORE_BG（VDIG2）、VDDS_PLL_DDR（VDIG2）、VDDS_OSC（VDIG2）、VDDS_PLL_CORE_LCD（VDIG2）、VDDA3P3V_USB0（VMMC）、VDDSHV5（VMMC）、VDDSHV4（VMMC）、VDDSHV3（VMMC）、VDDSHV2（VMMC）、VDDA3P3V_USB1（VMMC）、VDDA1P8V_USB0（VAUX1）、VDDA1P8V_USB1

(VAUX1)、VDDS _ RTC (VRTC)、VDDA _ ADC (VPLL)、VDDS (VDAC)、VDDSHV6(VAUX2)、VDDSHV1(VDIG1)。

其中,VDD_CORE 和 VDD_MPU 是 AM335x 处理器的内核电压,非常重要,直接决定了处理器能否正常工作。另外,VDD_CORE 和 VDD_MPU 的电流通常都非常大,在处理器不同的工作状态(如正常工作,待机等)和不同的工作频率下电压值可以做相应的调节(1.1 V 或 0.95 V,最大不超过 1.5 V),所以通常情况下都由 DC/DC 单独提供,并做动态调节。

VDDS_PLL_MPU、VDDS_SRAM_MPU_BB 等,即上述由"(VDIG2)"标识的因电流都很小,要求也不高,故可共用一路 LDO,通常为 1.8 V,最大不超过 2.1 V。

VDDA1P8V_USB0、VDDA1P8V_USB1 可以共用,通常也为 1.8 V,最大不超过 2.1 V。

VDDS_RTC 为 RTC 时钟,由单独的 LDO 提供,通常为 1.8 V,最大不超过 2.1 V。

VDDA_ADC 为模/数转换控制器的电源,并为模/数转换提供参考电压,由单独的 LDO 提供,通常为 1.8 V,最大不超过 2.1 V。

VDDS 由单独的 LDO 提供,通常为 1.8 V,最大不超过 2.1 V。

VDDSHV6 由单独的 LDO 提供,通常为 3.3 V,最大不超过 3.8 V。

VDDSHV1 由单独的 LDO 提供,通常为 3.3 V,最大不超过 3.8 V。

如图 2.32 所示为外部 5 V 主电源通过电源座输入、系列保护及滤波后连接到 VBAT 端。

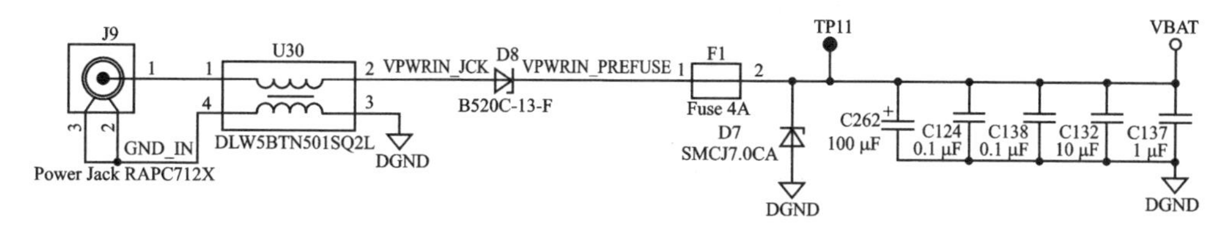

图 2.32　5 V 外部电源输入

如图 2.33 所示为整个板上的 PMIC 电源管理芯片 TPS65910A3,共 3 路 DC/DC 和 9 路 LDO,它提供了 TMDSSK3358 板上的大部分电源。

如图 2.34 所示为 RTC 提供的 LDO 电源管理芯片 TPS71718。

如图 2.35 所示为电源的启动按键,EXP_PB_POWERON 连接 PMIC 的 PWRON 引脚,用于启动电源。

如图 2.36 所示为板上的 1.8 V LDO 电源输出。

如图 2.37 所示为板上的 3.3 V LDO 电源输出。

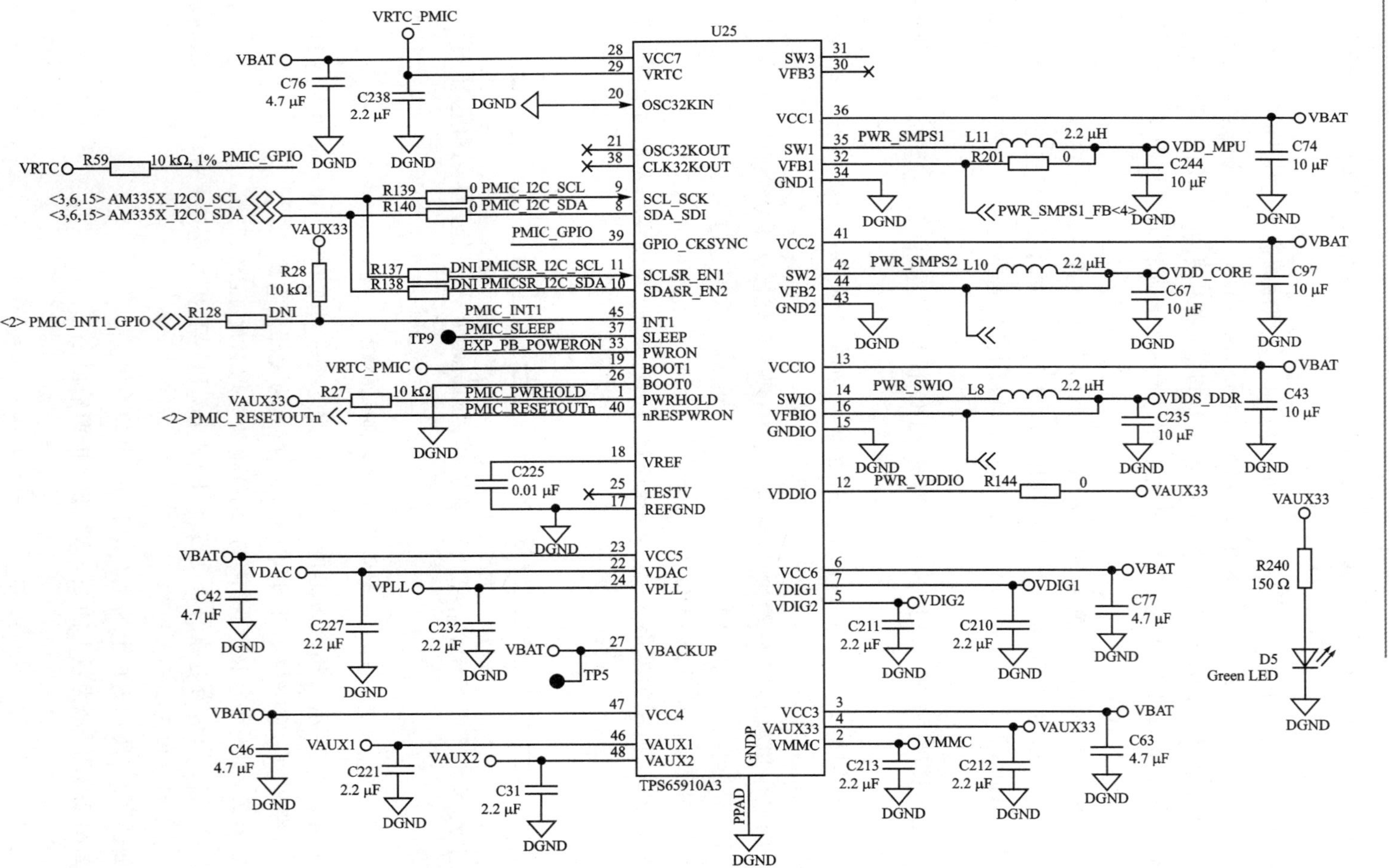

图2.33　PMIC电源管理

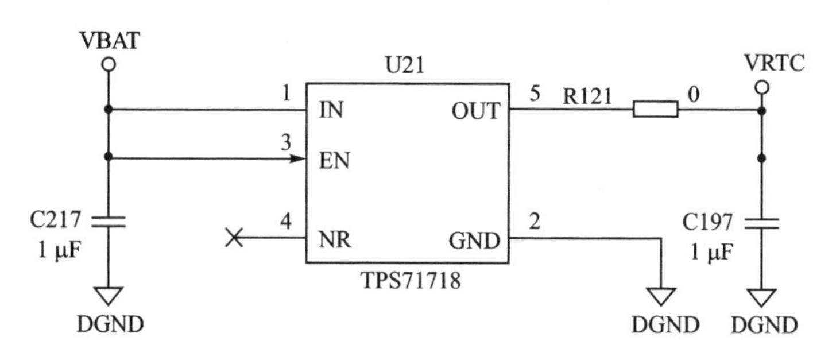

图 2.34　RTC LDO 电路

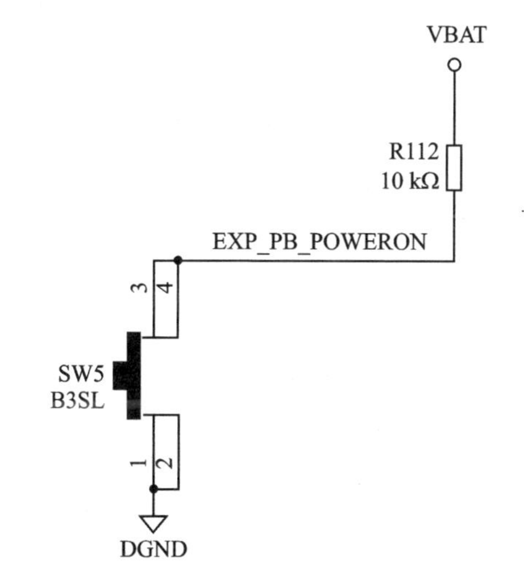

图 2.35　电源启动按键电路

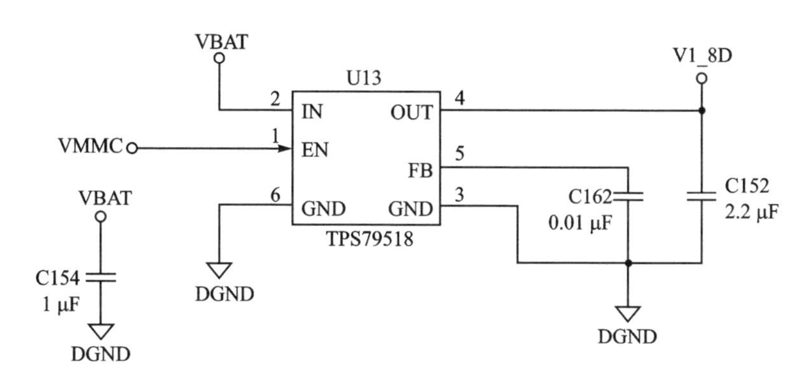

图 2.36　板上 1.8 V LDO 电路

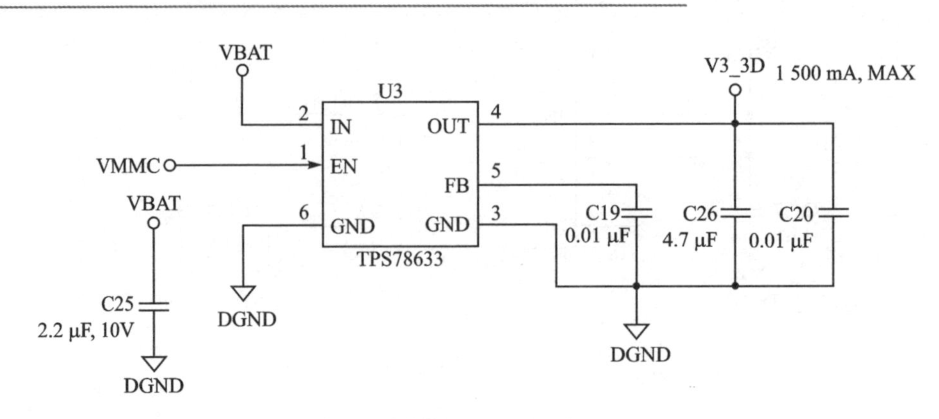

图 2.37　板上 3.3 V LDO 电路

2.4.15　Wi-Fi 和蓝牙模块

如图 2.38 所示为集成 Wi-Fi、蓝牙双功能的 LBEE5ZSTNC－523 模块接口电路。Y2 为输出 32.768 kHz 时钟的有源晶振，MM8130－2600 为 RF 连接器。

如图 2.39 所示为集成 Wi-Fi、蓝牙双功能的 LBEE5ZSTNC－523 模块的电源电路。

这里还需要特别注意，因为 LBEE5ZSTNC－523 模块的接口电压为 1.8 V，而 AM335x 的接口电压为 3.3 V，所以使用了 SN74AVC4T245 总线缓冲器件作为 1.8 V 和 3.3 V 接口电平的转换，如图 2.40 所示。

2.4.16　USB Host/Device

AM335x 集成了两个高速的 USB 2.0 OTG 端口，都可作为 HOST（主机或称主设备）和 DEVICE（设备或称从设备），如图 2.41 所示为它们的接口原理图。

图 2.41 中，J5 为 USB 接口座，TPD4S012 为具有电源钳位的 4 通道 USB ESD 器件，ACM2012 为专门用于高速差分信号线上的滤波器。TPS2051BD 为固定限流开关，由 AM335x 处理器 AM335X_USB1_DRVVBUS 信号输出控制 USB1 的电源 VUSB_VBUS1 是否输出，其中电源 V5_0D 如图 2.42 所示。

TPS63010 为具有 2 A 开关的高效率、单电感器、降压升压转换器。

2.4.17　IIS 音频电路

IIS（Inter-IC Sound Bus）又称 IIS，是飞利浦公司提出的串行数字音频总线协议，该总线专门用于音频设备之间的数据传输。AM335x 集成了 IIS 总线接口，可以直接连接一个外部 8/16 位立体声的音频编/解码器。如图 2.43 所示为 AM335x 与 TI 公司的音频 CODEC（多媒体数字信号编解码器）TLV320AIC3106 的接口电路，用于实现模拟音频信号的采集和数字音频信号的模拟输出，并彼此通过 IIS 数字音频接口实现音频信号的数字化处理。

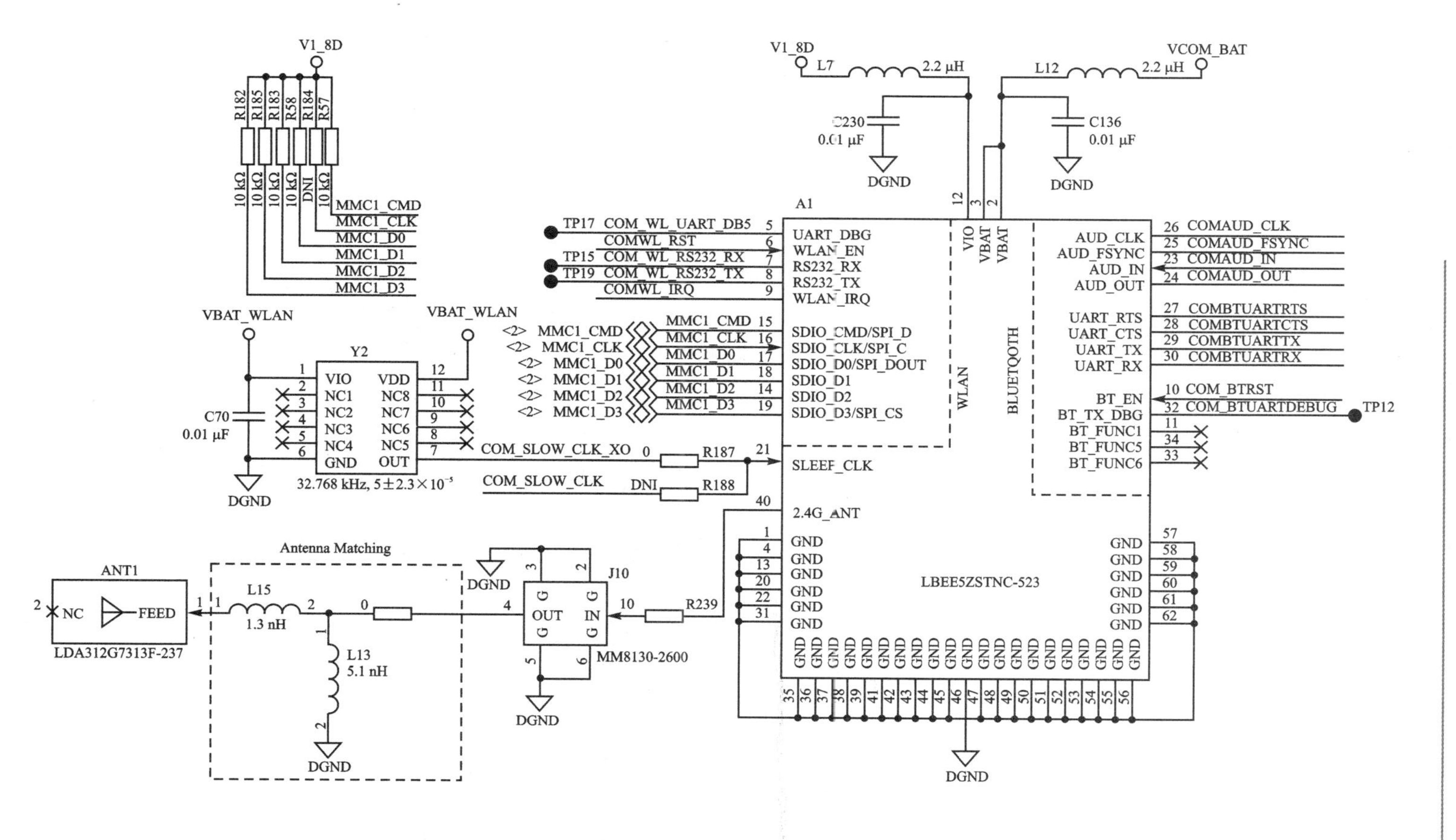

图2.38　Wi-Fi蓝牙模块

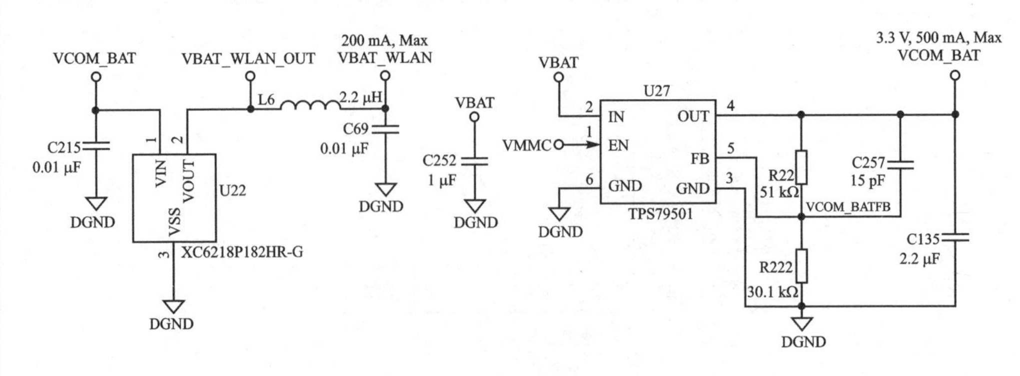

图 2.39　Wi-Fi 蓝牙双功能模块电源

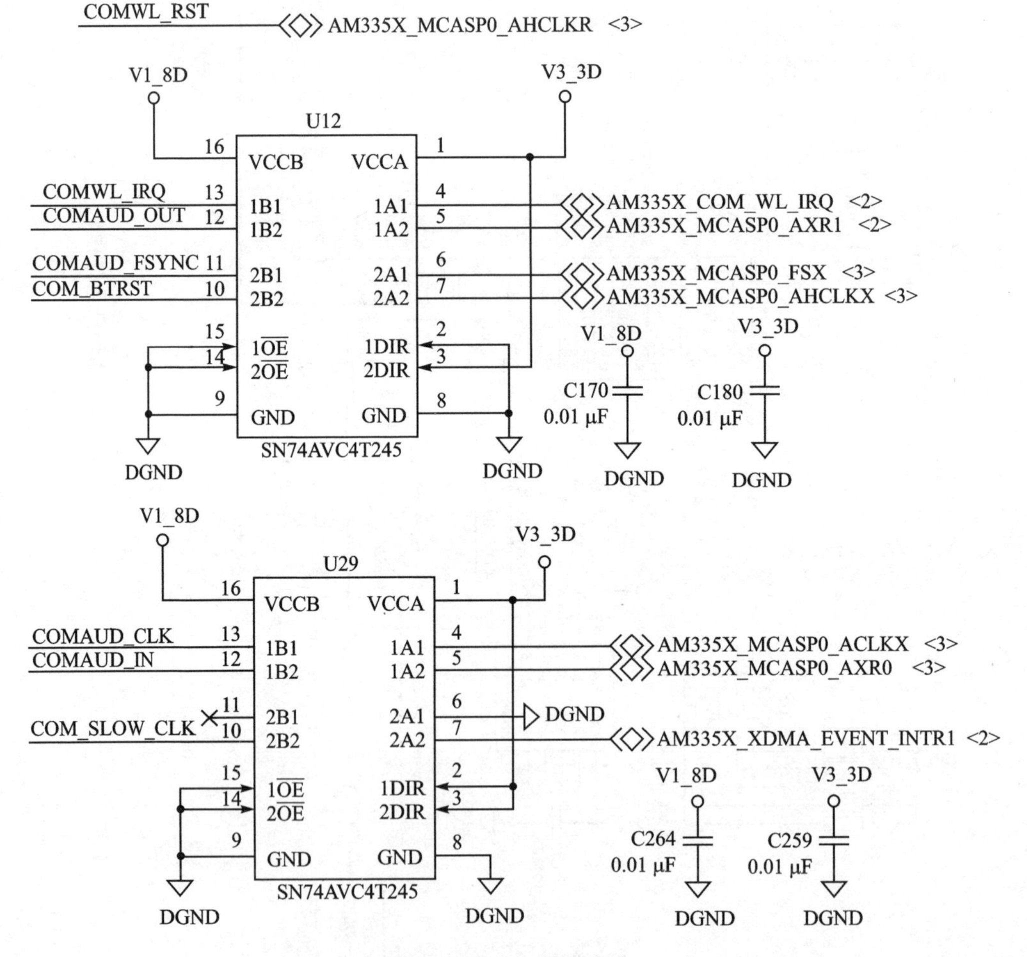

图 2.40　电平转换电路

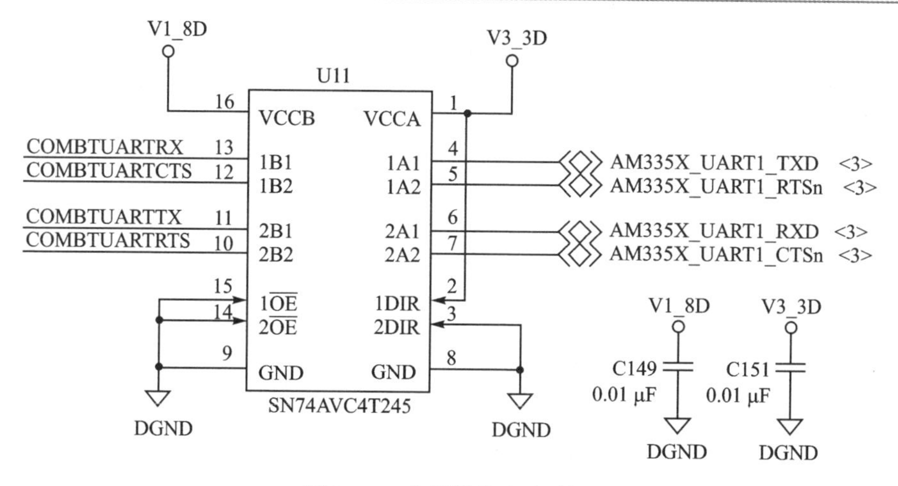

图 2.40 电平转换电路(续)

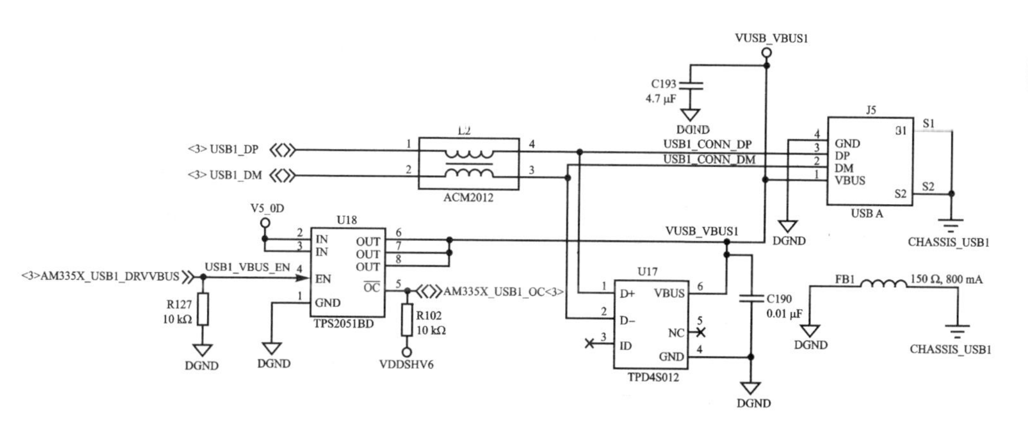

图 2.41 USB 接口原理图

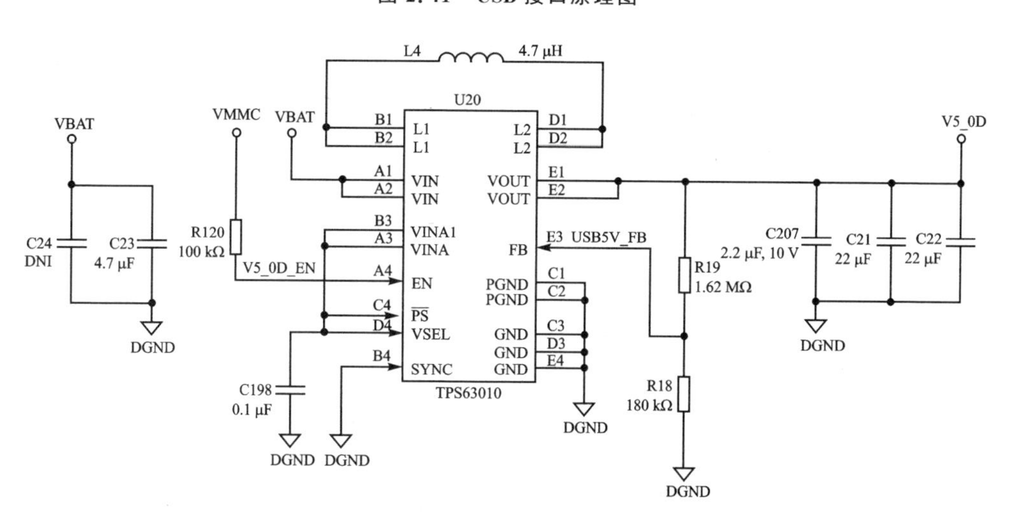

图 2.42 USB 电源(V5_0D)

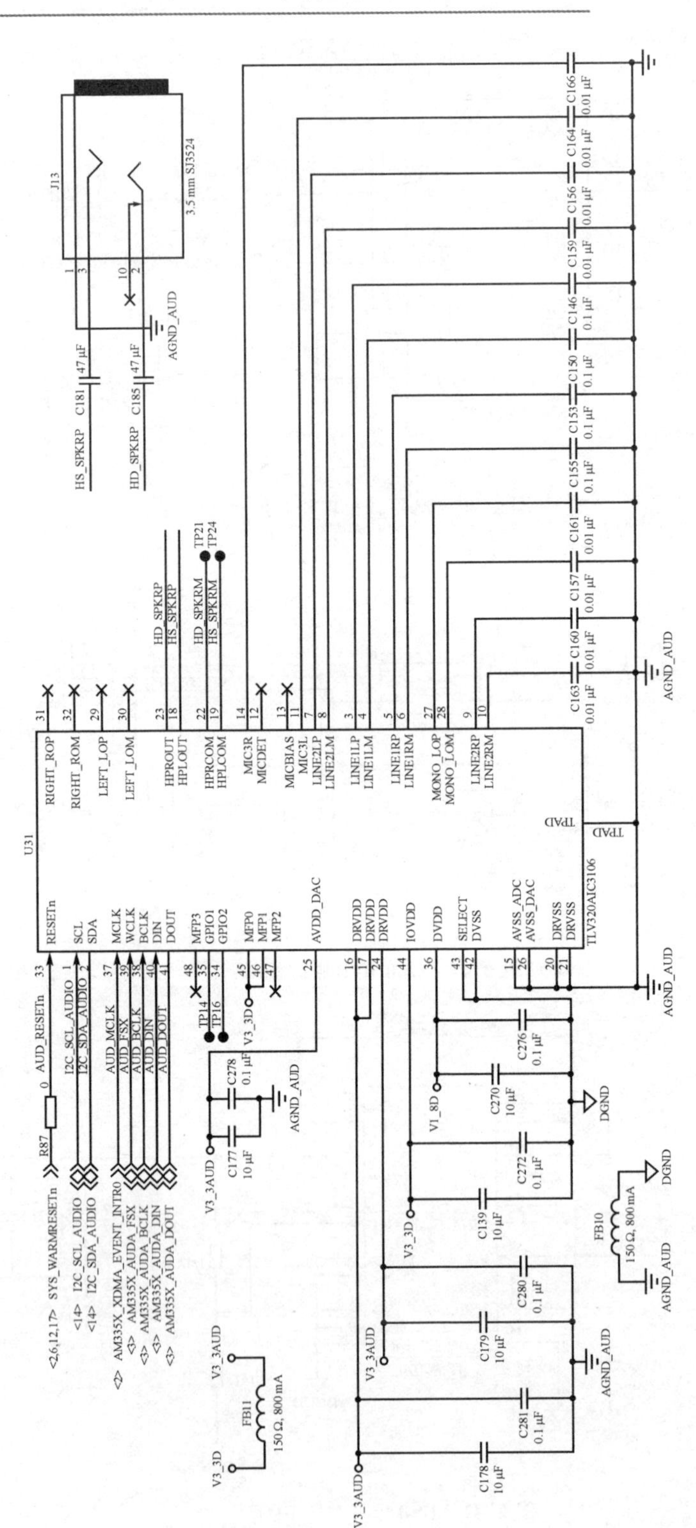

图2.43　IIS音频电路

在实际的项目中，如果不需要编码（录音）功能，可以选择成本更低、使用更加简便的音频 D/A 转换器，如 HT82V731 等。

TLV320AIC3106 共有 4 个 IIS 总线接口控制信号引脚，DI/DOUT（AUD_DI/AUD_DOUT）为音频数据输入/输出，分别与总线的 DI/DOUT（AM335X_AUDA_DIN/AM335X_AUDA_DIOUT）相连；BCLK（AUD_BCLK）为位时钟输入，与 IIS 总线时钟 BCLK（AM335X_AUDA_BCLK）相连，触发每位音频数据的串行移入/移出。WCLK 为字选择输入，与 IIS 的左右通道时钟 LRCK（AM335X_AUDA_FSX）相连，指定当前输入的数据为左声道还是右声道的数据。由于 TLV320AIC3106 工作在从模式下，还需要 AM335x 提供 CODEC 的系统时钟 MCLK，与 AM335x 的 AM335X_XDMA_EVENT_INTR0 相连。HS_SPKRP、HD_SPKRP 为音频的左右通道输出，也是通过隔直通交的电容后从耳机插孔输出。关于 TLV320AIC3106 更详细的引脚描述及介绍请读者参考它的数据手册。

2.4.18　LCD 显示

1. LCD 液晶显示屏接口电路

LCD 液晶显示屏通常按颜色可以分为灰度、黑白和彩色，彩色屏通常还会分成 STN 型和 TFT 型。STN 屏也称伪彩屏，读者经常听到的 256 色屏（或 256 色伪彩屏）指的就是这种，256 色是指每个像素点由 8 位颜色值构成的 256（$2^8=256$）种色彩。TFT 屏也称真彩屏，64K 真彩屏就是指这种，每个像素点由 16 位颜色值构成 64K（$2^{16}=65\,536=64K$）种色彩。AM335x 集成了 LCD 控制器，可支持 256 色和 4 096 色的 STN 屏，也可支持 64K 色和 16M 色的 TFT 屏。读者可能还会听到模拟屏和数字屏，它们的主要区别在与外部控制器的接口上。我们这里指的都是数字屏，是直接输入数字量来表示每个像素的颜色值。而模拟屏输入的是 RGB 三基色模拟信号，在屏的内部再将模拟信号转换成数字量来驱动每个像素的颜色值。

如图 2.44 所示为 AM335x 与 LCD 屏的接口电路。其中，在 AM335x 的各个 LCD 信号输出串接 33 Ω 的电阻。

2. 电容触摸屏接口电路

在《ARM Linux 自学体验》中，我们使用的是电阻式触摸屏，而随着智能手机的普及，触摸屏也从原来的电阻式被更易使用的电容屏所取代。TMDSSK3358 板使用的也是电容式触摸屏，接口电路如图 2.45 所示，该接口为可选项，因为图 2.44 中的 LCD 接口中已经带有电容式触摸屏。

目前电容式触摸屏不像电阻式触摸屏那样，需要由我们控制模拟信号的采集，而是由专门的电容式触摸屏厂家直接加 MCU 来控制实现，然后直接通过 IIC 接口并按照厂家提供的现成通信协议实现坐标值的读取。

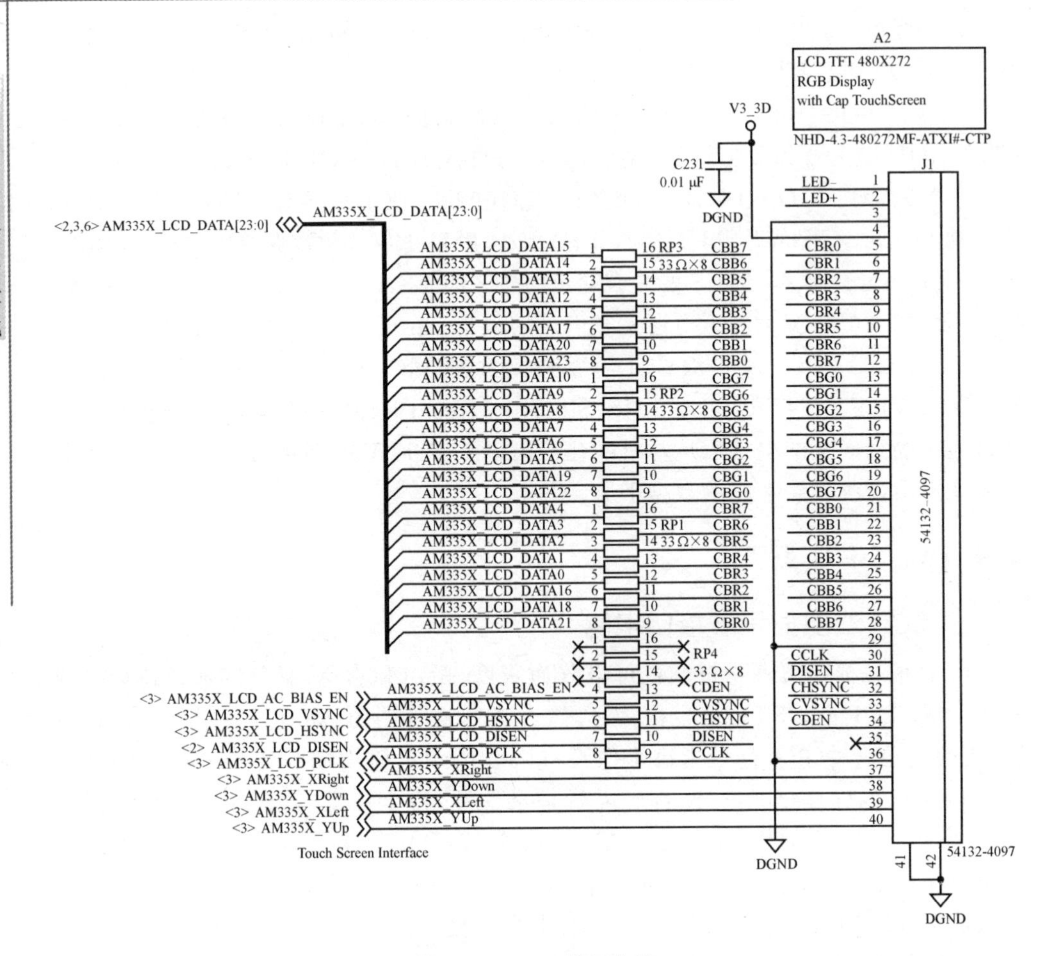

图 2.44　LCD 接口电路

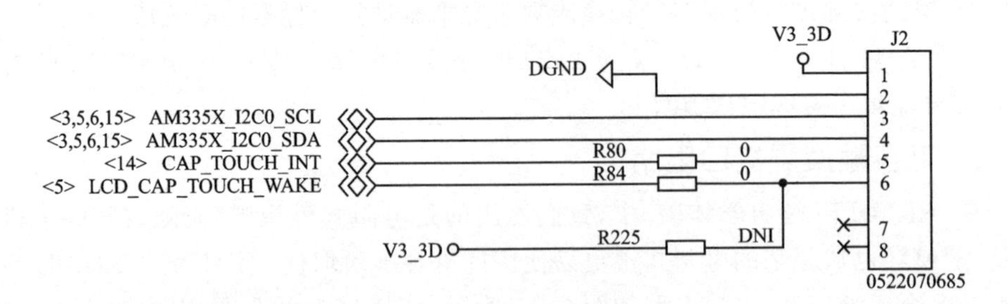

图 2.45　电容触摸屏接口电路

3. LCD 背光调节电路

目前的 TFT 型 LCD 屏中常用的背光有 CCFL 和白光 LED 两种。5 in 以上的大尺寸 LCD 背光普遍采用冷阴极荧光灯(CCFL)光源,而 5 in 以下(如 3.5 in 等)的普遍采用白光 LED 背光源,有些还需要多个 LED 并联或串联以提供足够的亮度。

这些背光电源可以由通用的升压型 DC/DC(或 DC/AC 逆变器)实现,但通常都由专用的背光电源 IC 解决方案实现。如图 2.46 所示为 LED 背光驱动电路芯片 TPS61081。实际上 TPS61081 就是一个升压型的 DC/DC 转换器,具有集成功率二极管的 27 V、1.2 A 开关,1.2 MHz 的升压转换器。

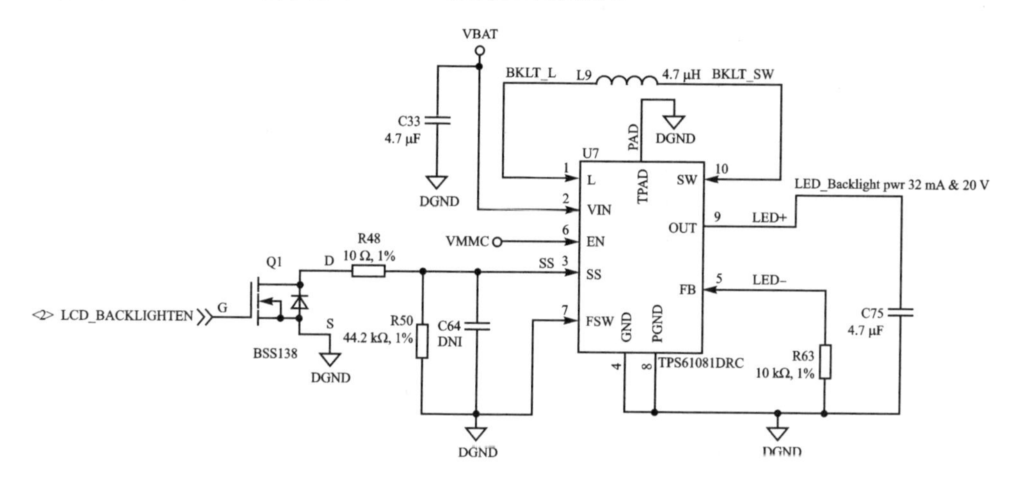

图 2.46　LED 背光驱动电路

TPS61081DRC 的 EN 引脚直接与 VMMC 电源相连,以控制 LCD 背光电源的使能。TPS61081DRC 的 SS(Soft Start programming)引脚由 AM335x 的 LCD_BACKLICHTEN 输出经 BSS138 场效应管控制。

第3章

AM335x Starter Kit 入门

在《ARM Linux 自学体验》中，这章直接介绍硬件调试，顺着硬件制作后的思路，自然要将焊接回来的硬件调试好，让处理器及电路顺利跑通。但本书的思路是直接针对现成的 AM335x Starter Kit 开发板，假设读者已经购买到 AM335x Starter Kit 或类似的开发板，所以本章先介绍 AM335x Starter Kit 开发板的快速入门指南（读者也可以直接参考随板子自带的入门手册*AM335x Starter Kit Quick Start Guide*），然后为了使读者思路顺畅，并考虑部分读者会自己 PCB 打样制作硬件等硬件调试的需要，还会介绍硬件调试的步骤等内容。

3.1 AM335x Starter Kit 快速入门指南

本节的内容基本上是翻译 TI 随板自带的入门指南，英文较好的读者，或者对本书中的介绍理解不清楚或想进一步理解的，可直接参考英语版入门指南，TI 官网应该也有其电子版。

随板附带的 AM335x Starter Kit(SK)快速入门指南是为了帮助用户通过 SK 的初始设置，允许用户去评估基于 Linux 和 Android 两大操作系统的 AM335x Cortex-A8 处理器，包括 3D 图像、集成外设等更多功能特性。AM335x SK 包含下述内容：

1. 硬件方面

- Sitra 系列 AM3358 Cortex-A8 处理器，主频最高达 1 GHz；
- TPS65910 电源管理芯片；
- 4.3 in 电阻触摸屏和 LCD；
- 256 MB DDR3 存储器；
- 1 000 Mbps 双网口；
- 板上自带 XDS100 USB JTAG 仿真器。

如图 3.1 所示为 AM335x Starter Kit 各接口元件位置标识图。

2. 印刷手册

AM335x SK 快速入门指南。

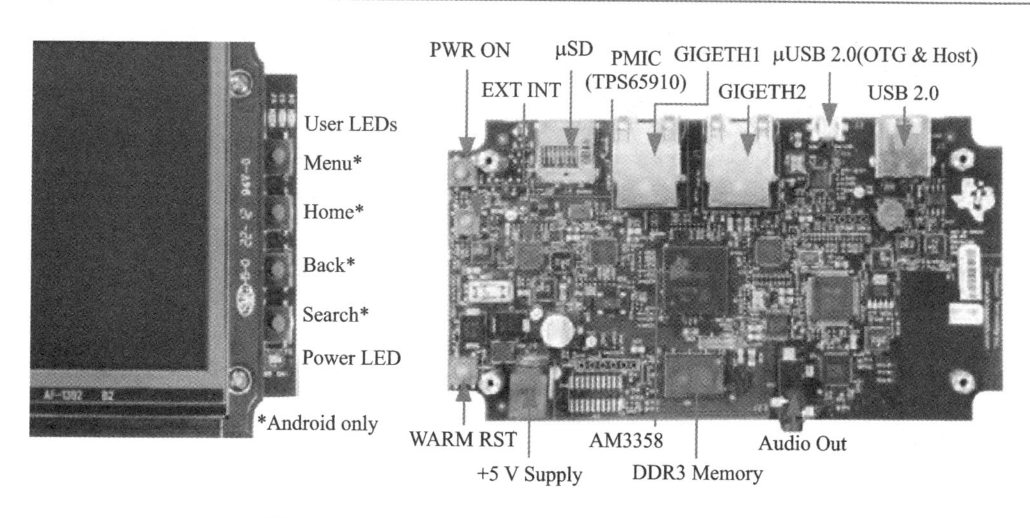

图 3.1　AM335x Starter Kit 接口元件位置标识图

3. 其　他

- 国际通用的电源适配器；
- 2 张 μSD 卡(Android ＋ Linux)；
- 1 张 μSD 卡转 SD 卡适配器；
- 1 根 USB 2.0 线缆。

4. 操作步骤

(1) 默认设置(从 μSD 卡引导操作系统)

① LCD 屏上贴有一张保护膜，可以撕掉以改善触摸特性或者保留起到保护作用，如图 3.2 所示。

图 3.2　AM335x Starter Kit 保护膜

② 选择随板附带的 Android 或 Linux 版 μSD 卡，并将其插入 AM335x Starter Kit 开发板的 μSD 卡槽，如图 3.3 所示。

③ 开发板电源插孔连接电源适配器，如图3.4所示。

图3.3 AM335x Starter Kit μSD卡槽

图3.4 AM335x Starter Kit 电源插孔

④ 按下电源按键，直到板上电源LED指示灯亮起，即给开发板上电，如图3.5所示。如果需要让开发板掉电关机，则常按电源按键约10 s，或一直到电源LED指示灯熄灭。

图3.5 AM335x Starter Kit 电源按键

⑤ 开机启动完毕后将看到对应的Android或Linux操作系统界面，如图3.6所示分别为Android和Linux启动后的界面。

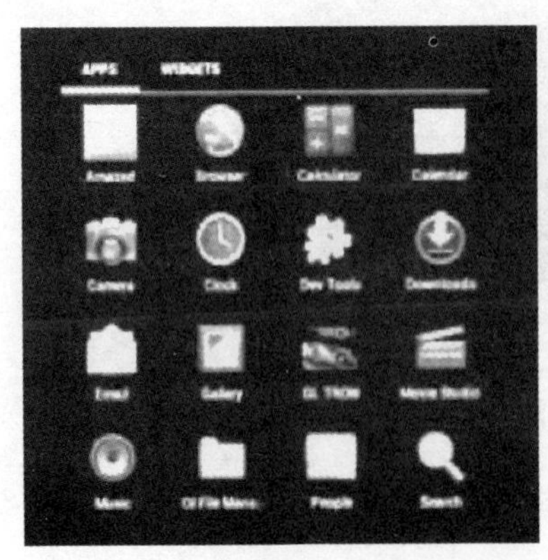

Android

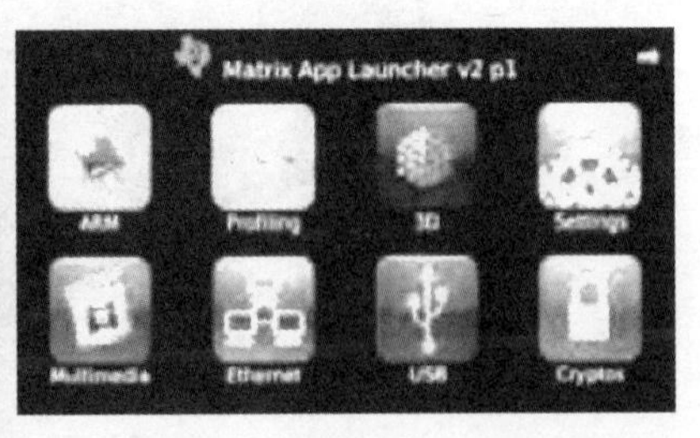

Linux

图3.6 AM335x Starter Kit 的Android和Linux界面

⑥ 如果是 Android SDK，则可以从应用菜单(application menu)中选择 Amazed demo 去展现 AM335x 处理器运算特性和包含的加速度计，如图 3.7 所示。

如果是 Linux SDK，则选择 QT Playground demo 去展现 AM335x 处理器运算特性，如图 3.8 所示。QT Playground 是位于 Matrix menu 下的 QT application。

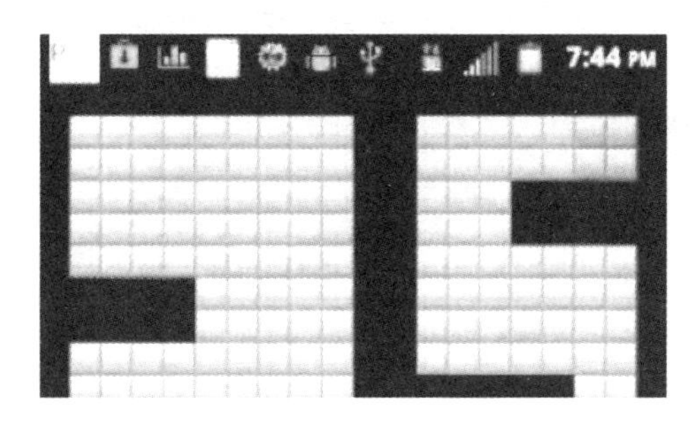

图 3.7 Android Amazed demo 界面

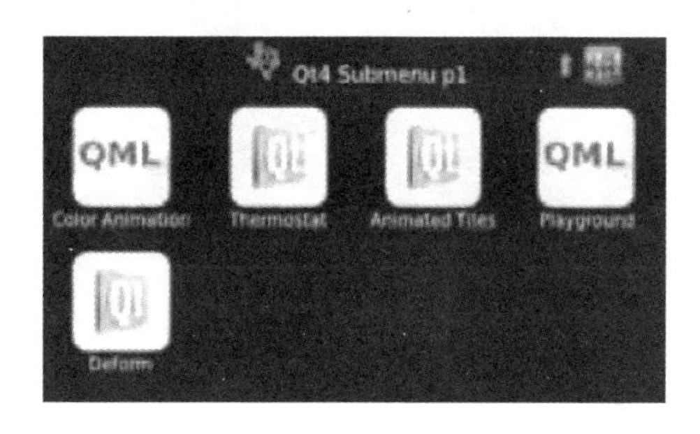

图 3.8 Linux QT Playground 界面

(2) 开始开发

① 准备软件开发环境，开发板关机，移出 μSD 卡，将 μSD 卡插入 SD 卡适配器并插入 PC 的 SD 卡槽，如图 3.9 所示。

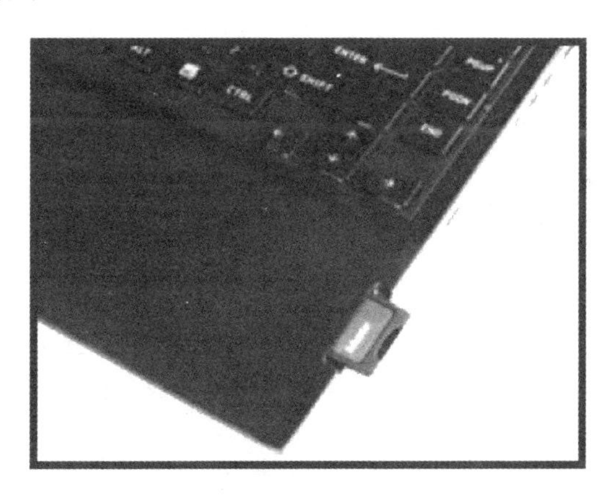

图 3.9 插入 PC 的 SD 卡槽

如果 PC 没有 SD 卡槽，则可由 USB 接口的 SD 卡读卡器替代。

下面针对不同的操作系统进行介绍。

Android 开发：插入 Android 版本的 SD 卡到 PC 主机，出现 START HERE 文件夹，执行其中的 setup.htm.

Linux 开发：在 Linux 环境(如 Ubuntu 12.04)下的 PC 主机上，将 Linux 版本的 SD 卡插入到 PC 主机，出现 START HERE 文件夹，执行 setup.htm.

如果在设置 Linux 主机 PC 时需要帮助，请访问 www.ti.com/startyoulinux.

② 使用 Micro USB 连接 AM335x Starter Kit 开发板和 PC 主机，如图 3.10 所示。

③ 使用网线连接 AM335x Starter Kit 开发板和 PC 主机，或直接接入以太网路由器或交换机等网络(注：PC 主机也需接入网络)，如图 3.11 所示。

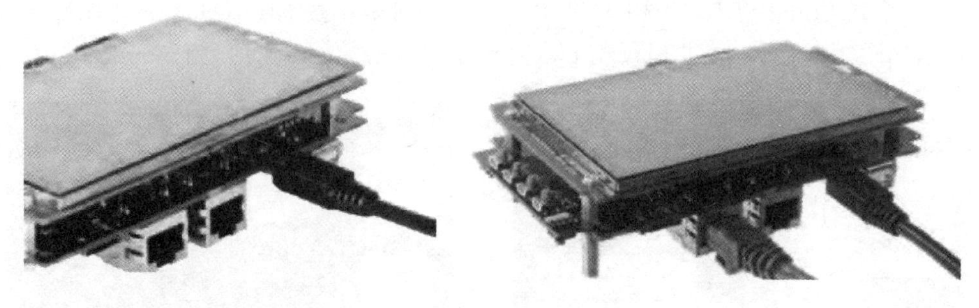

图 3.10 Micro USB　　　　图 3.11 以太网 RJ45

5. 附加资源

AM335x 处理器更多的信息，如 User Guide、Software、How Tos、Design File，请读者参考 www.ti.com/am335x。

3.2 硬件调试概述

AM335x Starter Kit(SK)快速入门指南主要是帮助大家快速地利用附带的 Android 和 Linux μSD 卡让 AM335x Starter Kit 开发板运行起来，且能通过系统中自带的 DEMO 程序评测处理器各方面的特性。但是，如果开发板硬件存在某些问题而无法顺利启动，或者是自己新打板制作回来的硬件，那该怎么办呢？接下来我们就简单介绍一下硬件调试的基本步骤。

ARM 系统板的一般调试步骤如下(每个人的习惯不一样，会略有不同)：

1. 电源调试

无论是 S3C2410，还是 AM335x，或是其他的 ARM 处理器系统板，当拿到一块刚焊接完成的板子后，首先应该测试它的电源是否正常，只有处理器所需的电源都稳定了，MPU 和其他的外设才能正常工作。

注意：第一次通电前，应该用万用表测量电源和地是否短路。

2. 复位电路和时钟

如果复位电路不正常，则下载到目标板上的程序在上电(或手动复位)后可能会不工作。时钟是处理器工作的基础，一般很少出现问题，除非是晶振坏了，或电源没有正常供电，通常用示波器测试晶振是否有起振，振荡频率是否为标称值。

3. 处理器内核的检测

如果朋友们有仿真器，那么一般要先检测目标处理器的内核，只有内核找到了，才能说明JTAG接口已经正常，处理器有反应了；如果没有找到，那么一般先用示波器测试JTAG的各种波形，然后查找电源是否正常。注：AM335x Starter Kit开发板已经将XDS100v2仿真器集成到板子上，直接使用USB将板子连接到PC即可连接仿真调试。

4. 仿真器仿真

如果可以使用仿真器控制目标处理器进入调试状态，那么这个板子基本上已经"活"了，我们的心也大可以放下来，因为最复杂的BGA部分都可以正常工作了，还有什么可以担心的呢？

有时虽然可以找到处理器的内核，但却进入不了仿真状态，这是最危险，也是最头痛的事情。遇到这种情况，通常是先检查处理器的供电电源、时钟，然后对着用户手册的信号引脚描述部分，检查处理器的关键信号等。

5. DDR SDRAM测试

当仿真器进入调试状态时，PC通常还没有指向DDR SDRAM区域，所以也不能装载映像文件进行源码级调试，必须先让DDR SDRAM正常读/写。由于没有向Flash下载有效代码，所以上电后或仿真器进入时都还没有执行过有效的代码，处理器也没有进行内存控制器及SDRAM的初始化工作，因此SDRAM不能正常访问是正常的。我们必须手动装载初始化文件（S3C2410在命令窗口obey初始化文件，而AM335x在CCS中有Tools和Scripts等菜单直接加载gel文件）对其初始化，初始化成功后可以读/写DDR SDRAM内存。如果读/写不正确，那么还要查找SDRAM部分电路。最后再装载映像文件，进入源码级的调试。

6. 串口测试

当可以装载程序进行调试后，要测试一下串口是否可以顺利地往PC终端打印信息及接收命令等，此时可以装载官方提供的现成的DEMO程序测试串口输出。AM335x Starter Kit板上集成了串口转USB芯片，需要注意串口转USB芯片的供电。

7. Flash或μSD的烧写与引导

像S3C2410及老一代的ARM系统，一般都会带有NAND Flash，通常利用像sjf2410等的烧写工具烧写后，再引导启动。而现在Cortex-A8/A9或更新的处理器往往都可以支持SD卡引导启动，所以更常见的是直接制作SD卡启动。AM335x Starter Kit板上没有带Flash，只能利用μSD卡启动，可以利用TF卡读卡器插入PC，将MLO和app文件复制到μSD卡后，再插入到板上，然后上电引导启动。

8. 外围电路的调试

当系统的核心部分都正常工作后，其他外设或接口都只是做具体的工作，有些外设接口可以直接利用现成的DEMO程序进行测试，有些需要编写测试代码调试，另外还可以让系统运行Linux再调试等，这可由读者自己决定。此外，读者也可以一边学习某个外设，一边编写它的测试程序，一边调试硬件。

3.3 XDS100v2 仿真器和 CCS 软件的使用

3.3.1 集成开发环境 CCS 的下载与安装

在连接 XDS100v2 仿真器与硬件开发板之前，需要先安装配套的集成开发环境 CCS 软件，其最新版本的下载地址：http://processors.wiki.ti.com/index.php/Download_CCS#Download_the_latest_CCS。

CCSv4、CCSv5、CCSv6 版本的下载地址：

http://processors.wiki.ti.com/index.php/CCSv4；

http://processors.wiki.ti.com/index.php/CCSv5；

http://processors.wiki.ti.com/index.php/CCSv6。

如果用户没有 TI 的网络账号，则需注册后才能下载。由于笔者使用的是 32 位的 Windows 系统，所以下载 Win32 版本：CCS6.0.0.00190_win32.zip 或者 CCS5.2.0.00069_win32.zip（笔者两个都下载了，而且上也都安装在了计算机上，虽然 CCS6.0 比较新，但 TI 的很多例程还是基于 CCS5.0，所以也经常使用 CCS5.0）。下载完成后，解压 CCS6.0.0.00190_win32.zip 进入该文件夹，如图 3.12 所示。

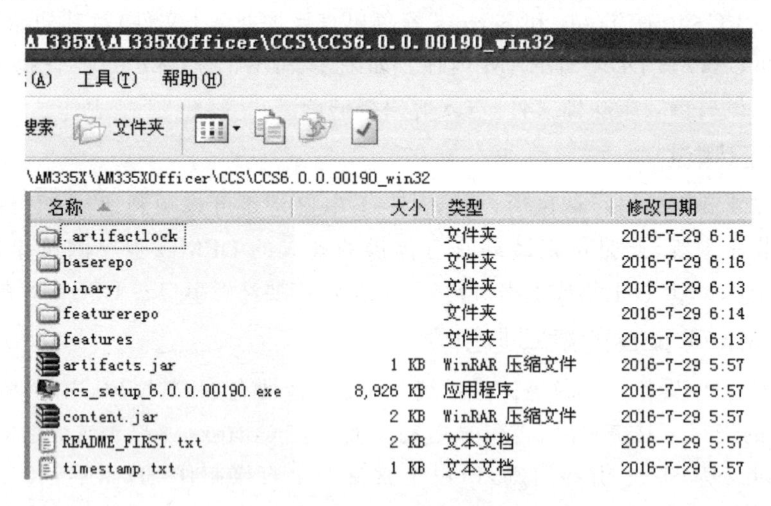

图 3.12 CCSv6 安装文件夹

双击 ccs_setup_6.0.0.00190.exe，如果弹出如图 3.13 所示的对话框，则单击“否”按钮。

出现上面这种情况可能是电脑打开了 360 安全卫士等软件，退出 360 安全卫士等软件后，再重新双击 ccs_setup_6.0.0.00190.exe，将出现如图 3.14 所示的License Agreement 对话框。

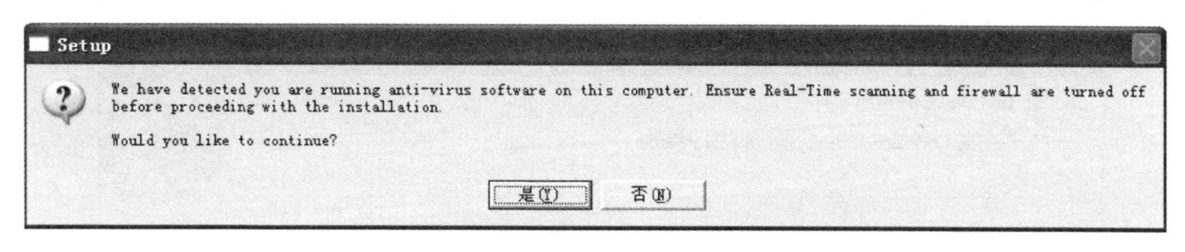

图 3.13 CCS 安装配置对话框——提示关闭防病毒软件

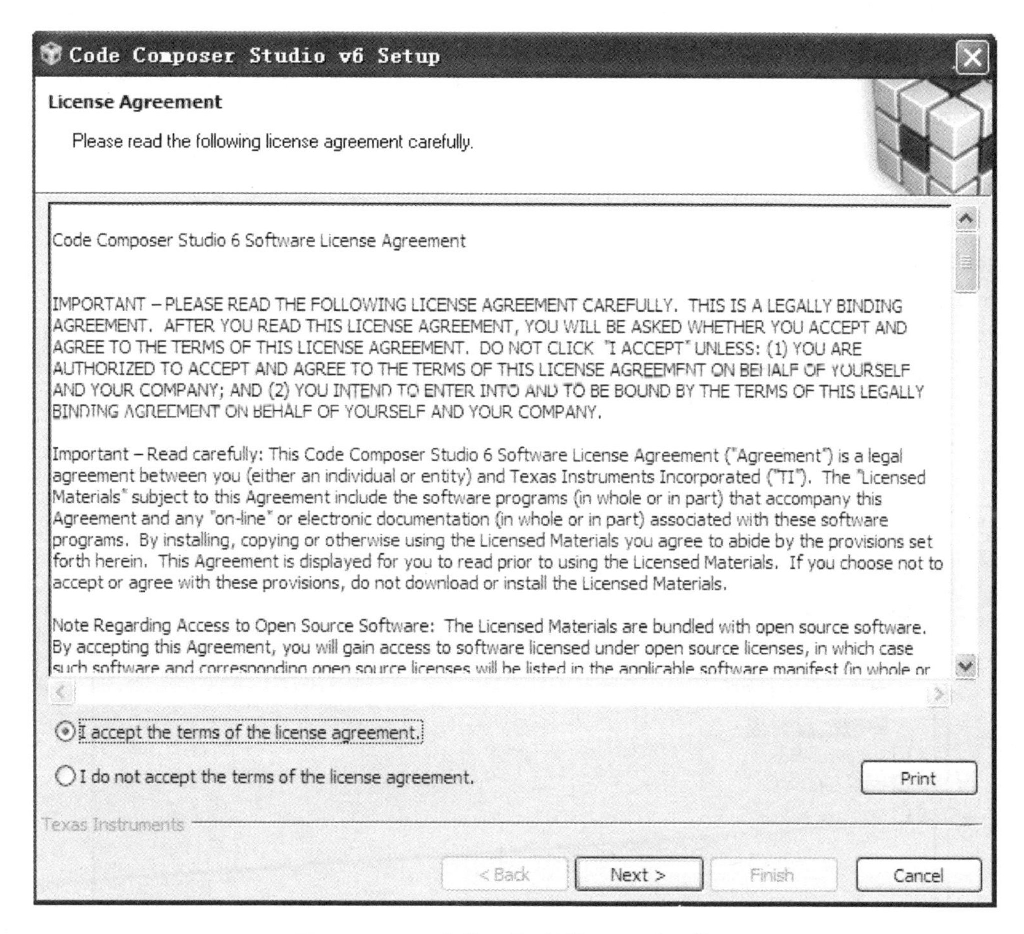

图 3.14 CCS 安装配置对话框——许可协议

单击 Next 按钮，弹出如图 3.15 所示的对话框用户可以选择安装路径。

一般保持默认路径 C:\ti 即可，当然用户也可以选择其他路径。单击 Next 按钮，在 Processor Support 对话框中选择需要安装的处理器（AM335x 属于 Sitara 32-bit ARM Processors），如图 3.16 所示为选择安装的 TI 处理器类别和对应的编译器。

这里勾选了 GCC ARM Compiler 和 TI ARM Compiler，其实两者可以只选择一项。单击 Next 按钮，在 Select Emulators 对话框中选择要支持的仿真器，如图 3.17

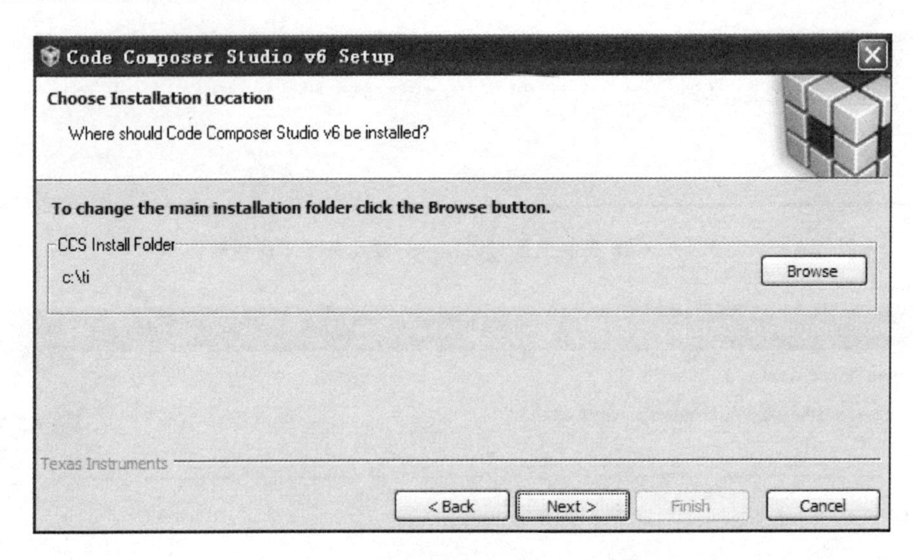

图 3.15　CCS 安装配置对话框——安装路径

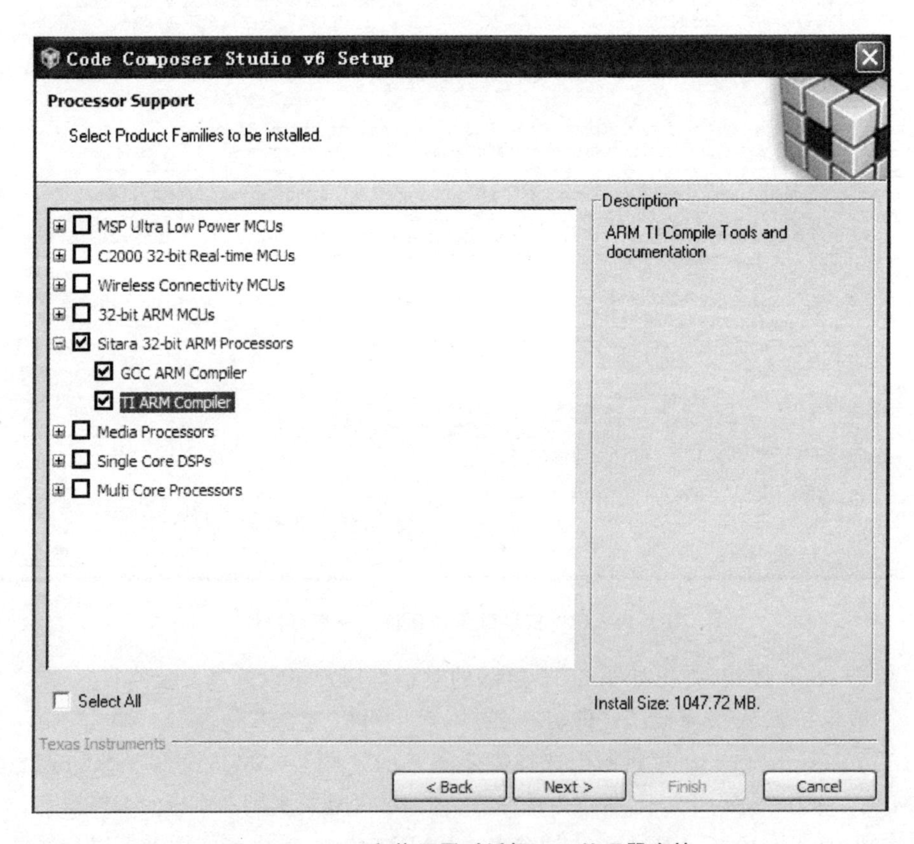

图 3.16　CCS 安装配置对话框——处理器支持

所示。

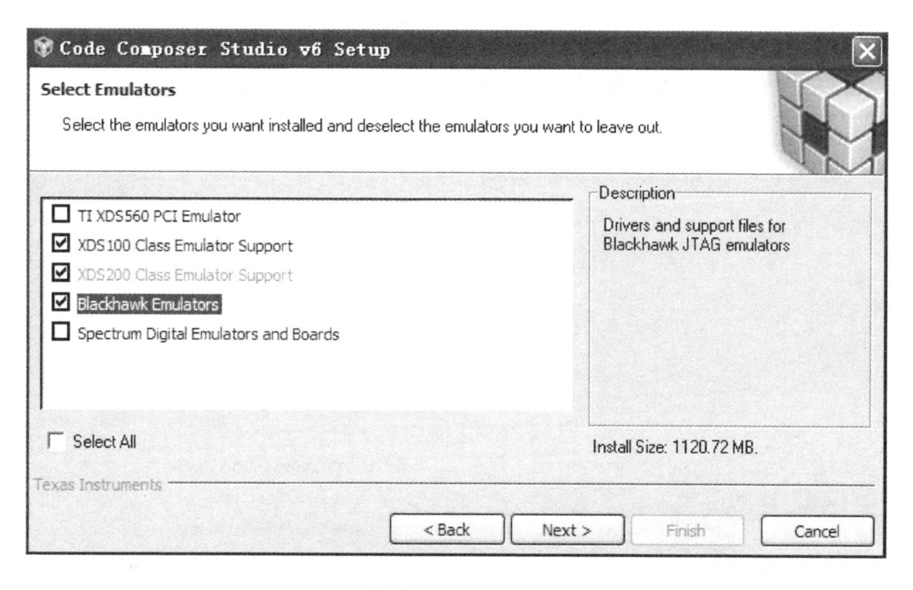

图 3.17　CCS 安装配置对话框——选择仿真器

我们购买的 XDS100v2 USB 仿真器是 Blackhawk 公司的产品，所以勾选 Blackhawk Emulators，单击 Next 按钮，弹出如图 3.18 所示的对话框。

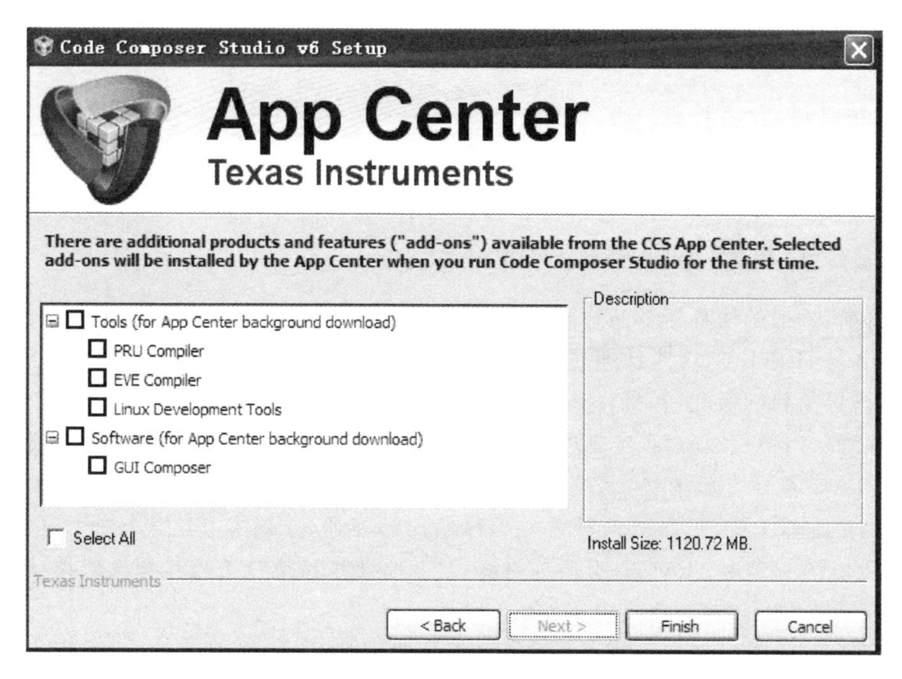

图 3.18　CCS 安装配置对话框——应用中心

单击 Finish 按钮，开始安装。安装完成后，如图 3.19 所示，再次单击 Finish 按钮完成。

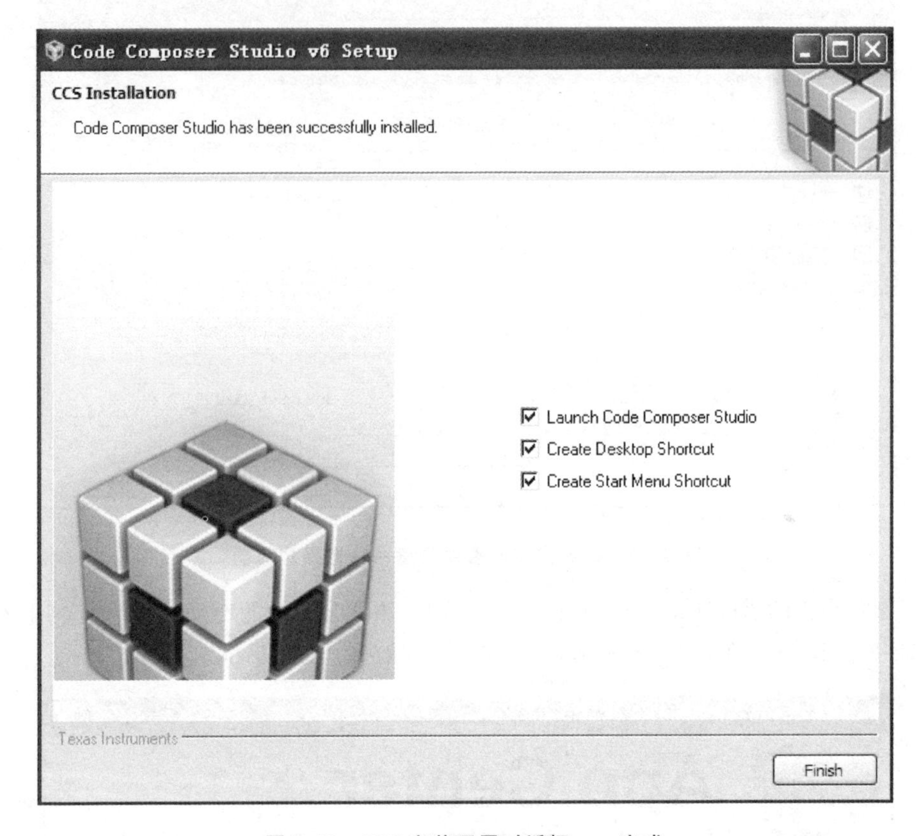

图 3.19 CCS 安装配置对话框——完成

3.3.2 仿真器与目标板的硬件安装

目前，一般的仿真器都通过标准的 20 Pin JTAG 接口与目标板相连（也有的如 STM32 与 JLINK 仿真器是通过四线的 SWD 接口相连的），或通过 USB 与 PC 相连，然后再给目标板和仿真器上电。

注意： TI AM335x 系列处理器有 TI 自己定义的 JTAG 接口，与传统 ARM 标准的 JTAG 接口有所不同，主要是 JTAG 接口座引脚信号的排序不同，而 XDS100 系列仿真器都支持，但在选购时需要选择不同的型号，分别有 TI 14 引脚、TI 20 引脚、ARM 10 引脚和ARM 20 引脚连接器，这里需要根据所使用的目标板的接口进行选择。

另外，AM335x Starter Kit 开发板已经内置了 XDS100v2 仿真器，只需要将开发板通过 USB 线连接 PC 的 USB 2.0 接口，再给开发板上电即可。

3.3.3 XDS100v2 USB 仿真器在 CCSv6 集成开发环境中的配置

双击桌面上的快捷图标或选择“开始”菜单中的 Code Composer Studio 6.0.0，首次启动时会提示设置工作目录，可以默认设置，也可以重新选定自己喜欢的其他目录。

启动 CCS 后会显示 Getting Started 界面，单击 Import Project，弹出 Import CCS Eclipse Projects 对话框，再单击 Browse 按钮选择导入现成的项目文件，如 StarterWare 下的 boot 例程，并单击“确定”按钮，如图 3.20 所示。

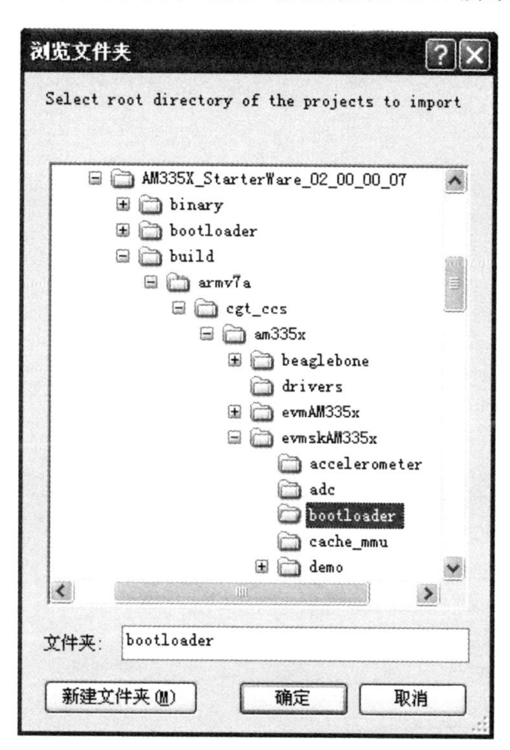

图 3.20 CCS 导入 StarterWare 例程

注：之前我们已经安装了 TI 官方自带的 StarterWare，安装方面的内容见相关章节。导入项目文件后，如图 3.21 所示，单击 Finish 按钮完成。

在 CCS 左边区域 Project Explorer 中右击刚导入的 demo 工程名，选择 New→Target Configuration File 命令，如图 3.22 所示。

在打开的新建配置对话框中输入配置文件名，如图 3.23 所示，勾选 Use shared location 后，单击 Finish 按钮。

在随后出现的 AM335xSKEVM.ccxml 中，选择仿真器和调试板或芯片，然后单击右边的 Save Configuration 下的 Save 按钮，如图 3.24 所示。

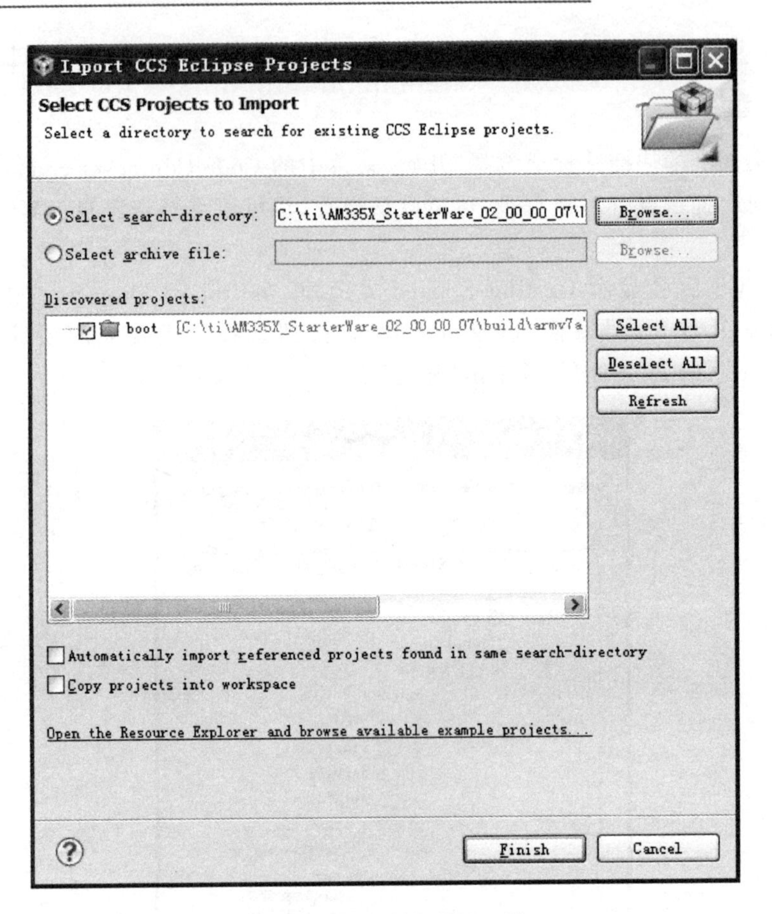

图 3.21　CCS 导入项目工程

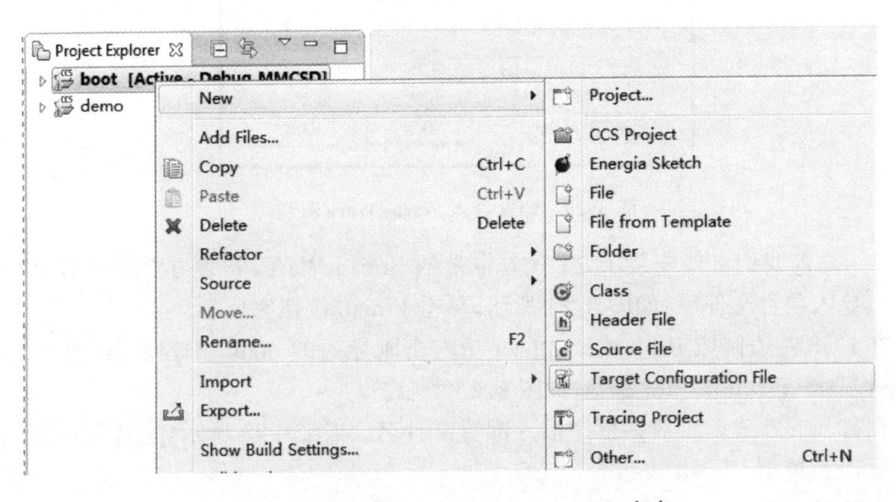

图 3.22　选择 Target Configuration File 命令

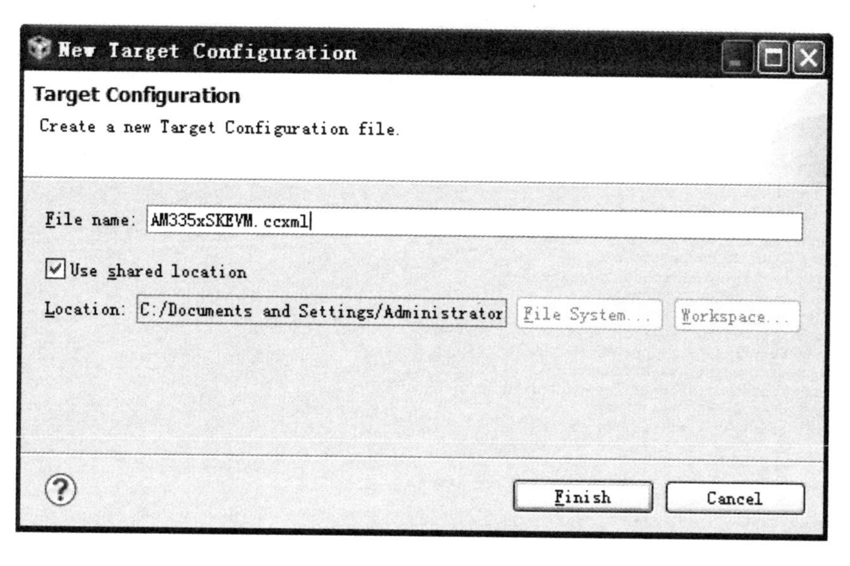

图 3.23　New Target Configuration 对话框

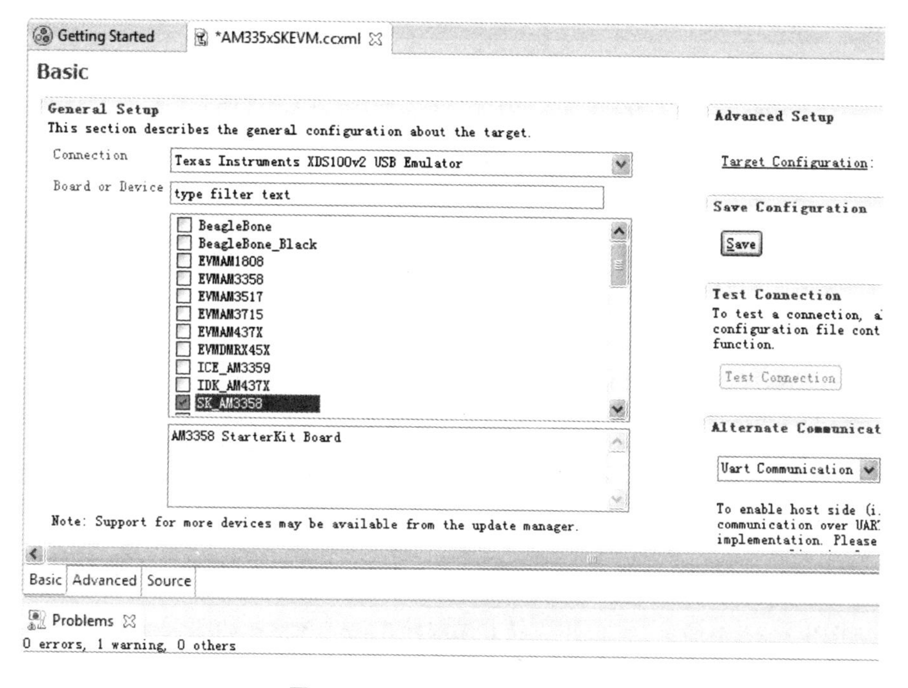

图 3.24　AM335xSKEVM.ccxml

完成以上操作且给 Starter Kit 上电后，可以单击图 3.24 右边的 Test Connection 按钮进行连接测试，测试 CCS 是否能够与仿真器相连接。如果连接正常，则会弹出 Test Connection 对话框，如图 3.25 所示。

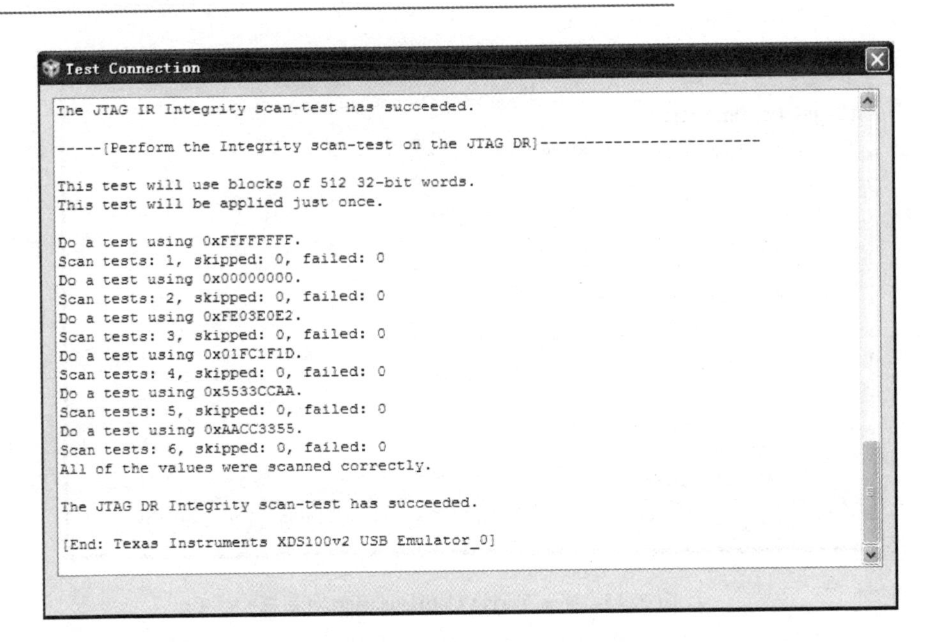

图 3.25 Test Connection 对话框

注：如果已经存在 *.ccxml 文件，则可以直接单击菜单工具栏中的 Debug 按钮，或右击 Project Explorer 窗口中的 demo 工程，选择 Debug As→Debug Configurations 命令打开对话框，在左边选择 *.ccxml 文件，然后单击右下方的 Debug 按钮。

单击工具栏中的 Debug 按钮，此时在 CCS 中会出现 Debug 窗口，如图 3.26 所示。

Debug
AM335xSKEVM.ccxml [Code Composer Studio - Device Debugging]
Texas Instruments XDS100v2 USB Emulator_0/M3_wakeupSS_0 (Disconnected : Unknown)
Texas Instruments XDS100v2 USB Emulator_0/CortxA8 (Disconnected : Unknown)
Texas Instruments XDS100v2 USB Emulator_0/PRU_0 (Disconnected : Unknown)
Texas Instruments XDS100v2 USB Emulator_0/PRU_1 (Disconnected : Unknown)

图 3.26 Debug 窗口

右击图 3.26 中的第二项，并选择 Connect Target 命令，如图 3.27 所示。

连接成功后，如图 3.28 所示。

此时还需要加载 gel 文件。单击 Tools→GEL Files，在 CCS 主窗口右下方会出现一个 GEL Files 新功能窗口，CCSv6 默认已经装载有 gel 文件，如图 3.29 所示。

如果没有默认装载，则如图 3.30 所示，在窗口内右击，选择 Load GEL。

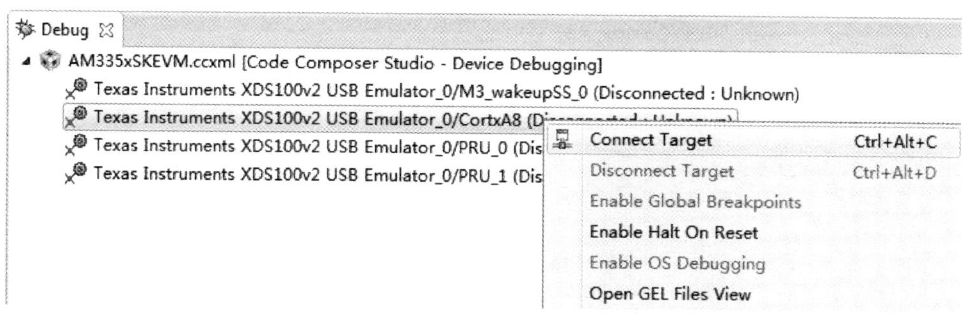

图 3.27 选择 Connect Target 命令

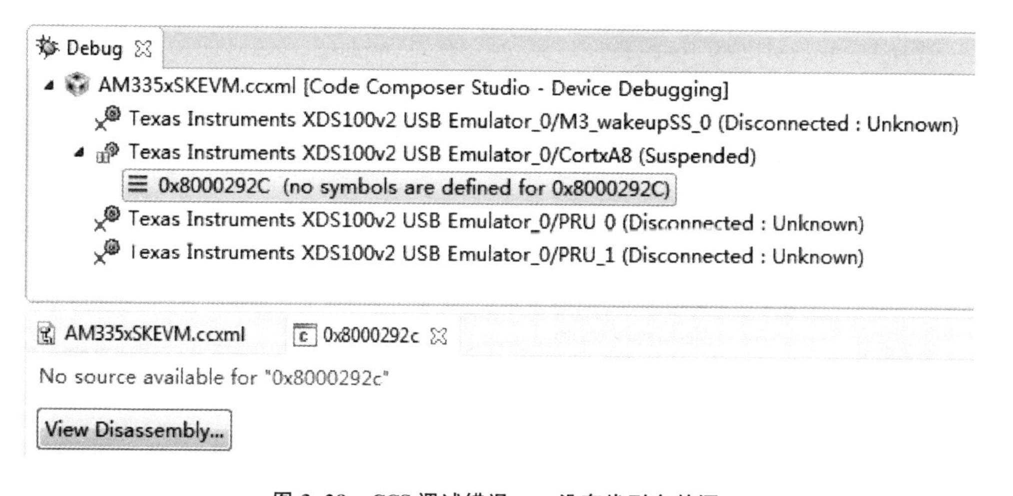

图 3.28 CCS 调试错误——没有找到有效源

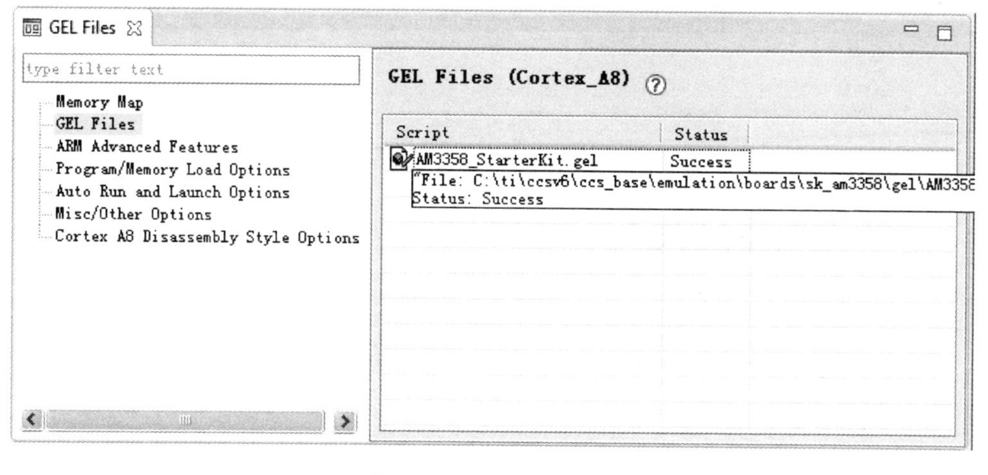

图 3.29 显示 GEL 文件

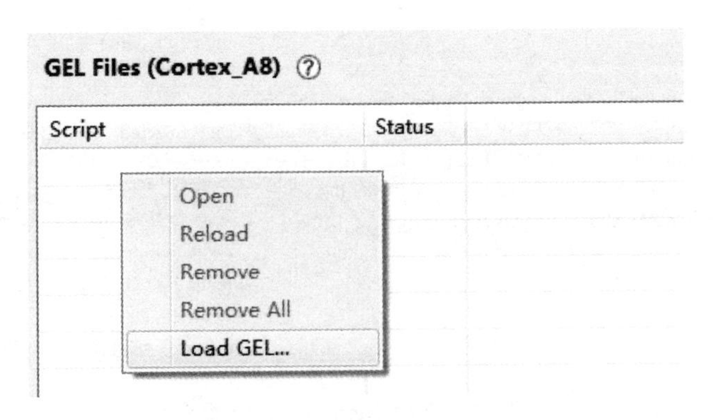

图 3.30 加载 GEL 文件(一)

选择 C:\ti\AM335X_StarterWare_02_00_01_01\tools\gel\ AM335x_SK_1.2.gel，如图 3.31 和图 3.32 所示。

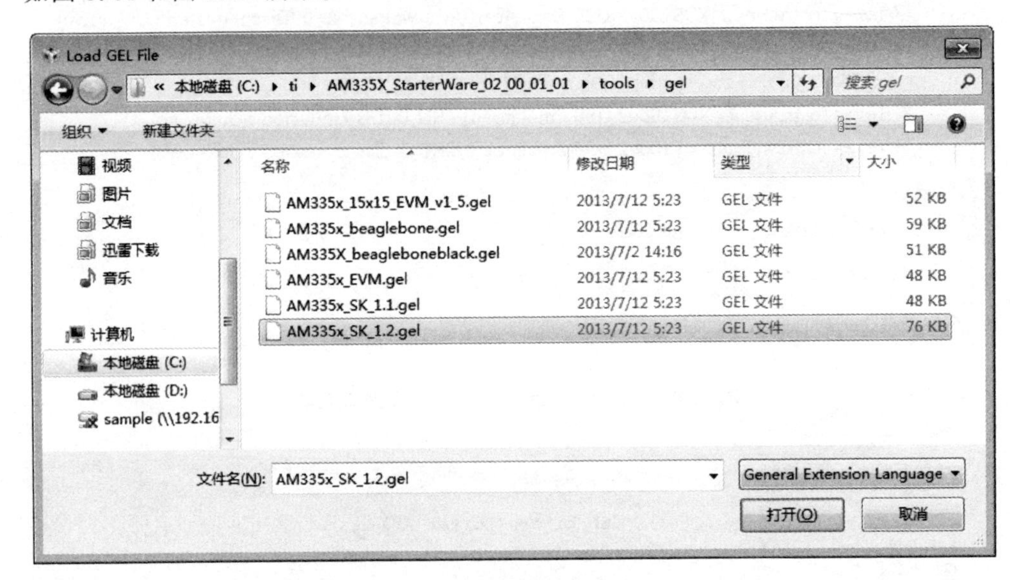

图 3.31 加载 GEL 文件(二)

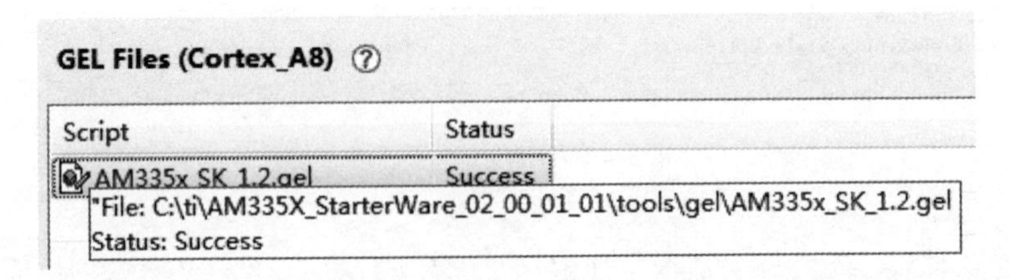

图 3.32 加载 GEL 文件(三)

然后选择 Scripts→AM335x System Initialization→AM3358_SK_Initialization，进行初始化操作，如图 3.33 所示。

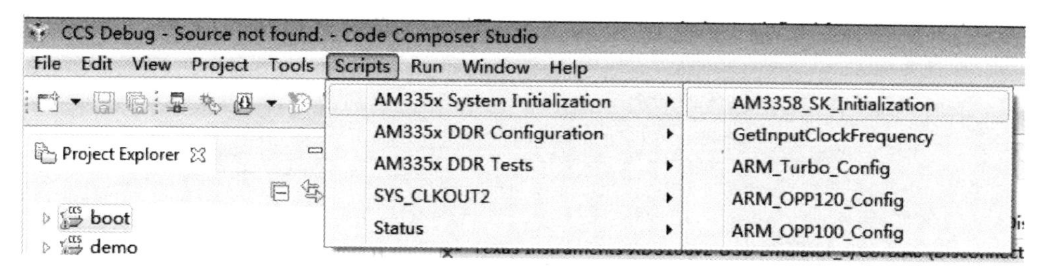

图 3.33　装载 AM3358 Starter Kit 板初始化脚本

选择 Run→Load→Load Program，装载程序，如图 3.34 所示。

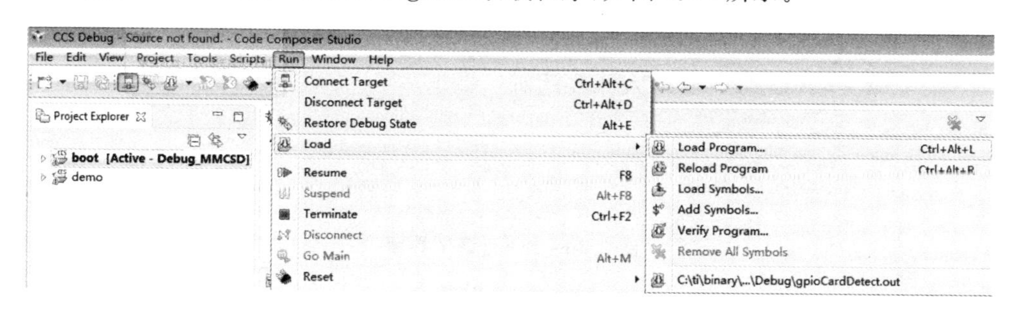

图 3.34　装载程序

在弹出的对话框中，单击 Browse 按钮，打开 C:\ti\AM335X_StarterWare_02_00_01_01\binary\armv7a\cgt_ccs\am335x\evmskAM335x\bootloader\Debug_MMCSD\boot，如图 3.35 所示。

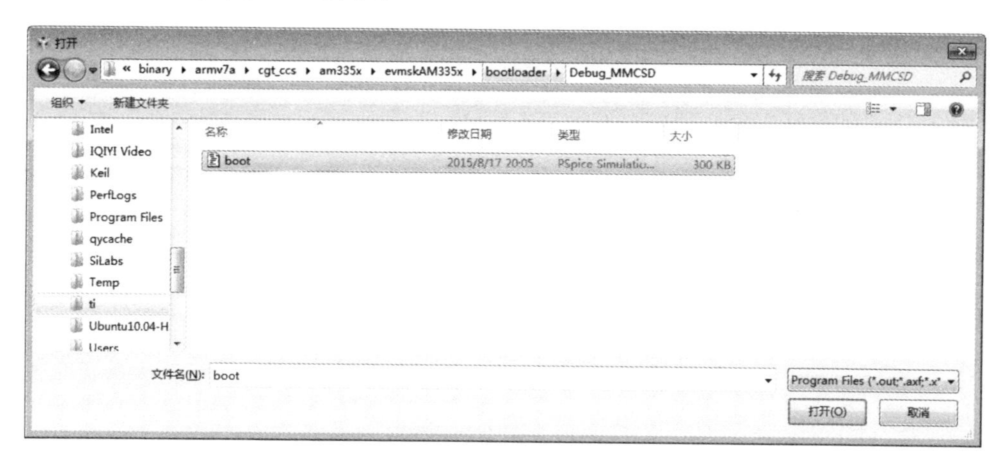

图 3.35　装载 boot 程序

完成装载后将进入调试界面，如图 3.36 所示。

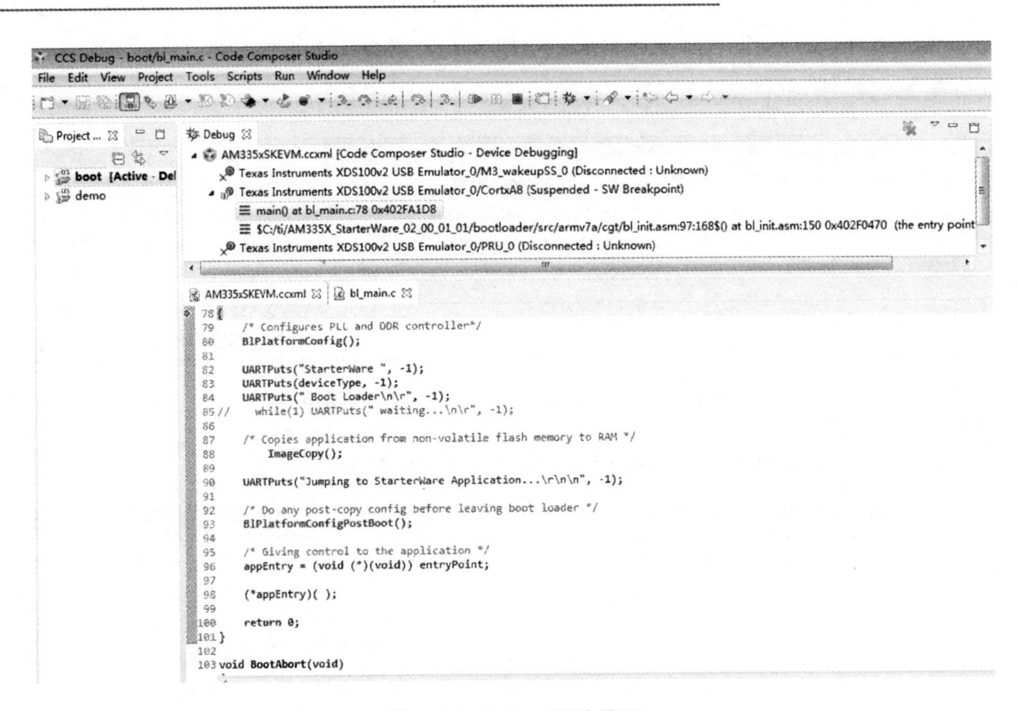

图 3.36 Debug 调试界面

现在我们就可以单步执行，或设置断点等各种调试手段了。

第二篇

ARM 前后台系统

第 4 章

无操作系统平台下的应用库——StarterWare

我们在前面章节，已经简单介绍了 TI 为 AM335x 提供的无操作系统平台下的应用库 StarterWare，本章将进行详细的分析及实践。

4.1 StarterWare 下载安装

StarterWare 最新官方下载地址：http://software-dl.ti.com/dsps/dsps_public_sw/am_bu/starterware/latest/index_FDS.html，主页如图 4.1 所示。

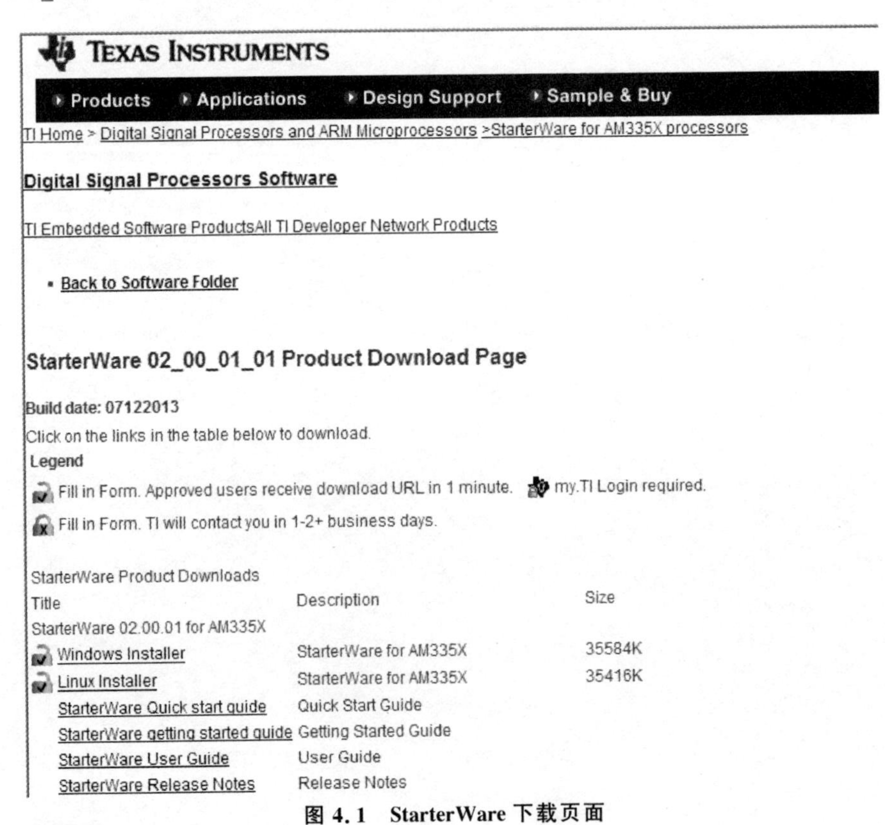

图 4.1　StarterWare 下载页面

选择 Windows Installer，下载 AM335X_StarterWare_02_00_01_01_Setup.exe 并安装。StarterWare 的安装非常简单，按提示操作即可，我们可以将其安装到默认的位置，如图 4.2 所示。

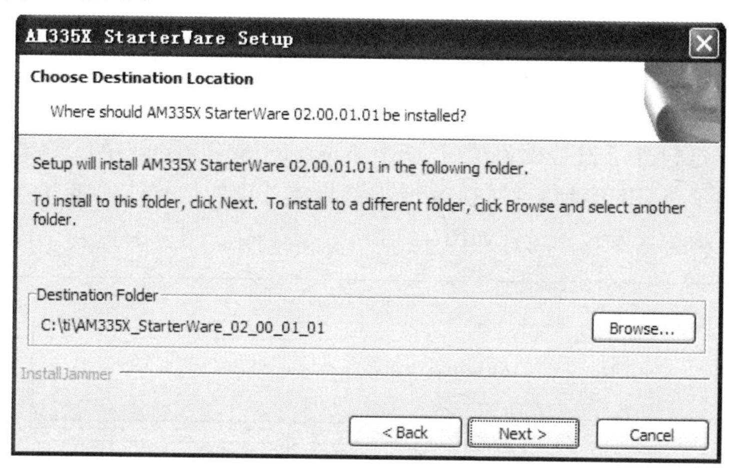

图 4.2　StarterWare 默认安装目录

安装完成后，将在 C:\ti 目录下生成 AM335X_StarterWare_02_00_01_01 文件夹，包含各个库及例程，如图 4.3 所示。其中 docs 文件夹下包含有 StarterWare 的使用手册 UserGuide_02_00_01_01.pdf。

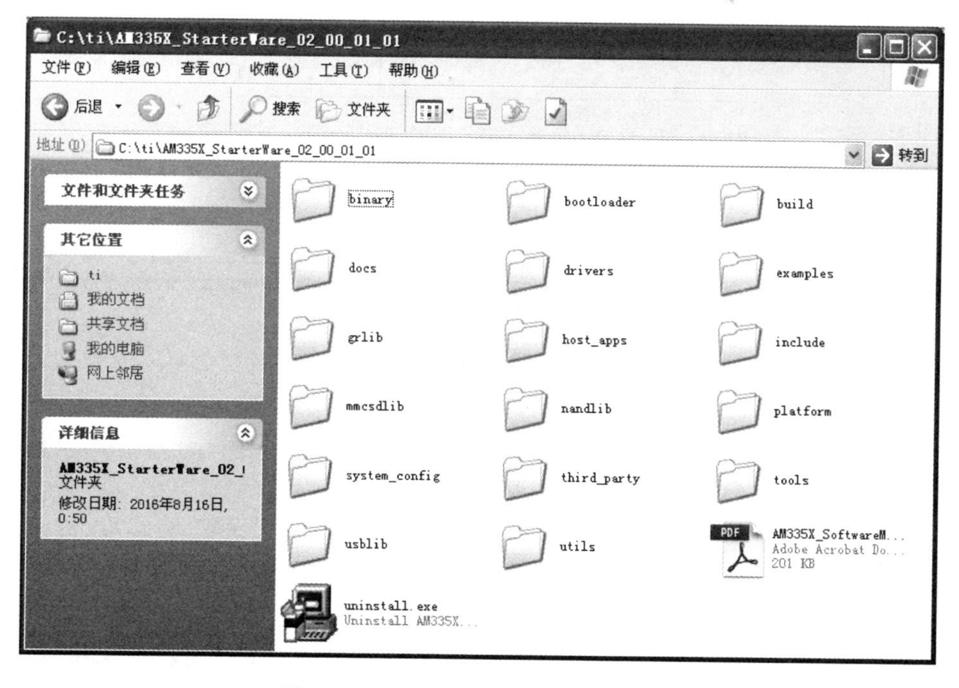

图 4.3　StarterWare_02_00_01_01 目录

4.2 StarterWare快速入门指南

4.2.1 StarterWare概述

StarterWare_02_00_01_01为AM335x提供了无操作系统平台下的应用库支持，整个开发包包含了设备抽象层（Device Abstraction Layer，DAL）库和外设/板级样例/评估例子等，用于评估AM335x及内部集成外设的运算处理能力。

StarterWare开发包的组件如图4.4所示。

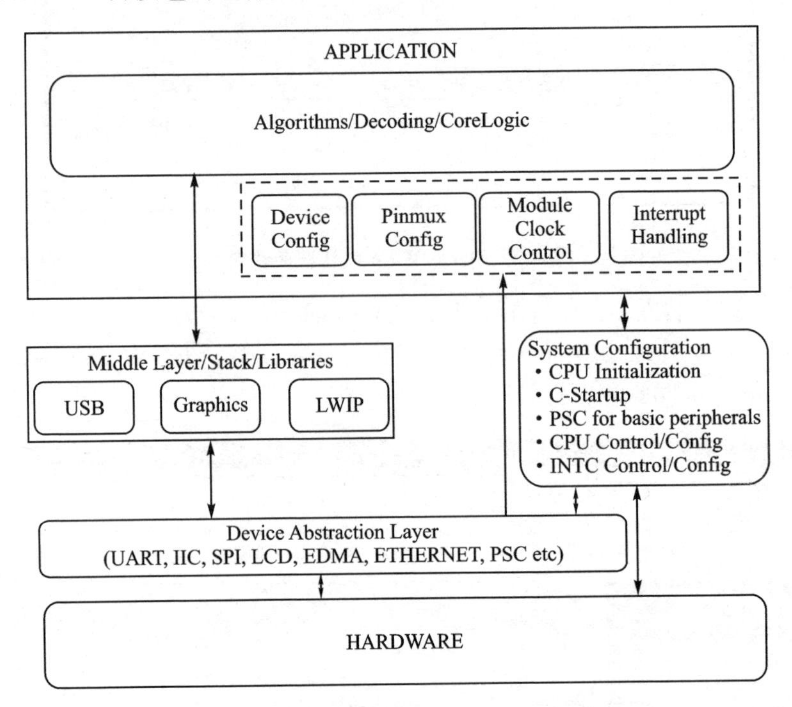

图4.4 StarterWare_02_00_01_01多层结构

Device Abstraction Layer Library：设备抽象层库，该层包含AM335x集成外设支持的设备抽象层APIs。

Peripheral Examples：集成外设的应用例子，包含通过使用设备抽象层库APIs评估AM335x集成外设处理能力的外设应用例子。

System Configuration Code：处理器系统配置代码，包含设置AM335x处理器内核和为应用例子的执行而准备的处理器级的主要配置代码。

Platform Code：平台级代码，包含AM335x Starter Kit开发板的初始化配置代码，其用于配置AM335x Starter Kit开发板，例如AM335x芯片引脚的复用、通信I/O的扩展等。

StarterWare 目录结构如图 4.5 所示，目录中的文件夹介绍如下：

```
|----StarterWare_#.#.#.#
|-- Software-manifest.pdf
|-- docs
|    |-- ReleaseNotes-#.#.#.#.pdf
|    |-- UserGuide-#.#.#.#.pdf
|-- drivers
|-- examples
|    |--evmAM335x
|        |-uart
|-- grlib
|-- usblib
|-- mmcsdlib
|-- nandlib
|-- host_apps
|-- build
|    |--armv7a
|        |--gcc
|            |--am335x
|                 |--drivers
|                 |--system_config
|                 |--evmAM335x
|                      |--uart
|                      |--platform
|                      |--bootloader
|-- binary
|    |--armv7a
|        |--gcc
|            |--am335x
|                 |--drivers
|                 |--system_config
|                 |--evmAM335x
|                      |--uart
|                      |--platform
|                      |--bootloader
|-- binary
|    |--armv7a
|        |--gcc
|            |--am335x
|                 |--drivers
|                 |--system_config
|                 |--evmAM335x
|                      |--uart
|                      |--platform
|                      |--bootloader
|-- include
|    |-- hw
|    |-- armv7a
|         |--am335x
|-- platform
|    |-- evmAM335x
|    |-- beaglebone
|-- system_config
|    |--armv7a
|        |-- am335x
|        |-- gcc
|-- bootloader
|    |--include
|    |--src
|-- third_party
|-- tools
|-- utils
|-- test_bench
```

图 4.5　StarterWare_02_00_01_01 目录结构

- drivers：这个文件夹包含驱动库 APIs 的源文件。
- examples：Examples 例程作为 StarterWare 开发包提供的一部分，展示了 AM335x 集成外设的一些（不是所有）功能。
- grlib：这个文件夹包含图形库的源文件。
- mmcsdlib：这个文件夹包含 MMCSD 库源文件。
- nandlib：这个文件夹包含 NAND 库源文件。
- usblib：这个文件夹包含 USB 库源文件。
- host_apps：这个文件包含用于执行以太网例子的文件。
- build：这个文件夹包含 makefile 文件，用于编译各种库（drivers、system_config、platform、utils、usblib、grlib 等）和例子应用程序（uart、mcspi、hsi2c 等）。
- binary：所有可执行的目标镜像文件都放在这个文件夹里。
- include：这里包括所有的头文件，如用户接口驱动头文件、处理器家族和 SoC 配置头文件、处理器的寄存器级头文件。
- platform：所有支持 StarterWare 开发包的硬件平台。
- system_config：系统配置和初始化代码，如启动代码、中断向量表初始化、底层 CPU 配置代码等都在这里提供。
- bootloader：包含 bootloader 引导程序的配置文件。
- third_party：包含第三方的一些软件。
- tools：一些软件工具。
- utils：包含一些用户可利用的代码小工具。
- test_bench：这个文件夹包含 StarterWare 自动测试 framework 框架和测试 scripts 脚本。

编译和运行 StarterWare 的计算机主机可以是 Windows 系统，也可以是 Linux 系统，需要安装 CCSv5 或 CCSv6 等版本的集成开发环境。

4.2.2 在 AM335x Starter Kit 开发板上运行 StarterWare 应用

下面介绍如何将 StarterWare 的 Bootloader 和系统级评估例程复制到 μSD 卡并在 AM335x Starter Kit 开发板上装载并启动，以快速查看 StarterWare 的各项功能及样例。

1. AM335x Starter Kit 板的设置

使用 USB 线将开发板（J3 Micro USB 端口）连接到计算机，并确认 USB 虚拟串口的驱动已经安装（驱动可以从 TI 官网下载）。选择串行端口号，并启动超级终端或其他的串口调试助手，设置串口波特率为 115 200，无奇偶校验位，1 位停止位，无流控。

使用以太网网线连接开发板（J6 GIGETH1 以太网端口 1）与计算机。

准备一张格式化完成的μSD卡，作为AM335x Starter Kit开发板的引导启动卡。

2. 将StarterWare的Bootloader和Demo目标二进制镜像复制到μSD卡

将C:\ti\AM335X_StarterWare_02_00_01_01\binary\armv7a\cgt_ccs\am335x\evmskAM335x\bootloader\Release_MMCSD\目录下的boot.bin复制到μSD卡中，并重命名为MLO。

将C:\ti\AM335X_StarterWare_02_00_01_01\binary\armv7a\cgt_ccs\am335x\evmskAM335x\demo\Release目录下的demo.bin也复制到μSD卡中，并重命名为app。

因为StarterWare_02_00_01_01的Bootloader和Demo目录下已经分别包含有boot.bin和demo.bin重命名好的MLO和app文件，所以可以直接将这两个目标二进制文件镜像复制到μSD卡。

将C:\ti\AM335X_StarterWare_02_00_01_01\binary\armv7a\cgt_ccs\am335x\evmskAM335x\bootloader\Release_MMCSD\MLO复制到μSD卡中。

将C:\ti\AM335X_StarterWare_02_00_01_01\binary\armv7a\cgt_ccs\am335x\evmskAM335x\demo\Release\app复制到μSD卡中。

3. 装载和执行

将μSD卡插入AM335x Starter Kit开发板的μSD卡槽(J4 MICROSD)后，按电源键(SW5 PWRON)上电启动开发板，或按复位键(SW7 WARMRST)复位重新启动开发板。开发板上的AM335x处理器会先装载μSD上的Bootloader并执行，而Bootloader又将装载Demo程序并执行，此时可以在UART终端上查看到启动信息，LCD也将显示Demo运行后的界面。在计算机上打开网络浏览器，当输入AM335x Starter Kit执行Demo程序自带默认的IP地址时，也会有相应的显示界面，如图4.6所示。

注：这里需要特别注意的是由于串口转USB芯片的供电是由板子提供的，而不是由计算机提供的，所以每次上电都会和计算机连接在一起，此时串口终端将不会收到数据，只有当键盘上有任意键被按下后才会在终端上显示串口发送的数据，所以我们插入TF卡后，最好不要按电源按键重新下电再上电，而应使用Starter Kit的复位按键重新复位，这样终端会立即打印串口的信息。当然也有另外一个解决方法，就是将串口转USB芯片的电源改由计算机通过USB线提供，这样一旦接入计算机后，无论Starter Kit是否上电都一直保持连接。

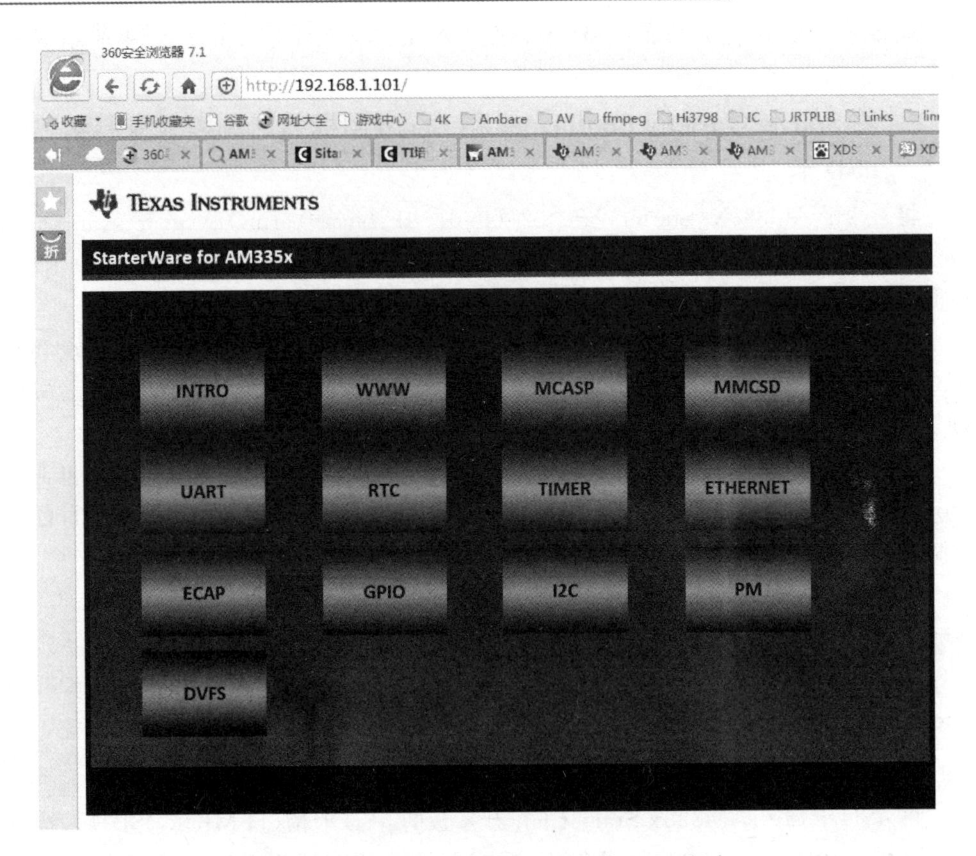

图 4.6　StarterWare_02_00_01_01 的 Demo 程序

4.2.3　Windows 下开发环境的搭建

在 Windows 计算机上下载安装 CCSv5 或 CCSv6，相关内容可参考之前的章节。

1. 导入 StarterWare 工程

StarterWare 已经安装在 C:\ti 目录下。

启动 Code Composer Studio 6.0.0，选择 File→Import 命令，弹出如图 4.7 所示的对话框。

选中 Code Composer Studio→CCS Projects，再单击 Next 按钮。

单击 Select search-directory 后的 Browse 按钮，选择 C:\ti\AM335X_StarterWare_02_00_01_01\build\armv7a\cgt_ccs\am335x\evmskAM335x 目录，如图 4.8 所示。

选择所需的例子，或者选择全部导入，单击 Finish 按钮，如图 4.9 所示。

2. 编译工程

导入之后，右击需要编译的工程名，选择 Rebuild Project 命令进行编译，如图 4.10 所示。

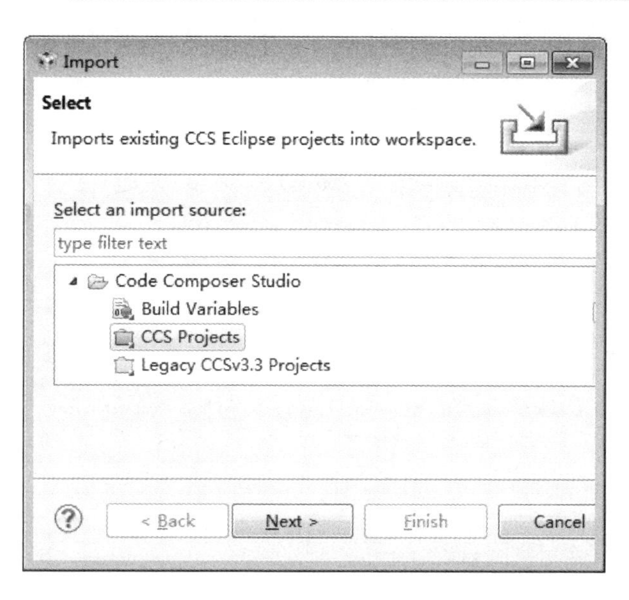

图 4.7　CCS 导入对话框

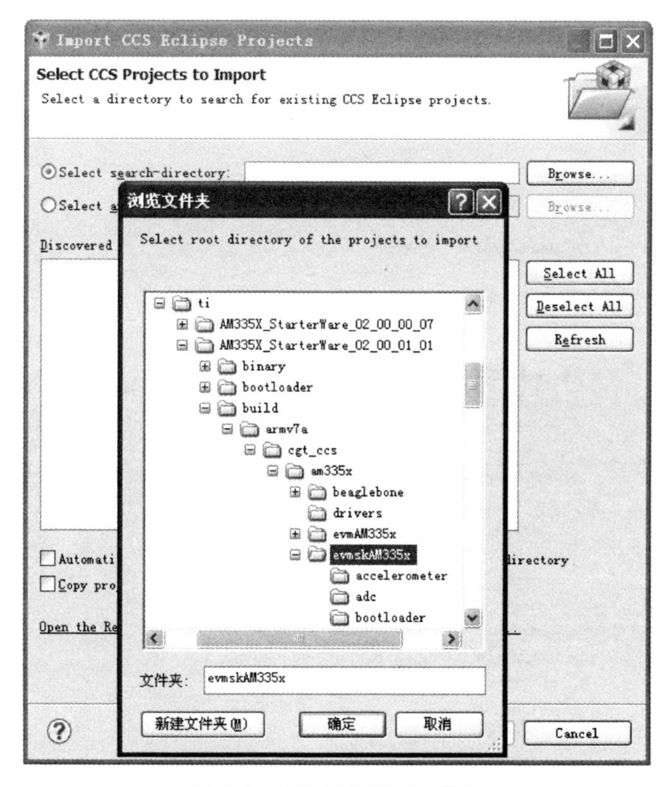

图 4.8　CCS 导入浏览文件夹

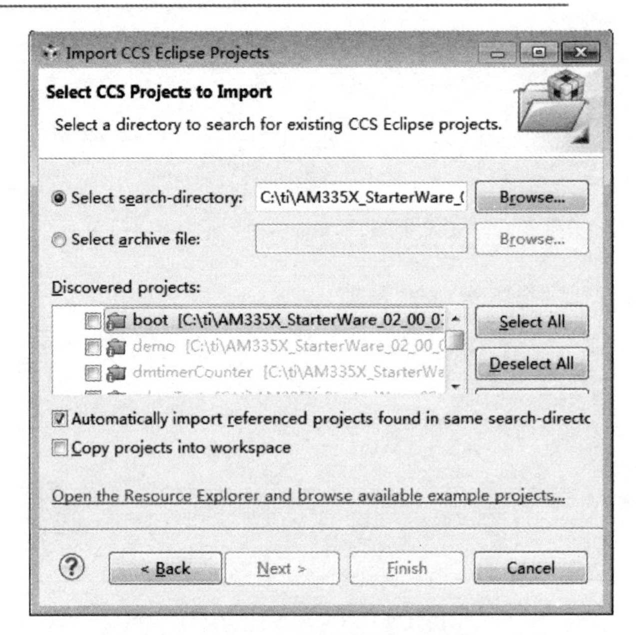

图 4.9　CCS 导入完成对话框

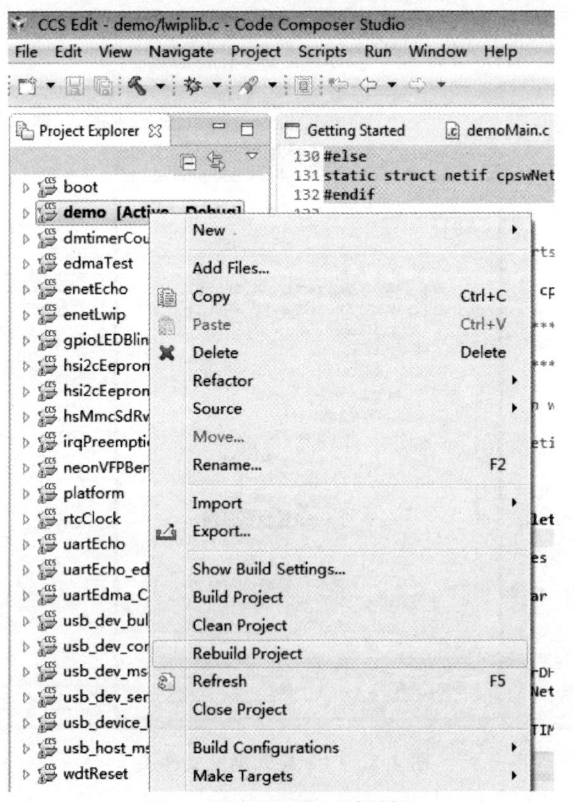

图 4.10　CCS 编译对话框

注：此时编译可能会出现错误，如图 4.11 所示。

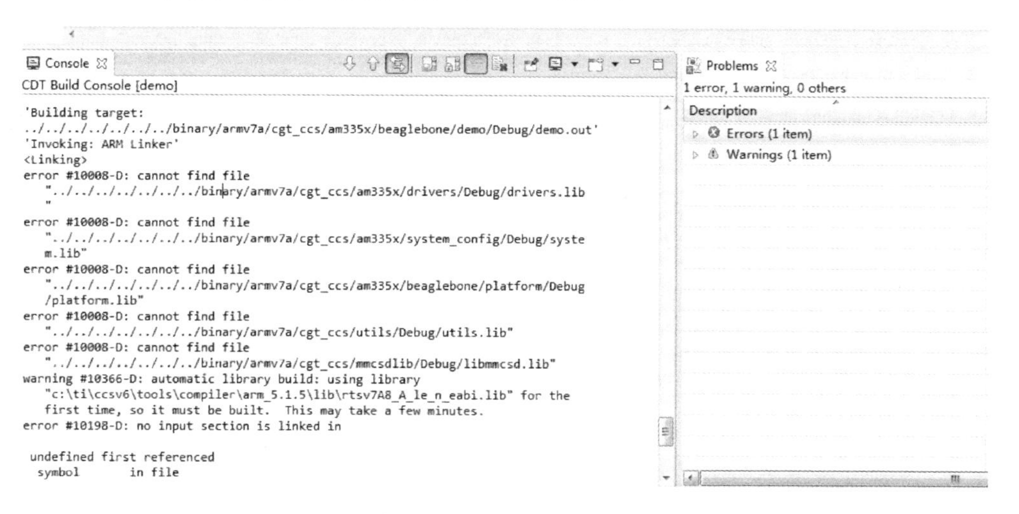

图 4.11　StarterWare 编译出错

出现这种错误的原因是没有找到 Debug 下的各个库文件，这时需要进入提示出错的目录下手动创建 Debug 文件夹，然后从 Release 文件夹中将 *.lib 文件复制到 Debug 文件夹中，再重新编译就可以通过了。生成的目标文件位于：C:\ti\AM335X_StarterWare_02_00_01_01\binary\armv7a\cgt_ccs\am335x\evmskAM335x\demo\Debug，如图 4.12 所示。

图 4.12　StarterWare 编译生成的目标文件

3. 认识目标文件及 ti_image 工具

生成的目标文件有 demo.bin 和 demo_ti.bin，开始笔者不知道这两个文件有何区别，就直接将 demo.bin 重命名为 app，然后将其和 MLO 一同放入 μSD 卡，但运行出错。参考 C:\ti\AM335X_StarterWare_02_00_01_01\docs\UserGuide_02_00_01_01.pdf 发现，它需要通过 ti_image 转化成 app 才能运行，原文解释如图 4.13 所示。

打开 cmd，并进入 ti_image 工具目录，如图 4.14 所示。

我们再用 Hex 比较工具打开，转换前后两个文件间的区别如图 4.15 所示。

Usage of ti_image Tool

TI image tool, available in "tools" directory of the package can be used to add the header information to the boot loader or application binary images. It has separate executable for Windows and Linux.

Usage of the same is explained below.

For Windows

1. Open command prompt
2. Go to /tools/ti_image
3. If the tiimage.exe is not available, generate one with Cygwin or MingW [3]environvment using the command "gcc tiimage.c -o tiimage"
4. Execute the image converter by giving proper inputs in the format, tiimage.exe <load address> <boot mode> <input image path/name> <output image path/name>

- Examples
  - tiimage.exe 0x402F0400 MMCSD boot.bin MLO
  - tiimage.exe 0x80000000 NONE uartEcho.bin app

图 4.13　ti_image 工具说明

```
管理员: C:\Windows\system32\cmd.exe
Microsoft Windows [版本 6.1.7601]
版权所有 (c) 2009 Microsoft Corporation。保留所有权利。

C:\Users\Administrator>cd C:\ti\AM335X_StarterWare_02_00_01_01\tools\ti_image

C:\ti\AM335X_StarterWare_02_00_01_01\tools\ti_image>dir
 驱动器 C 中的卷没有标签。
 卷的序列号是 BC62-BE39

 C:\ti\AM335X_StarterWare_02_00_01_01\tools\ti_image 的目录

2015/08/05  11:03    <DIR>          .
2015/08/05  11:03    <DIR>          ..
2015/08/03  21:30            44,932 boot.bin
2015/08/04  16:06           311,817 demo.bin
2015/08/03  20:19             4,660 gpioLEDBlink.bin
2013/07/12  14:31             6,675 tiimage
2013/07/12  05:23             4,619 tiimage.c
2013/07/12  05:23            51,618 tiimage.exe
               6 个文件        424,321 字节
               2 个目录 35,425,869,824 可用字节

C:\ti\AM335X_StarterWare_02_00_01_01\tools\ti_image>tiimage.exe 0x402F0400 MMCSD
 boot.bin MLO


C:\ti\AM335X_StarterWare_02_00_01_01\tools\ti_image>tiimage.exe 0x80000000 NONE
demo.bin app


C:\ti\AM335X_StarterWare_02_00_01_01\tools\ti_image>dir
 驱动器 C 中的卷没有标签。
 卷的序列号是 BC62-BE39

 C:\ti\AM335X_StarterWare_02_00_01_01\tools\ti_image 的目录

2015/08/05  11:07    <DIR>          .
2015/08/05  11:07    <DIR>          ..
2015/08/05  11:07           311,825 app
2015/08/03  21:30            44,932 boot.bin
2015/08/04  16:06           311,817 demo.bin
2015/08/03  20:19             4,660 gpioLEDBlink.bin
2015/08/05  11:06            44,940 MLO
2013/07/12  14:31             6,675 tiimage
2013/07/12  05:23             4,619 tiimage.c
2013/07/12  05:23            51,618 tiimage.exe
               8 个文件        781,086 字节
               2 个目录 35,423,891,456 可用字节

C:\ti\AM335X_StarterWare_02_00_01_01\tools\ti_image>_
```

图 4.14　ti_image 工具操作

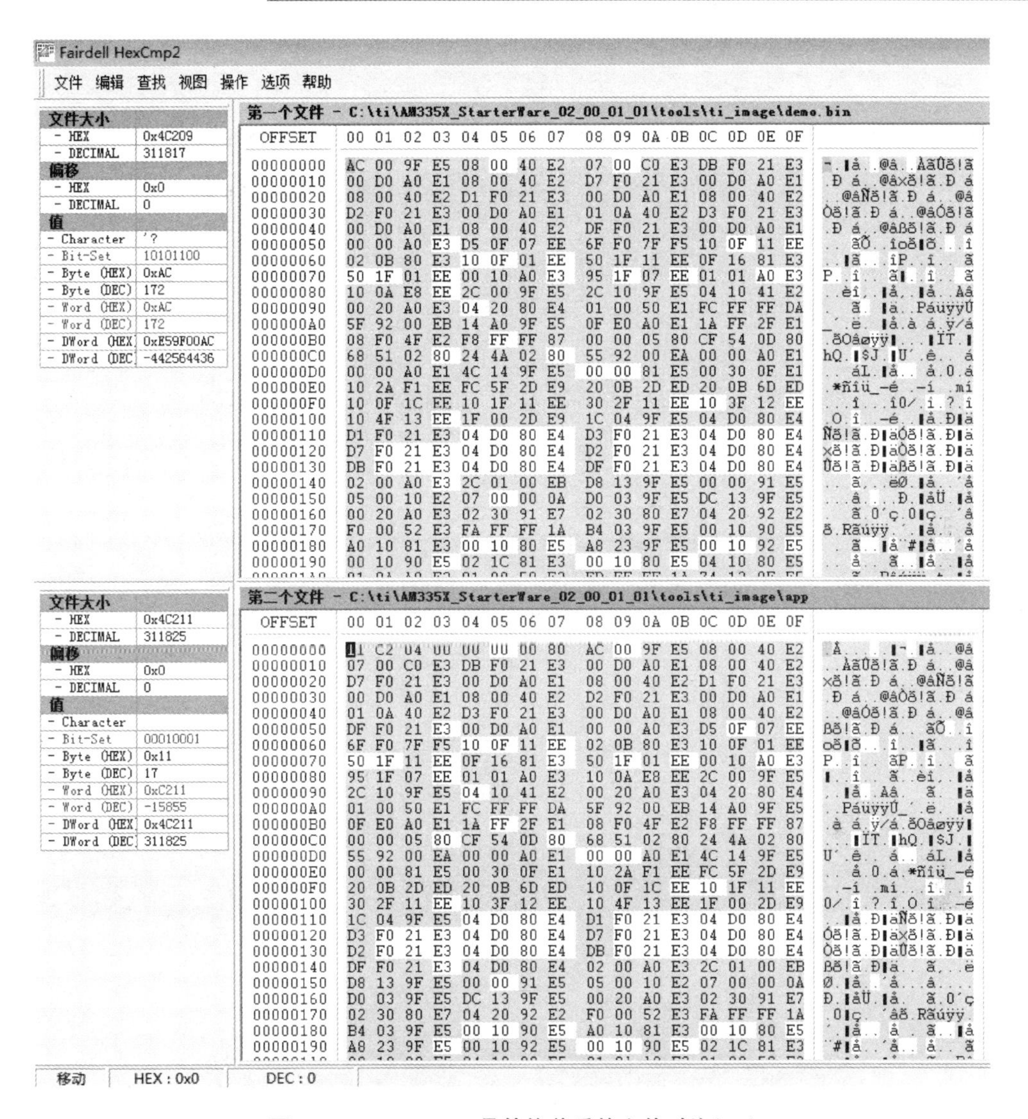

图 4.15　ti_image 工具转换前后的文件对比(一)

实际上，ti_image 只是在原文件 demo.bin 的开始加入了 8 个字节头，其中有 4 字节共 32 位为装载的地址，另 4 字节应该是代表 app 的标志。我们再打开原生成目录下的 demo.bin 和 demo_ti.bin，如图 4.16 所示。

原来 demo_ti.bin 已经增加了 8 个字节头，也就是说可以直接通过将 demo_ti.bin 重命名为 app，而不需要再调用 ti_image 工具进行转换。

4. 下载验证

将 MLO 和 app 复制到 μSD 卡并插入到 AM335x Starter Kit 开发板上，按住 SW5 按键再上电，终端打印信息如图 4.17 所示。

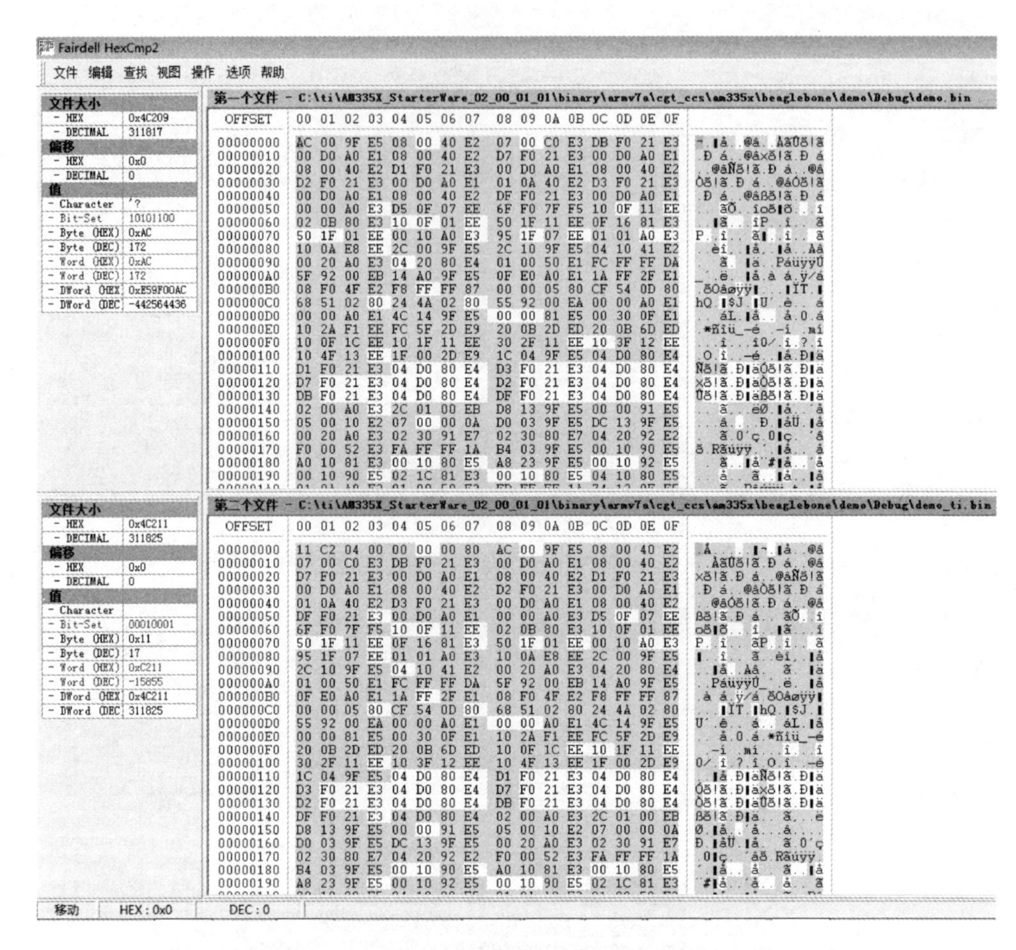

图 4.16　ti_image 工具转换前后的文件对比(二)

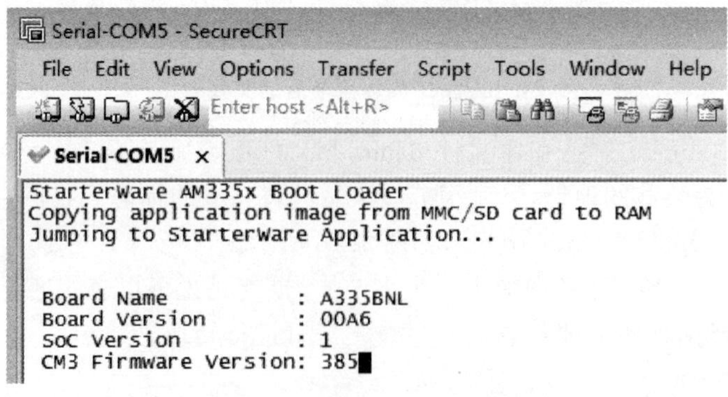

图 4.17　StarterWare 验证信息

4.3 AM335x内存映射和启动过程

在第2章硬件分析部分已经介绍过AM335x处理器内存映射的内容，这里我们用软件的思维再次介绍内存映射的相关内容，由于会存在重复性，还请各位读者见谅！

4.3.1 AM335x处理器内存映射

在分析AM335x启动之前，我们有必要对AM335x的各种存储器及地址映射关系作进一步强调，因为它保存着程序代码，以及承载运行中的代码和数据。

内存映射相关的详细内容请读者参考官方 *Technical Reference Manual.pdf* 的第177页，如图4.18所示。

L3 Memory Map

Block Name	Start_address (hex)	End_address (hex)	Size	Description
GPMC (External Memory)	0x0000_0000(1)	0x1FFF_FFFF	512MB	8-/16-bit External Memory (Ex/R/W)(2)
Reserved	0x2000_0000	0x3FFF_FFFF	512MB	Reserved
Boot ROM	0x4000_0000	0x4001_FFFF	128KB	
	0x4002_0000	0x4002_BFFF	48KB	32-bit Ex/R(2) – Public
Reserved	0x4002_C000	0x400F_FFFF	848KB	Reserved
Reserved	0x4010_0000	0x401F_FFFF	1MB	Reserved
Reserved	0x4020_0000	0x402E_FFFF	960KB	Reserved
Reserved	0x402F_0000	0x402F_03FF	64KB	Reserved
SRAM internal	0x402F_0400	0x402F_FFFF		32-bit Ex/R/W(2)
L3 OCMC0	0x4030_0000	0x4030_FFFF	64KB	32-bit Ex/R/W(2) OCMC SRAM
•	•	•	•	•
EMIF0 SDRAM	0x8000_0000	0xBFFF_FFFF	1GB	8-/16-bit External Memory (Ex/R/W)(3)

图4.18　AM335x L3 Memory Map(截图)

GPMC为通用内存控制器，用于外扩NOR Flash、NAND Flash等，起始地址为0x00000000，我们通常会将程序代码保存在此处。

Boot ROM为处理器内部的ROM，用于存放处理器自身的引导代码(出厂后已经固化，StarterWare文档里称它为RBL，即Read Only Memory BootLoader)，也是处理器上电及复位后最先执行的一段引导代码。

SRAM internal和L3 OCMC0是处理器内部的RAM，RBL会将例如SarterWare中的MLO从存储介质中装载到该区域执行。

0x402F0400～0x4030B800共109 KB如图4.19所示，RBL将MLO装载到此区域，所以MLO最大不能超过109 KB。

EMIF0 SDRAM外部存储器接口SDRAM控制器，用于扩展如DDR2或DDR3

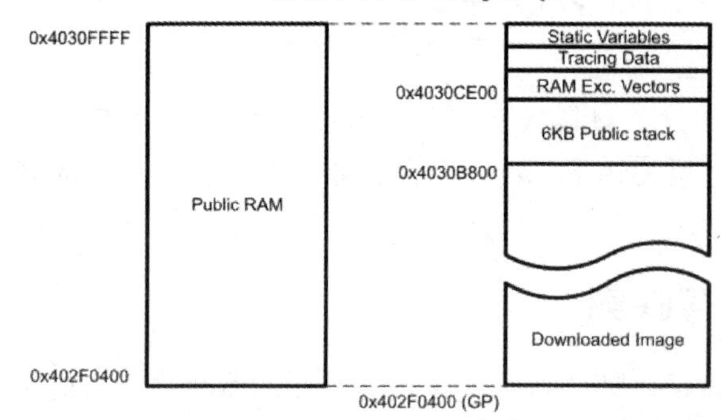

RAM Exception Vectors

Address	Exception	Content
4030CE00h	Reserved	Reserved
4030CE04h	Undefined	PC = [4030CE24h]
4030CE08h	SWI	PC = [4030CE28h]
4030CE0Ch	Pre-fetch abort	PC = [4030CE2Ch]
4030CE10h	Data abort	PC = [4030CE30h]
4030CE14h	Unused	PC = [4030CE34h]
4030CE18h	IRQ	PC = [4030CE38h]
4030CE1Ch	FIQ	PC = [4030CE3Ch]

图 4.19　AM335x Public RAM 内存映射

SDRAM 存储器，起始地址为 0x80000000，如 StarterWare 是由 MLO 将 app 装载到此开始区域执行的。

4.3.2　AM335x 处理器启动过程

当目标板上电或复位时，处理器开始执行位于内部 Boot ROM 区域的 RBL，启动顺序如图 4.20 所示。

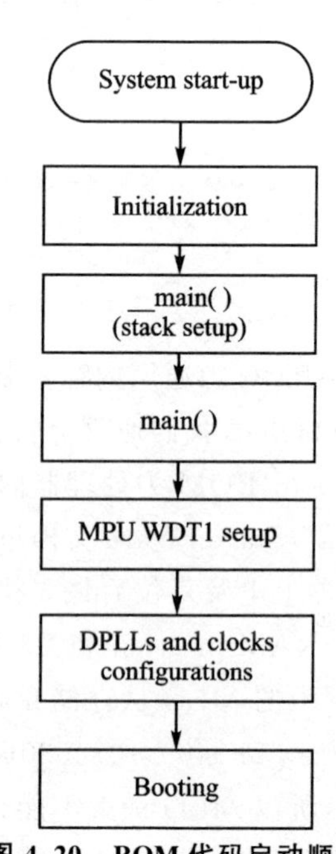

图 4.20　ROM 代码启动顺序

RBL 对自身的运行环境做初始化，并设置好堆栈、看门狗、锁相环 PLL 和系统时钟后进入引导过程，如图 4.21 所示。

在引导过程中会根据软件引导配置或处理器外部 SYSBOOT 引脚来设置引导器件列表，列表中会指明先查找哪个引导设备，再查找那个引导设备。这里的引导设备分为存储器（Memory）和外部设备接口（Peripheral）两种，前者有 NOR、NAND、MMC 和 SPI-EEPROM，后者有 Ethernet、USB 和 UART。

根据引导列表的先后，当检查到对应设备中有有效代码后就将其装载到 SRAM internal 和 L3 OCMC0 的内部 RAM 中执行，如果没有检查到就循环检查，一

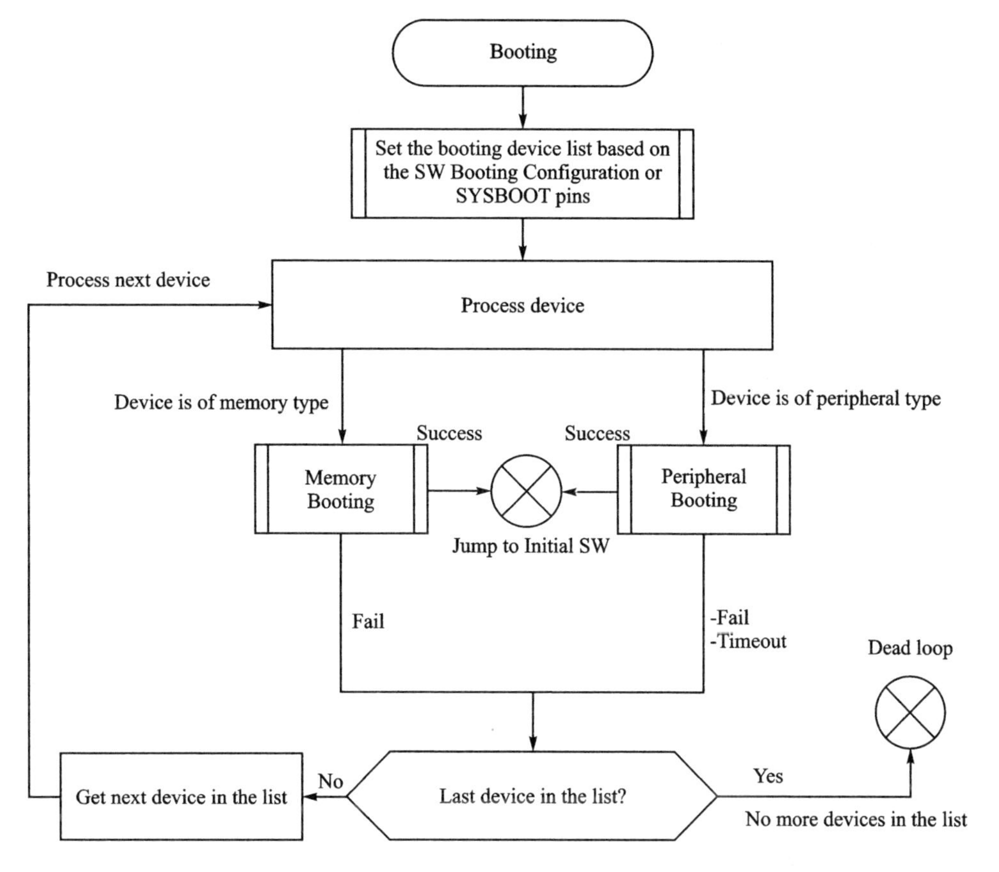

图 4.21　ROM 代码引导过程

直到最后，没有有效的代码后进入 Dead loop。

如图 4.22 所示为 AM335x Starter Kit 中 SYSBOOT 的引脚定义，定义的启动检查设备顺序为 MMC0→SPI0→UART0→USB0。

在电源上电复位完成时，LCD_DATA[15:0]引脚分别作为 SYSBOOT[15:0]功能被锁定为配置定义。

当我们将 StarterWare 的引导程序 MLO 和应用程序 app 复制到 μSD 卡，并插入到 AM335x Starter Kit 的 μSD 卡槽时，此时 RBL 在扫描 MMC0 接口时就会读取 μSD 卡中的 MLO 文件，并将其装载到内部 RAM 中运行。而 MLO 执行后，则会将 app 装载到 0x80000000 开始的 DDR 中执行。

如图 4.23 所示为 StarterWare 的 UserGuide_02_00_01_01.pdf 中对从 μSD 卡引导步骤的描述。

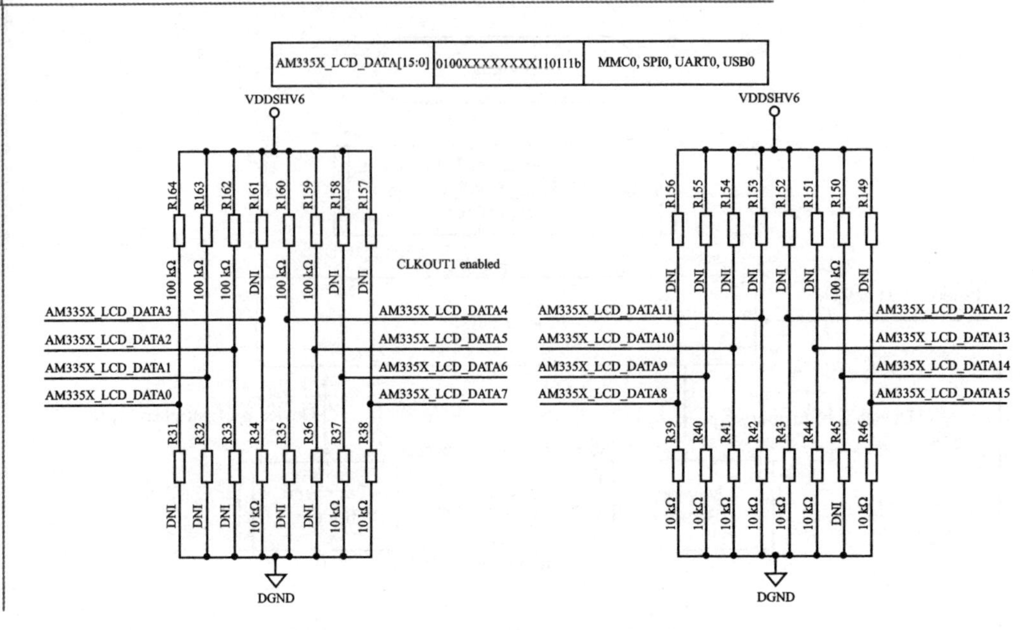

图 4.22　AM335x 启动配置定义

Stages in SD Booting

- Stage 1 : On System reset, the ROM Code copies the bootloader by reading the MLO file from the SD to the address embedded on its header, Then ROM Code handsover the execution control to Bootloader.
- Stage 2 : The Bootloader copies the StarterWare application by reading the file named app to the addresss embedded on its header. After copying, the bootloader jumps to start address of the application to handover the control to the application.
- Stage 3 : StarterWare application executes.

图 4.23　AM335x StarterWare μSD 卡引导步骤

第5章

启动代码分析

在第一篇ARM硬件部分，我们主要通过对AM335x各个信号引脚，到系统的原理图，再到仿真测试等基本都是基于硬件的内容进行讲解的，帮助读者从外部去理解、学习ARM处理器AM335x。有了这些硬件基础后，就可以通过底层软件去理解AM335x。

为什么有些朋友说ARM比单片机更容易学习，这是有一定道理的。因为单片机什么都要自己做，自己写代码去控制每个外设，而32位处理器（包括ARM）由于本身比单片机更复杂、外设更多等各方面原因，官方或第三方都会提供DEMO板和完整的测试程序，包括启动代码、各种外设的库函数、操作系统及BSP包等，让用户花更短的时间去学习，将精力都放在自己的应用上。以前的ARM9 S3C2410是，现在的AM335x更是如此。因此，我们也要充分利用这个特点和资源，学会去理解官方提供的代码；学会如何将这些代码应用在自己的产品上，将它们变成自己的东西；学会辨认哪些是重要的，哪些是不需要去深究的；学会在出现问题后如何去调试、去修改官方代码中的BUG等。此时，读者应该具有C语言的基础，参考ARM指令表能够理解汇编语言的能力（不要求记住汇编指令，也不要求汇编学得有多好，只需要能够读懂、理解启动代码，理解官方提供的汇编程序就可以了，其他地方基本用不到。阿南经常会收到网友及读者的邮件，阐述学汇编、ARM的困惑，其实这些担忧是没有必要的），最好还要有一些单片机的基础，再了解一些ARM体系结构方面的知识，如ARM工作模式、寄存器、异常中断等。有英语基础的读者，建议看IDE开发环境（S3C2410是ADS1.2，AM335x是CCSv5或CCSv6）安装后自带的在线文档，那里什么都有；不想看英文的就买一本清华大学出版社出版的《ARM体系结构与编程》一书。

从某种意义上说，如果读者已经掌握单片机的应用，那么只需要理解好所用ARM处理器的启动代码，清楚ARM处理器的启动、中断等底层工作过程，就基本可以将ARM处理器当作超级单片机来使用了，其他外围的应用和读者接触一种新的单片机的外围应用相差不大。所以阿南将启动代码作为单独的一章，希望读者要特别重视对启动代码的理解，因为只有这样才能理解这个处理器的启动工作过程、中断的处理过程，才能真正掌握这种处理器，清楚它与8位单片机的不同之处。

5.1 启动代码和 Bootloader 的区别

很多初学的朋友总是将启动代码和 Bootloader 混为一谈，这是错误的。

通常我们将处理器复位开始执行指令到进入 C 语言的 main 函数之前执行的那段汇编代码称为启动代码，这是由于 C 语言程序的运行需要具备一定的条件，如分配好外部数据空间、堆栈空间和中断入口等。另外，汇编代码可以更直接地对硬件进行操作，效率更高，这在启动和中断等对时间要求更高的情况下是非常必要的。在以前的 S3C2410 中，几乎每个 S3C2410A 的应用工程都会包含启动代码，通常是在 2410init.s 汇编文件中，特殊功能寄存器定义在 2410addr.s 中，Memory bank 配置在 memcfg.s 中，还有系统的选项等在 option.s 文件中。2410init.s 不仅包括复位后执行的代码，还包括 CPU 进入掉电模式、产生中断等与处理器直接相关的、用汇编实现的代码。而现在的 AM335x 更加精减，只在 boot 代码中包含启动代码 bl_init.asm，而其他应用则由 boot 来引导装载，boot 也就是 Bootloader。

Bootloader 不是一段代码，它是一个具有引导装载功能的完整程序，如可以引导装载 Linux 或 WinCE 的 vivi、U-boot 等，以及通常 PC 上的 BIOS 程序等。而以前基于 S3C2410A 的 Bootloader，也包含了 S3C2410A 的启动代码，也就是说 S3C2410A 的启动代码是基于 S3C2410A 的 Bootloader 的一部分，是所有基于 S3C2410A 应用程序的一部分。现在的 AM335x 的启动代码 bl_init.asm 也是其 boot 的一部分。

5.2 汇编基础

这里只介绍启动代码中经常使用到的 ARM 汇编指令、伪操作等，其他的汇编知识及详细的解释请参考其 IDE 开发环境在线文档中的 *Assembler Guide* 或《ARM 体系结构与编程》等。

5.2.1 伪操作

1. GET 及 INCLUDE

GET 伪操作类似于 C 语言中的“include ＊.h”，是将一个源文件包含到当前源文件中，并将被包含的文件在其当前位置进行处理。INCLUDE 是 GET 的同义词。

语法格式：

```
GET filename
```

其中，filename 为被包含的源文件的名称，可以使用路径信息。

2. GBLA、GBLL 及 GBLS

GBLA、GBLL 及 GBLS 伪操作用于定义一个 ARM 程序中的全局的变量，并将其初始化。

GBLA 伪操作定义一个全局的算术变量，并将其初始化为 0。

GBLL 伪操作定义一个全局的逻辑变量，并将其初始化为{FALSE}。

GBLS 伪操作定义一个全局的串变量，并将其初始化为空串“”。

语法格式：

```
<gblx> variable
```

其中，<gblx>为上述 3 种伪操作之一，variable 是所定义的全局变量的名称。

3. SETA、SETL 及 SETS

SETA、SETL 及 SETS 伪操作分别用于给一个 ARM 程序中的算术、逻辑和串变量赋值。

语法格式：

```
variable <setx> expr
```

其中，<setx>是上述 3 种伪操作之一，variable 是相应的<gblx>定义的变量，expr 为赋予的值。

4. IMPORT

IMPORT 伪操作相当于 C 语言中的 extern 声明，它告诉编译器当前的符号不在源文件中定义，而是在其他源文件中定义，在本源文件中可能引用该符号。

语法格式：

```
IMPORT symbol
```

其中，symbol 为声明的符号名称，区分大小写。

5. EXPORT 及 GLOBAL

EXPORT 声明一个符号可以被其他文件引用，相当于声明了一个全局变量。GLOBAL 是 EXPORT 的同义词。

```
EXPORT symbol
```

其中，symbol 为声明的符号名称，区分大小写。

6. LTORG

LTORG 伪操作用于声明一个数据缓冲池（也称为文字池）的开始。在使用伪指令 LDR 时，常常需要在适当的地方加入 LTORG 声明数据缓冲池，这样 LDR 加载的数据暂时放于数据缓冲池内，再用 ARM 的加载指令读出数据。如果没有用 LTORG 声明，则汇编器会在程序末尾自动声明。

语法格式：

```
LTORG
```

当程序中使用LDR之类的指令时，数据缓冲池的使用可能越界。为防止越界发生，可使用LTORG伪操作定义数据缓冲池。通常大的代码段可以使用多个数据缓冲池。ARM汇编编译器一般把数据缓冲池放在代码段的最后面，即下一代码段开始之前，或者END伪操作之前。LTORG伪操作通常放在无条件跳转指令之后，或者子程序返回指令之后，这样处理器就不会错误地将数据缓冲池中的数据当作指令来执行。例子如下：

```
……
ldr r1, = GSTATUS3      ;GSTATUS3 has the start address just after POWER_OFF wake - up
ldr r0,[r1]
mov pc,r0
LTORG
……
```


7. DATA

DATA伪操作声明在代码段中使用数据。

阿南在*Assembler Guide*中只能找到这样的描述："The DATA directive is no longer needed. It is ignored by the assembler."，在其他书籍和网上也没能找到它的具体语法格式，所以还请读者谅解。

8. DCD及DCDU

DCD用于分配一段字内存单元(分配的内存都是字对齐的)，并用伪操作中的expr初始化。& 是DCD的同义词。DCDU与DCD的不同之处在于分配的内存单元并不严格字对齐。

语法格式：

{label} DCD expr{,expr}

其中，{label}为可选，expr可以为数字表达式或者程序的标号。

9. AREA

AREA伪操作用于定义一个代码段或者数据段。

语法格式：

AREA sectionname {,attr}{,attr}…

其中，sectionname为所定义的代码段或者数据段的名称。attr是该代码段(或者程序段)的属性，在AREA伪操作中，各属性间用逗号隔开。下面列举2410init.s中用到的属性：

- CODE定义代码段，默认属性为READONLY；
- DATA定义数据段，默认属性为READWRITE；
- READONLY指定本段为只读，代码段默认属性为READONLY；

➢ READWRITE 指定本段为可读可写，数据段的默认属性为 READWRITE。

10. ENTRY

ENTRY 伪操作指定程序的入口点。

语法格式：

ENTRY

一个程序(可以包含多个源文件)中至少要有一个 ENTRY(可以有多个 ENTRY)，但一个源文件中最多只能有一个 ENTRY(可以没有 ENTRY)。

11. END

END 伪操作告诉编译器已经到了源程序的结尾。

语法格式：

END

每一个汇编源程序都包含 END 伪操作，以告诉本源程序的结束。

12. ASSERT

在汇编编译器对汇编程序的扫描中，如果 ASSERT 中的条件不成立，则 ASSERT 伪操作将报告该错误信息。

语法：

ASSERT logical expression

其中，logical expression 为一个逻辑表达式或运算符。如：

```
ASSERT    :DEF:ENTRY_BUS_WIDTH
```

DEF 为运算符。DEF 运算符判断是否定义某个符号。

语法：

:DEF :X

如果 X 已经定义，则结果为真，否则为假。

13. IF、ELSE 及 ENDIF

IF、ELSE 及 ENDIF 伪操作类似于 C 语言中的条件编译，根据条件把一段源代码包括在汇编语言程序内或者将其排除在程序之外。为了书写方便，在实际的程序中往往用符号表示，"["是 IF 伪操作的同义词，"|"是 ELSE 伪操作的同义词，"]"是 ENDIF 伪操作的同义词。

语法格式：

IF logical expression

Instructions or derectives

```
{ELSE
Instructions or derectives
}
ENDIF
```

其中，ELSE 伪操作为可选，logical expression 为控制的逻辑表达式，“[”表示 IF，它与表达式之间要有空格。

14. MACRO 及 MEND

MACRO 伪操作标识宏定义的开始，MEND 标识宏定义的结束。用 MACRO 及 MEND 定义一段代码，称为宏定义体，这样在程序中就可以通过宏指令多次调用该代码段。

语法格式：

```
MACRO
{ $ label} macroname { $ parameter{, $ parameter}…}
; code
…
; code
MEND
```

其中，$ label 用在宏指令被展开时，label 可被替换成相应的符号，通常是一个标号。在一个符号前使用 $ 表示程序被汇编时将使用相应的值来替代 $ 后的符号。

15. EQU

EQU 伪操作为数字常量，基于寄存器的值和程序中的标号定义一个字符名称。

语法格式：

```
name EQU expr{,type}
```

其中，name 为定义的字符名称；expr 为基于寄存器的地址值、程序中的标号、32 位的地址常量或者 32 位的常量；type 是当 expr 为 32 位常量时，用来指示 expr 表示的数据类型，可以为 CODE16、CODE32 和 DATA。

16. MAP

MAP 用于定义一个结构化的内存表的首地址。^是 MAP 的同义词。

语法格式：

```
MAP expr{,base-register}
```

其中，expr 为数字表达式或者是程序中的标号；base-register 为一个寄存器，当指令中包含它时，结构化内存表的首地址为 expr 和 base-register 寄存器值的和。

17. FIELD

FIELD 用于定义一个结构化内存表中的数据域。# 是 FIELD 的同义词。

语法格式：

{label} FIELD expr

其中，{label}为可选项，当指令中包含这一项时，label 的值为当前内存表的位置计数器{VAR}的值；expr 表示本数据域在内存表中所占的字节数，汇编编译器处理了 FIELD 伪操作后，内存表计数器的值将加上 expr。

MAP 伪操作和 FIELD 伪操作配合使用来定义结构化的内存表结构。MAP 定义内存表的首地址，FIELD 定义内存表中各数据域的字节长度，并可以为每一个数据域指定一个标号，其他指令可以引用该标号。

5.2.2 CCS 支持的伪操作

5.2.1 小节介绍的伪指令是 ADS1.2 及后续版本都支持的，但在 CCS 中还会有自己的一些伪指令，我们可以在 CCS 的 Help 里搜索查看解释。

如图 5.1 所示为 def、ref 和 global 的解释。

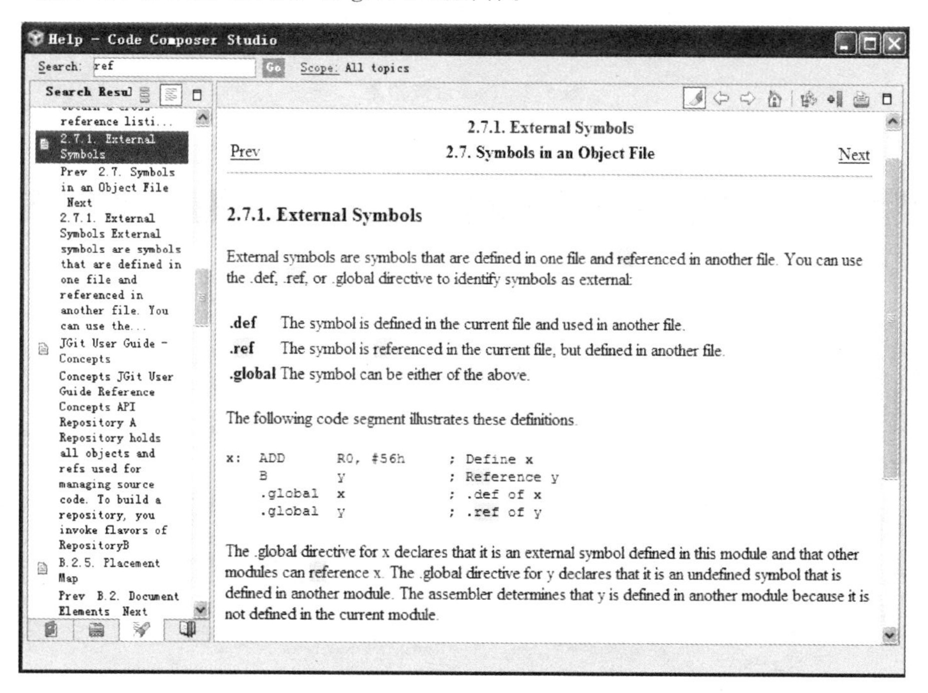

图 5.1 def、ref 和 global 的解释

.def 是在当前文件中定义被其他文件引用的变量；.ref 是在当前文件中引用其他文件定义的变量；.global 是定义一个全局变量，它既可以被当前文件引用，也可以

被其他文件引用。

如图5.2所示为.set设置常量。

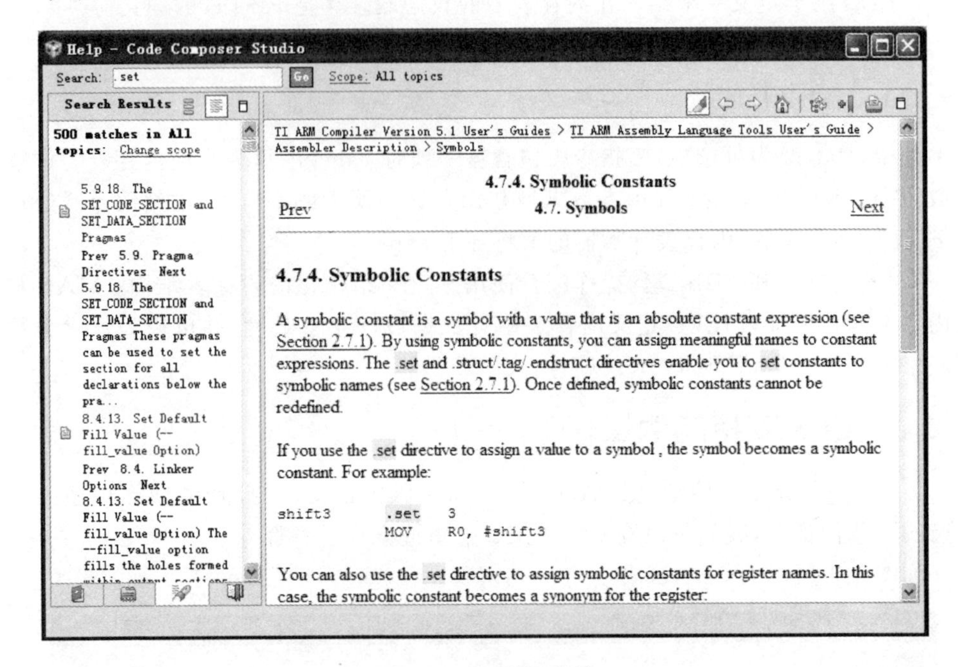

图5.2 .set设置常量

如图5.3所示为.text声明代码段入口。

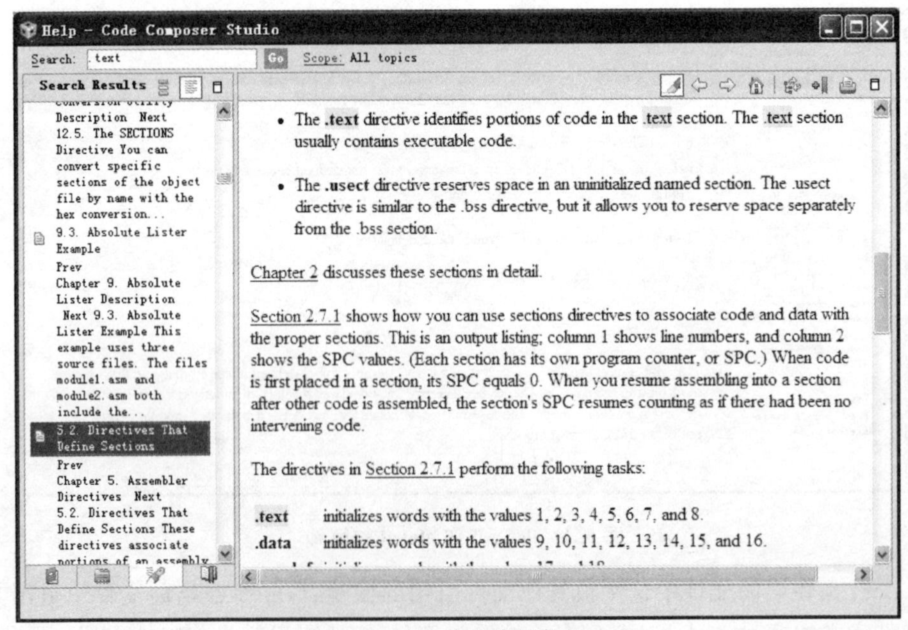

图5.3 .text声明代码段入口

如图 5.4 所示为.state32 声明 32 bit 指令。

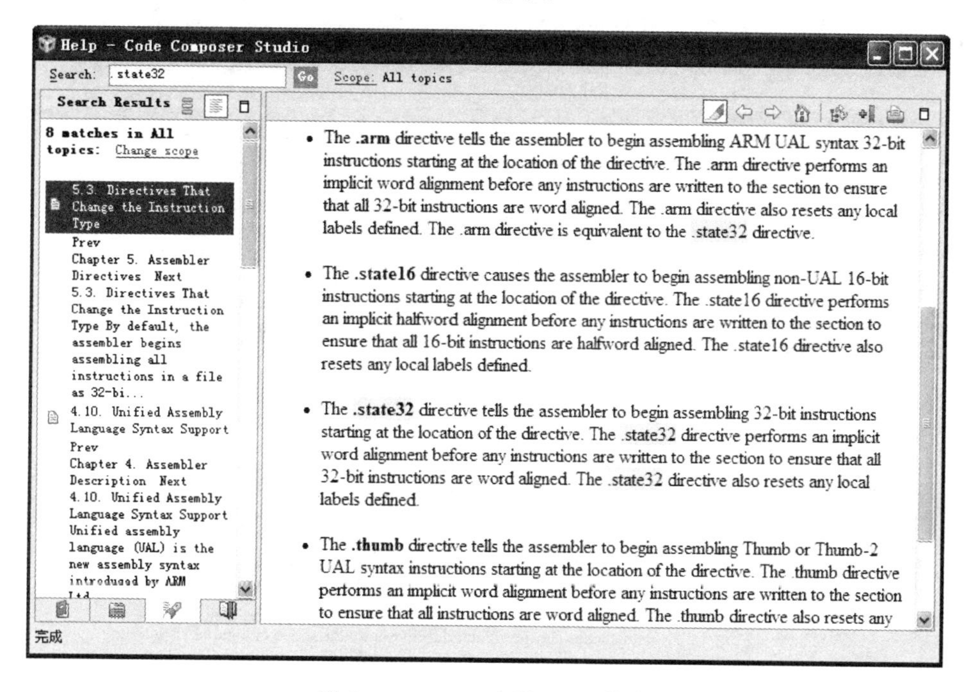

图 5.4 .state32 声明 32 bit 指令

如图 5.5 所示为.word 指定 4 字节。

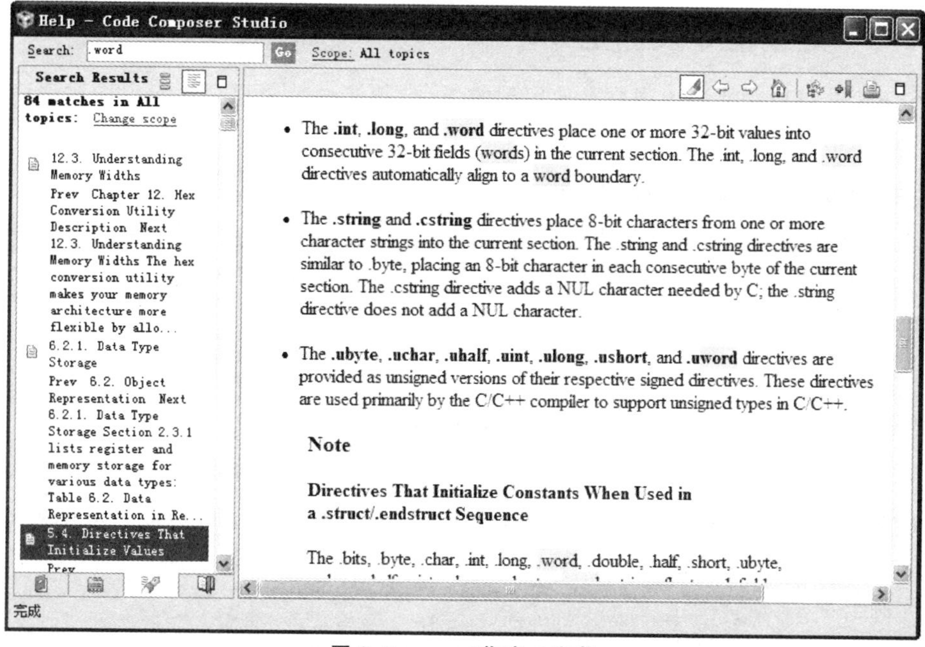

图 5.5 .word 指定 4 字节

如图5.6所示为.end指定汇编语言的结束。

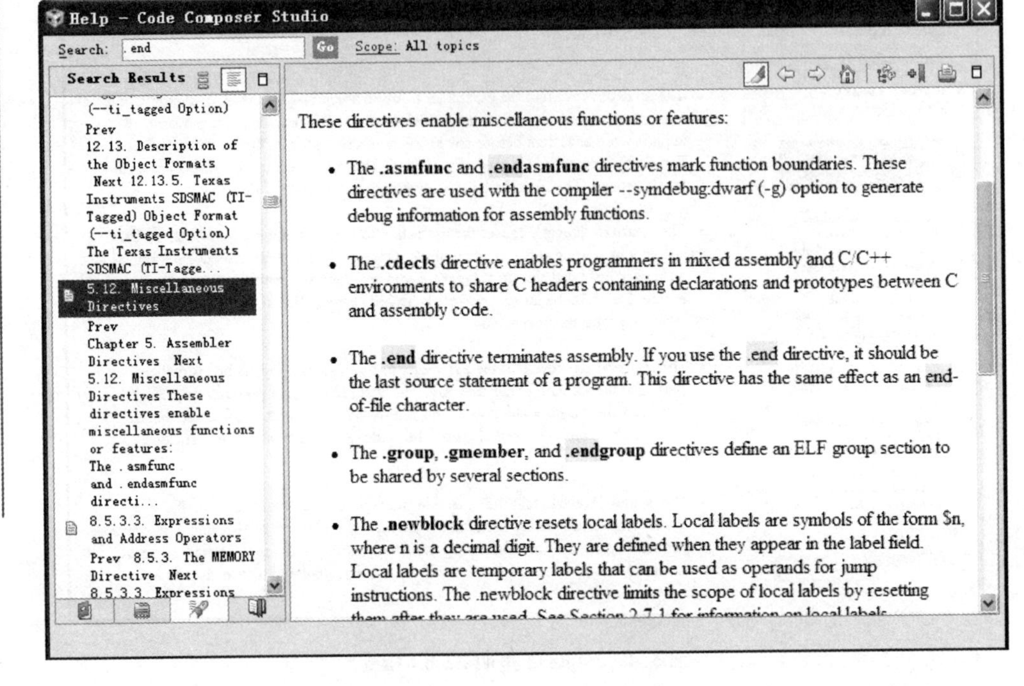

图5.6　.end指定编译语言的结束

5.2.3　汇编指令及伪指令

下面介绍启动代码中通常会使用到的指令，其他没有介绍到的指令描述请读者参考《ARM体系结构与编程》一书。

1. Load/Store内存访问指令

Load指令用于从内存中读取数据放入寄存器中；Store指令用于将寄存器中的数据保存到内存中。这里的装载和存储都是相对寄存器的，即数据装载到寄存器，将寄存器的数据保存。

语法格式：

LDR{＜cond＞} ＜Rd＞,＜addressing_mode＞

其中，＜cond＞为指令执行的条件码，可选项，当忽略时指令为无条件执行，本节的其他指令与此相同；＜Rd＞为目标寄存器，本节其他指令与此相同；＜addressing_mode＞为指令的内存地址。

例子：

```
ldr r1,[r0]          ;r0保存内存地址，该指令将内存的内容装载到r1寄存器中
```

```
str r1,[r0]          ;该指令将 r1 寄存器的值保存到 r0 指向的内存中
```

2. LDR 大范围的地址读取伪指令

LDR 伪指令将一个 32 位的常数或一个地址值读取到寄存器中。

语法格式：

LDR{<cond>} <Rd>,=[expr|label-expr]

其中，expr 为 32 位常量；label-expr 为基于 PC 的地址表达式或者是外部表达式。

例子：

```
ldr r0, = WTCON ;  WTCON 为特殊功能寄存器地址 0x53000000,该指令使 r0 = 0x53000000
```

3. MOV 传送指令

MOV 指令将<shifter_operand>表示的数值传送到目标寄存器<Rd>中，并根据操作结果更新 CPSR 中相应的条件标志位。

语法格式：

MOV{<cond>}{S} <Rd>,<shifter_operand>

其中，S 决定指令的操作是否影响 CPSR 中条件标志位的值；<shifter_operand>为向目标寄存器传送的数据常数或保存数据的源寄存器，本节其他指令与此相同。

例子：

```
mov r1,r3            ;将 r3 的内容传送给 r1,此时 r1 与 r3 的值相同
mov r1,#16           ;将立即数传给 r1,此时 r1 的值为 16
```

4. TST 位测试指令

TST 指令将<shifter_operand>表示的数值与寄存器<Rn>的值按位做逻辑与操作，根据操作结果更新 CPSR 中相应的条件标志位。它通常用于测试寄存器中某个位是 1 还是 0。

语法格式：

TST{<cond>} <Rn>,<shifter_operand>

其中，<Rn>为第一个源操作数所在的寄存器，本节其他指令与此相同。

例子：

```
ldr    r1, = GSTATUS2
ldr    r0,[r1]
tst    r0,#0x2        ;测试 GSTATUS2 寄存器的 bit1 是否为 1,以确定下条指令的跳转
bne    WAKEUP_POWER_OFF
```

5. CMP 比较指令

CMP 指令从寄存器<Rn>中减去<shifter_operand>表示的数值，根据操作的结果更新 CPSR 中相应的条件标志位。它与 SUBS 减法指令的区别在于 CMP 不保存操作结果。

语法格式：

CMP{<cond>} <Rn>,<shifter_operand>

例子：

```
    add    r2,r0,#52
0
    ldr    r3,[r0],#4
    str    r3,[r1],#4
    cmp    r2,r0          ;r2 减去 r0,以确定 r0 是否到达 r2 的值
    bne    %B0            ;不相等跳转
```

6. 跳转指令

指令跳转到指定的目标地址。

语法格式：

B{L}{<cond>} <target_address>

其中，L 决定是否保存返回地址。当有 L 时，当前 PC 寄存器的值(该跳转指令的下条指令的地址)将保存到 LR 寄存器中；当无 L 时，仅执行跳转。<target_address>为目标地址。

例子：

```
    bl     InitStacks       ;跳转到 InitStacks 标号指定的地址处执行
……
InitStacks
    ……
    mov    pc,lr            ;返回
```

7. ADD 加法指令

ADD 指令将<shifter_operand>表示的数据与寄存器<Rn>中的值相加，并把结果保存到目标寄存器<Rd>中，同时根据操作结果更新 CPSR 中相应的条件标志位。

语法格式：

ADD{<cond>}{S} <Rd>,<Rn>,<shifter_operand>

例子：

```
add    r2,r0,#52     ;将r0值与立即数52相加,结果保存在r2中
```

8. SUB 减法指令

SUB 指令从寄存器<Rn>中减去<shifter_operand>表示的数值,并把结果保存到目标寄存器<Rd>中,同时根据操作的结果更新 CPSR 中相应的条件标志位。

语法格式:

SUB{<cond>}{S} <Rd>,<Rn>,<shifter_operand>

例子:

```
    mov r1,#16
0   subs r1,r1,#1      ;将r1内容减1,再将结果保存到r1中,结果更新CPSR标志位
    bne %B0
```

9. BIC 位清除指令

BIC 指令将<shifter_operand>表示的数值取反后再与<Rn>的值按位做逻辑与操作,并把结果保存到目标寄存器<Rd>中,同时根据操作的结果更新 CPSR 中相应的条件标志位。

语法格式:

BIC{<cond>}{S} <Rd>,<Rn>,<shifter_operand>

10. ORR 逻辑或操作指令

ORR 指令将<shifter_operand>表示的数据与寄存器<Rn>中的值按位做逻辑或操作,并把结果保存到目标寄存器 Rd 中,同时更新 CPSR 相应的条件标志位。

语法格式:

ORR{<cond>}{S} <Rd>,<Rn>,<shifter_operand>

11. MRS 和 MSR 状态寄存器访问指令

MRS 指令用于将状态寄存器的内容传送到通用寄存器中;MSR 指令用于将通用寄存器的内容或一个立即数传送到状态寄存器中。

语法格式:

MRS{<cond>} <Rd>,CPSR

MSR{<cond>} CPSR_<fields>,#<immediate>

MSR{<cond>} CPSR_<fields>,<Rm>

其中,<fields>设置状态寄存器中需要操作的位,状态寄存器的 32 位可以分为 4 个 8 位的域:bits[31:24]为条件标志位域,用 f 表示;bits[23:16]为状态位域,用 s 表示;bits[15:8]为扩展位域,用 x 表示;bits[7:0]为控制位域,用 c 表示。<imme-

diate>为将要传送到状态寄存器中的立即数。<Rm>寄存器包含将要传送到状态寄存器中的数据。

例子：

```
mrs r0,cpsr                        ;将状态寄存器中的内容传送到 r0 中
bic r0,r0,#MODEMASK                ;将模式位都清 0,再保存到 r0 中
orr r1,r0,#UNDEFMODE|NOINT         ;将 r0 设置成 UndefMode,且关中断,结果保存到 r1
msr cpsr_cxsf,r1                   ;将 r1 传给状态寄存器,即设置成 UndefMode,关中断
```

12. STMFD 和 LDMFD 堆栈寻址指令

LDM 和 STM 分别为批量 Load/Store 内存访问指令，FD(Full Descending)为满递减数据栈。

例子：

```
stmfd    sp!,{r8 - r9}             ;将 r8、r9 压入到堆栈,并更新堆栈指针
                                   ;SP 先递减,再将 r9 压栈,SP 再递减,r8 再压栈
……
ldmfd    sp!,{r8 - r9,pc}          ;从堆栈中恢复 r8、r9 和 PC 寄存器,并更新堆栈指针
                                   ;先将 SP 指向的地址内容出栈给 r8,然后递增,再出栈
                                   ;给 r9,再递增,再出栈给 PC
```


5.2.4 ARM 程序状态寄存器和段

如图 5.7 所示为 ARM 程序状态寄存器 CPSR。

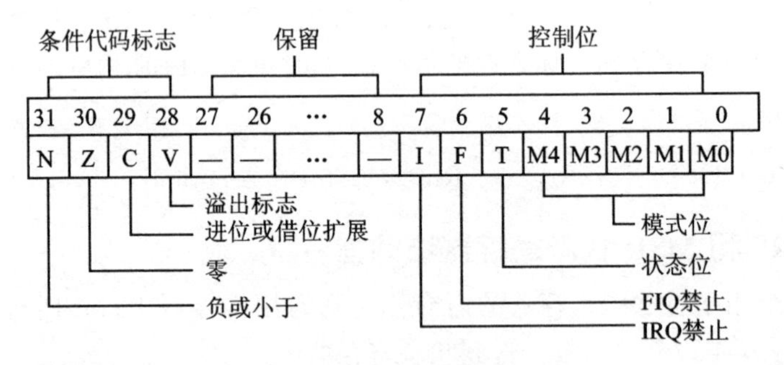

不同工作模式对应的二进制

模式名	用户	快中断	中断	管理	中止	未定义	系统
M[4:0]	10000	10001	10010	10011	10111	11011	11111

图 5.7　ARM 程序状态寄存器

CPSR 是 ARM 处理器的当前程序状态寄存器，共 32 位，其中 cpsr_c 代表的是这 32 位中的低 8 位，也就是控制位。Bit7、Bit6 分别为 FIQ 和 IRQ 的中断禁止位，当设置为 1 时，相应的中断被禁止。M[4:0]为 ARM 各种工作模式的设定位。该寄

存器只能通过 MRS 和 MSR 指令进行操作，前者为读取 CPSR 寄存器值保存到通用寄存器中，后者为写 CPSR 寄存器值。

另外，关于段和堆栈的描述如下。

BSS 段：BSS 段(bss segment)通常是指用来存放程序中未初始化的全局变量的一块内存区域。BSS 是英文 Block Started Symbol 的简称。BSS 段属于静态内存分配。

DATA 段：数据段(data segment)通常是指用来存放程序中已经初始化的全局变量的一块内存区域。数据段属于静态内存分配。

CODE 段：代码段(code segment/text segment)通常是指用来存放程序执行代码的一块内存区域。这部分区域的大小在程序运行前就已经确定，并且内存区域通常属于只读，某些架构也允许代码段为可写，即允许修改程序。在代码段中，也有可能包含一些只读的常数变量，例如字符串常量等。

堆(heap)：堆是用于存放进程运行中被动态分配的内存段，它的大小并不固定，可动态扩张或缩减。当进程调用 malloc 等函数分配内存时，新分配的内存就被动态添加到堆上(堆被扩张)；当利用 free 等函数释放内存时，被释放的内存从堆中被剔除(堆被缩减)。

栈(stack)：栈又称堆栈，是用户存放程序临时创建的局部变量，也就是我们函数括号“{}”中定义的变量(但不包括 static 声明的变量，static 意味着数据段中存放变量)。除此之外，在函数被调用时，其参数也会被压入发起调用的进程栈中，函数的返回值也会被存放回堆栈中。由于堆栈的先进先出特点，所以栈非常适于用来保存/恢复调用现场。从这个意义上讲，我们可以把堆栈看成一个寄存、交流临时数据的内存区。

5.3　启动代码 bl_init.asm 及功能模块分解

在下面的内容中，读者需要启动 CCSv5，且加载 StarterWare 库中的 Boot 工程，并打开 bl_init.asm 文件。在 S3C2410 中，基本每个前后台系统工程都会包含启动代码，它们的各个工程是相互独立的。但在 AM335x 的 StarterWare 库中，只有 Boot 工程中有一个启动代码(即 bl_init.asm)，而其他的应用工程都是依靠 Boot 加载的，所以除了 Boot 外，其他的应用工程都是没有启动代码的。

5.3.1　全局变量、内部符号等的定义

如图 5.8 所示为全局变量定义、引入变量声明和定义常数值等。

```
*bl_init.asm    bl_main.c
40 ;******************************* Global Symbols*********************************
41         .global Entry               ;定义全局变量，即可以被当前文件引用，也可以被其它文件引用
42         .global start_boot
43         .global __TI_auto_init
44
45         .ref __stack                ;引用其它文件定义的变量
46         .ref bss_start
47         .ref bss_end
48         .ref start_boot
49         .ref main
50
51 ;************************ Internal Definitions ******************************
52 ;
53 ; Define the stack sizes for different modes. The user/system mode will use
54 ; the rest of the total stack size
55 ;
56
57 UND_STACK_SIZE .set 0x8              ;定义常量
58 ABT_STACK_SIZE .set 0x8
59 FIQ_STACK_SIZE .set 0x8
60 IRQ_STACK_SIZE .set 0x100
61 SVC_STACK_SIZE .set 0x8
62
63 ;
64 ; to set the mode bits in CPSR for different modes
65 ;
66
67 MODE_USR .set 0x10
68 MODE_FIQ .set 0x11
69 MODE_IRQ .set 0x12
70 MODE_SVC .set 0x13
71 MODE_ABT .set 0x17
72 MODE_UND .set 0x1B
73 MODE_SYS .set 0x1F
74
75 I_F_BIT .set 0xC0
76
77 ;**************************** Code Seection ***********************************
78         .text                       ;声明代码段入口
79
80 ;
81 ; This code is assembled for ARM instructions
82 ;
83         .state32                    ;声明32-bit指令
84
85 ;****************************************************************************
86 ;
87 ;****************************************************************************
88 ;
89 ; The reset handler sets up the stack pointers for all the modes. The FIQ and
```

图 5.8　变量定义

5.3.2　程序入口及各种模式的堆栈初始化

如图 5.9 所示为程序入口 Entry(芯片内部的 ROM 代码加载运行的入口点)，且初始化了各种模式下的堆栈指针，并禁止 FIQ、IRQ 中断。

```
 90 ; IRQ shall be disabled during this. Then, clearthe BSS sections, switch to the
 91 ;  main() function.
 92 ;
 93 Entry:
 94 ;
 95 ; Set up the Stack for Undefined mode
 96 ;
 97          LDR    r0, _stackptr                   ; Read the stack address
 98          MSR    cpsr_c, #MODE_UND|I_F_BIT       ; switch to undef  mode
 99          MOV    sp,r0                           ; write the stack pointer
100          SUB    r0, r0, #UND_STACK_SIZE         ; give stack space
101 ;
102 ; Set up the Stack for abort mode
103 ;
104          MSR    cpsr_c, #MODE_ABT|I_F_BIT       ; Change to abort mode
105          MOV    sp, r0                          ; write the stack pointer
106          SUB    r0,r0, #ABT_STACK_SIZE          ; give stack space
107 ;
108 ; Set up the Stack for FIQ mode
109 ;
110          MSR    cpsr_c, #MODE_FIQ|I_F_BIT       ; change to FIQ mode
111          MOV    sp,r0                           ; write the stack pointer
112          SUB    r0,r0, #FIQ_STACK_SIZE          ; give stack space
113 ;
114 ; Set up the Stack for IRQ mode
115 ;
116          MSR    cpsr_c, #MODE_IRQ|I_F_BIT       ; change to IRQ mode
117          MOV    sp,r0                           ; write the stack pointer
118          SUB    r0,r0, #IRQ_STACK_SIZE          ; give stack space
119 ;
120 ; Set up the Stack for SVC mode
121 ;
122          MSR    cpsr_c, #MODE_SVC|I_F_BIT       ; change to SVC mode
123          MOV    sp,r0                           ; write the stack pointer
124          SUB    r0,r0, #SVC_STACK_SIZE          ; give stack space
125 ;
126 ; Set up the Stack for USer/System mode
127 ;
128          MSR    cpsr_c, #MODE_SYS|I_F_BIT       ; change to system mode
129          MOV    sp,r0                           ; write the stack pointer
```

图 5.9　堆栈初始化

5.3.3　BBS 段初始化

如图 5.10 所示为循环将 BBS 段内容清零。

```
130 ;
131 ; Clear the BSS section here
132 ;
133 Clear_Bss_Section:
134
135          LDR    r0, _bss_start                  ; Start address of BSS
136          LDR    r1, _bss_end                    ; End address of BSS
137          SUB    r1,r1,#4
138          MOV    r2, #0
139 Loop:
140          STR    r2, [r0], #4                    ; Clear one word in BSS
141          CMP    r0, r1
142          BLE    Loop                            ; Clear till BSS end
```

图 5.10　BBS 段清除

5.3.4 进入C语言程序

如图5.11所示为进入C语言的main函数，以及将进入前的PC值存入lr寄存器。进入C语言的main()函数，也就意味着启动代码的结束。

```
143
144 ;
145 ; Enter the start_boot function. The execution still happens in system mode
146 ;
147         LDR   r10, _main                  ; Get the address of main
148         MOV   lr,pc                            ; Dummy return to main
149         BX    r10                              ; Branch to main
150         SUB   pc, pc, #0x08                    ; looping
151
152 ;        MSR   cpsr_c, #MODE_SVC|I_F_BIT        ; change to SVC mode
153 ;        BX    lr
154 ;
155 ; End of the file
```

图5.11 C语言入口

5.3.5 bl_init.asm汇编结束

将汇编代码用到的变量指定为4字节，以及用.end指明汇编程序的结束，如图5.12所示。

```
156 ;
157
158 _stackptr:
159     .word __stack                              ;指定为4字节
160 _bss_start:
161     .word bss_start
162 _bss_end:
163     .word bss_end
164 _main:
165     .word main
166 _data_auto_init:
167     .word __TI_auto_init
168         .end                                   ;指定汇编语言的结束
```

图5.12 汇编程序结束

5.3.6 bl_init.asm总结

bl_init.asm共做3件事，或称3步。

Step1：初始化ARM各工作模式下的堆栈(共7种，但MODE_USR和MODE_SYS不需要设置堆栈)，并禁止IRQ和FIO中断使能位，最后切换到MODE_SYS模式运行。

Step2：将 BSS 段数据清除为 0。

Step3：跳转到 main()函数。

由上述内容可知，TI 将 AM335x 的启动代码汇编程序做了最大限度的精减，其只对处理器基本运行环境做了初始化，而处理器的各个时钟、外围设备等都移到 C 语言中做初始化设置，这也降低了开发理解的难度。

第6章

Boot 源代码分析

6.1 Boot 源代码目录结构

在 CCS 中导入 boot 工程如图 6.1 所示。

boot [Active - Debug_MMCSD]
- Binaries
- Includes
- Debug_MMCSD
- bl_copy.c
- bl_hsmmcsd.c
- bl_init.asm
- bl_main.c
- bl_platform.c
- boot.cmd
- cache.c
- cp15.asm
- device.c
- fat_mmcsd.c
- ff.c
- mmu.c
- bl_uart.c
- crc16.c
- xmodem.c

图 6.1 boot 代码目录结构

我们可以右击某个文件,选择 Properties 可以查看该文件所在路径。上述文件基本位于 C:\ti\AM335X_StarterWare_02_00_01_01\bootloader\src 和 C:\ti\AM335X_StarterWare_02_00_01_01\system_config\目录下,或 C:\ti\AM335X_StarterWare_02_00_01_01\third_party 目录。

6.2 启动代码 bl_init.asm 分析

bl_init.asm 共做以下三件事情。

Step1：初始化 ARM 各工作模式下的堆栈(共 7 种，但 MODE_USR 和 MODE_SYS 不需要设置堆栈)，并禁止 IRQ 和 FIO 中断使能位，最后切换到 MODE_SYS 模式运行。

Step2：将 BSS 段数据清除为 0。

Step3：跳转到 main()函数。

更详细的内容请参考之前启动代码的相关章节。

6.3　bl_main. c 主函数分析

如图 6.2 所示为 bl_main. c 源文件。

```
47 **                        Local Function Declararion
48 ****************************************************************************/
49 static void (*appEntry)();
50 /****************************************************************************
51 **                        Global Variable Definitions
52 ****************************************************************************/
53 unsigned int entryPoint = 0;
54 unsigned int DspEntryPoint = 0;
55 /****************************************************************************
56 **                        Global Function Definitions
57 ****************************************************************************/
58 /*
59  * \brief This function initializes the system and copies the image.
60  *
61  * \param  none
62  *
63  * \return none
64 */
65 int main(void)
66 {
67     /* Configures PLL and DDR controller*/
68     BlPlatformConfig();
69
70     UARTPuts("StarterWare ", -1);
71     UARTPuts(deviceType, -1);
72     UARTPuts(" Boot Loader\n\r", -1);
73
74     /* Copies application from non-volatile flash memory to RAM */
75         ImageCopy();
76
77     UARTPuts("Jumping to StarterWare Application...\r\n\n", -1);
78
79     /* Do any post-copy config before leaving boot loader */
80     BlPlatformConfigPostBoot();
81
82     /* Giving control to the application */
83     appEntry = (void (*)(void)) entryPoint;
84
85     (*appEntry)( );
86
87     return 0;
88 }
89 void BootAbort(void)
90 {
91     while(1);
92 }
93 /****************************************************************************
94 **                              END OF FILE
95 ****************************************************************************/
```

图 6.2　bl_main. c 的 main( )函数

从main()函数可知，主要做了3件事：BlPlatformConfig()对平台进行配置；ImageCopy()从Flash或μSD卡中将app映像文件复制到RAM中；(*appEntry)()从RAM中执行app。

6.4 bl_platform.c平台配置及硬件初始化分析

如图6.3所示为bl_platform.c源代码文件下的BlPlatformConfig()平台配置函数。

```
bl_init.asm    bl_main.c    bl_platform.c
1908 void BlPlatformConfig(void)
1909 {
1910     BoardInfoInit();//从EEPROM中读取板子信息，包括板的名称、版本和ID
1911 #ifdef beaglebone
1912     if(!strcmp(boardName,BNL_BOARD_NAME))
1913     {
1914         isBBB = TRUE;//确定为beaglebone black
1915     }
1916 #endif
1917     deviceVersion = DeviceVersionGet();//读取器件版本
1918     ConfigVddOpVoltage();//控制PMIC设定电源Vdd电压值
1919     oppMaxIdx = BootMaxOppGet();//获取处理器所支持的最大OPP(工作性能点，OPP是处理器电压和频率的组合，用户可对其进行控制以实现最佳处理器功耗，从而满足所有给定性能要求)
1920     SetVdd1OpVoltage(oppTable[oppMaxIdx].pmicVolt);//设置电源Vdd1电压值
1921
1922     HWREG(SOC_WDT_1_REGS + WDT_WSPR) = 0xAAAAu;//关闭看门狗
1923     while(HWREG(SOC_WDT_1_REGS + WDT_WWPS) != 0x00);
1924
1925     HWREG(SOC_WDT_1_REGS + WDT_WSPR) = 0x5555u;
1926     while(HWREG(SOC_WDT_1_REGS + WDT_WWPS) != 0x00);
1927
1928     /* Configure DDR frequency */
1929 #ifdef evmskAM335x
1930     freqMultDDR = DDRPLL_M_DDR3;
1931 #elif evmAM335x
1932     if(BOARD_ID_EVM_DDR3 == BoardIdGet())
1933     {
1934         freqMultDDR = DDRPLL_M_DDR3;
1935     }
1936     else if(BOARD_ID_EVM_DDR2 == BoardIdGet())
1937     {
1938         freqMultDDR = DDRPLL_M_DDR2;
1939     }
1940 #else
1941     if(isBBB)
1942     {
1943         freqMultDDR = DDRPLL_M_DDR3;
1944     }
1945     else
1946     {
1947         freqMultDDR = DDRPLL_M_DDR2;
1948     }
1949 #endif
1950
1951     /* Set the PLL0 to generate 300MHz for ARM */
1952     PLLInit();
1953
1954     /* Enable the control module */
1955     HWREG(SOC_CM_WKUP_REGS + CM_WKUP_CONTROL_CLKCTRL) =
1956             CM_WKUP_CONTROL_CLKCTRL_MODULEMODE_ENABLE;//使能软件唤醒
1957
```

图6.3 BlPlatformConfig()函数(一)

如图6.4所示也是bl_platform.c源代码文件下的BlPlatformConfig()平台配置函数，接图6.3。

BoardInfoInit()主要是先从板上的EEPROM中读取板的名称、版本和ID等板子的信息，然后再根据不同的板做相应的初始化工作。

DeviceVersionGet()读取板上主处理器的器件版本。

```
1958     /* EMIF Initialization */
1959     EMIFInit();
1960
1961     /* DDR Initialization */
1962
1963 #ifdef evmskAM335x
1964     /* Enable DDR_VTT */
1965     DDRVTTEnable(); //使能DDR的VTT，VTT主要为DDR的地址、控制线等信号的信号完整性而提供的终端电阻电源
1966     DDR3Init();
1967 #elif evmAM335x
1968     if(BOARD_ID_EVM_DDR3 == BoardIdGet())
1969     {
1970         DDR3Init();
1971     }
1972     else if(BOARD_ID_EVM_DDR2 == BoardIdGet())
1973     {
1974         DDR2Init();
1975     }
1976 #else
1977     if(isBBB)
1978     {
1979         DDRVTTEnable();
1980         DDR3Init();
1981     }
1982     else
1983     {
1984         DDR2Init();
1985     }
1986 #endif
1987     UARTSetup();
1988 }
```

图 6.4　BlPlatformConfig()函数(二)

ConfigVddOpVoltage()控制板上的PMIC，设定电源Vdd电压值。

BootMaxOppGet()获取处理器所支持的最大OPP(工作性能点，OPP是处理器电压和频率的组合，用户可对其进行控制以实现最佳处理器功耗，从而满足所有给定性能要求)。

SetVdd1OpVoltage(oppTable[oppMaxIdx].pmicVolt)设置电源Vdd1的电压值。

如图6.5所示为关闭看门狗定时器。

```
1922     HWREG(SOC_WDT_1_REGS + WDT_WSPR) = 0xAAAAu;//关闭看门狗
1923     while(HWREG(SOC_WDT_1_REGS + WDT_WWPS) != 0x00);
1924
1925     HWREG(SOC_WDT_1_REGS + WDT_WSPR) = 0x5555u;
1926     while(HWREG(SOC_WDT_1_REGS + WDT_WWPS) != 0x00);
```

图 6.5　关闭看门狗定时器操作语句

另外，也可以从AM335x技术手册上找到打开/关闭看门狗定时器的操作方法，如图6.6所示。

PLLInit()为锁相环时钟初始化函数。

如图6.7所示为使能sleep时软件唤醒的控制模块。

EMIFInit()为EMIF存储接口初始化。

DDRVTTEnable()使能DDR的VTT，VTT主要为DDR的地址、控制线等信号的信号完整性提供终端电阻电源。

DDR3Init()为DDR3存储器初始化。

UARTSetup()设置UART0为调试端口。

20.4.3.8 Start/Stop Sequence for Watchdog Timers (Using the WDT_WSPR Register)

To start and stop a watchdog timer, access must be made through the start/stop register (WDT_WSPR) using a specific sequence.

To disable the timer, follow this sequence:

1. Write XXXX AAAAh in WDT_WSPR.
2. Poll for posted write to complete using WDT_WWPS.W_PEND_WSPR.
3. Write XXXX 5555h in WDT_WSPR.
4. Poll for posted write to complete using WDT_WWPS.W_PEND_WSPR.

To enable the timer, follow this sequence:

1. Write XXXX BBBBh in WDT_WSPR.
2. Poll for posted write to complete using WDT_WWPS.W_PEND_WSPR.
3. Write XXXX 4444h in WDT_WSPR.
4. Poll for posted write to complete using WDT_WWPS.W_PEND_WSPR.

All other write sequences on the WDT_WSPR register have no effect on the start/stop feature of the module.

图 6.6　打开/关闭看门狗定时器说明


```
1954    /* Enable the control module */
1955    HWREG(SOC_CM_WKUP_REGS + CM_WKUP_CONTROL_CLKCTRL) =
1956            CM_WKUP_CONTROL_CLKCTRL_MODULEMODE_ENABLE;//使能软件唤醒
```

图 6.7　软件唤醒控制模块

6.5　bl_copy.c 映像复制分析

如图 6.8 所示为 bl_copy.c 源代码文件的 ImageCopy()函数。

```
bl_init.asm    bl_main.c    bl_copy.c
 87 ******************************************************************************/
 88 /*
 89  * \brief This function copies Image
 90  *
 91  * \param  none
 92  *
 93  * \return none
 94  */
 95 void ImageCopy(void)
 96 {
 97 #if defined(SPI)
 98     if (SPIBootCopy( ) != true)
 99         BootAbort();
100 #elif defined(MMCSD)
101         MMCSDBootCopy();
102 #elif defined(UART)
103     if (UARTBootCopy() != true)
104         BootAbort();
105 #elif defined(NAND)
106     if (NANDBootCopy() != true)
107         BootAbort();
108 #else
109     #error Unsupported boot mode !!
110 #endif
111 }
```

图 6.8　ImageCopy()函数

我们使用的是 AM335x Starter Kit，用 MMCSD 接口引导，故调用 MMCSD-BootCopy()函数，如图 6.9 所示。

```
140 /*
141  * \brief This function Initializes and configures MMCSD and copies
142  *         data from MMCSD.
143  *
144  * \param  none
145  *
146  * \return unsigned int: Status (Success or Failure)
147 */
148 #ifdef MMCSD
149 static unsigned int MMCSDBootCopy(void)
150 {
151     unsigned int retVal;
152 
153     BlPlatformMMCSDSetup();
154 
155     retVal = BlPlatformMMCSDImageCopy();
156 
157     return retVal;
158 }
159 #endif
```

图 6.9　MMCSDBootCopy()函数

如图 6.10 所示为 BlPlatformMMCSDSetup()函数。

```
1841 void BlPlatformMMCSDSetup(void)
1842 {
1843     /* Enable clock for MMCSD and Do the PINMUXing */
1844     HSMMCSDPinMuxSetup();
1845     HSMMCSDModuleClkConfig();
1846 }
```

图 6.10　BlPlatformMMCSDSetup()函数

对 MMCSD 接口进行一系列初始化后，最终调用图 6.11 中的 HSMMCSD-ImageCopy()函数，将 app 映像文件复制到 DDR 中运行，如图 6.11 和图 6.12 所示。

这里先打开 app 文件，再读取 8 字节的文件头，文件头中有 4 字节的文件大小，还有 4 字节的装载地址；然后再将文件中的映像读到缓冲区；再从缓冲区复制到装载地址区，读完后返回。

```
bl_init.asm   bl_main.c   bl_copy.c   bl_platform.c   bl_hsmmcsd.c
452 unsigned int HSMMCSDImageCopy(void)
453 {
454     FRESULT fresult;
455     unsigned short usBytesRead = 0;
456     ti_header imageHdr;
457     unsigned char *destAddr;
458     char *fname = "/app";
459     /*
460     ** Open the file for reading.
461     */
462     fresult = f_open(&g_sFileObject, fname, FA_READ);
463     /*
464     ** If there was some problem opening the file, then return an error.
465     */
466     if(fresult != FR_OK)
467     {
468         UARTPuts("\r\n Unable to open application file\r\n", -1);
469         return 0;
470     }
471     else
472     {
473         UARTPuts("Copying application image from MMC/SD card to RAM\r\n", -1);
474         fresult = f_read(&g_sFileObject, (unsigned char *)&imageHdr, 8,
475                          &usBytesRead);
476         if(fresult != FR_OK)
477         {
478             UARTPuts("\r\n Error reading application file\r\n", -1);
479             return 0;
480         }
481         if(usBytesRead != 8)
482         {
483             return 0;
484         }
485         destAddr = (unsigned char*)imageHdr.load_addr;
486         entryPoint = imageHdr.load_addr;
487     }
488     /*
489     ** Enter a loop to repeatedly read data from the file and display it, until
490     ** the end of the file is reached.
491     */
492     do
493     {
494         /*
495         ** Read a block of data from the file.  Read as much as can fit in the
496         ** temporary buffer, including a space for the trailing null.
497         */
498         fresult = f_read(&g_sFileObject, g_cTmpBuf, sizeof(g_cTmpBuf) - 1,
499                          &usBytesRead);
500         /*
501         ** If there was an error reading, then print a newline and return the
```

图 6.11 HSMMCSDImageCopy()函数(一)

```
502         ** error to the user.
503         */
504         if(fresult != FR_OK)
505         {
506             UARTPuts("\r\n Error reading application file\r\n", -1);
507             return 0;
508         }
509         if(usBytesRead >= sizeof(g_cTmpBuf))
510         {
511             return 0;
512         }
513         /*
514         ** Null terminate the last block that was read to make it a null
515         ** terminated string that can be used with printf.
516         */
517         g_cTmpBuf[usBytesRead] = 0;
518         /*
519         ** Read the last chunk of the file that was received.
520         */
521         memcpy(destAddr, g_cTmpBuf, (sizeof(g_cTmpBuf) - 1));
522         destAddr += (sizeof(g_cTmpBuf) - 1);
523         /*
524         ** Continue reading until less than the full number of bytes are read.
525         ** That means the end of the buffer was reached.
526         */
527     }
528     while(usBytesRead == sizeof(g_cTmpBuf) - 1);
529     /*
530     ** Close the file.
531     */
532     fresult = f_close(&g_sFileObject);
533     /*
534     ** Return success.
535     */
536     return 1;
537 }
```

图 6.12 HSMMCSDImageCopy()函数(二)

6.6 跳转到 app 运行

main()函数最后是跳转到 DDR 执行 app 程序，如图 6.13 和图 6.14 所示。

```
entryPoint = imageHdr.load_addr;
```

图 6.13 entryPoint 赋值

```
82      /* Giving control to the application */
83      appEntry = (void (*)(void)) entryPoint;
84
85      (*appEntry)( );
```

图 6.14 跳转到 app 运行

至此，Boot 引导结束。

第7章

LCD 例程源代码分析

7.1 LCD 例程源代码目录结构

在 CCS 左边 Project Explorer 中，右击选择 Import→CCS Projects 打开 Import CCS Eclipse Projects 对话框，单击 Browse 按钮选择目录 C:\ti\AM335X_StarterWare_02_00_01_01\build\armv7a\cgt_ccs\am335x\evmskAM335x\raster，导入 rasterDisplay 项目例程，如图 7.1 所示。

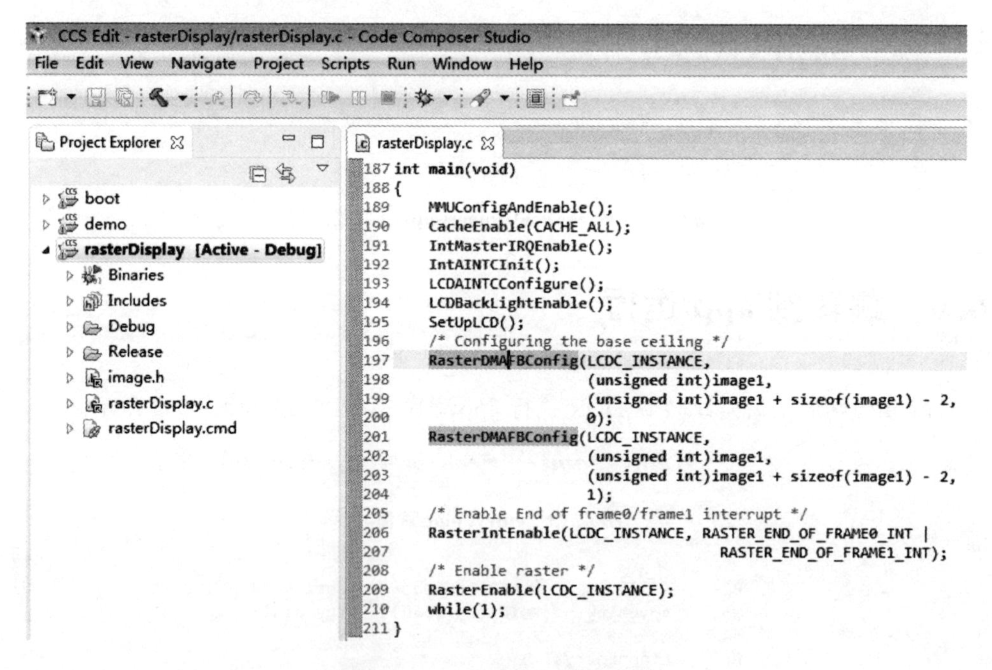

图 7.1　rasterDisplay 项目工程

虽然 rasterDisplay 项目工程中只有 rasterDisplay.c 源文件，但会调用如 system_config、platform 等目录下的一些文件。

7.2 rasterDisplay.c 文件分析

在如图7.1所示的rasterDisplay.c文件的main()主函数中，基本都是初始化调用。

7.2.1 内存管理和高速缓存的配置

MMUConfigAndEnable()函数为配置使能MMU内存管理单元。

CacheEnable()函数使能Cache高速缓冲区。

7.2.2 中断相关的配置分析

IntMasterIRQEnable()函数使能ARM处理器的IRQ中断。

IntAINTCInit()函数为初始化AM335x ARM处理器的中断控制器，而且通常在使用中断控制器之前完成调用，如图7.2所示。

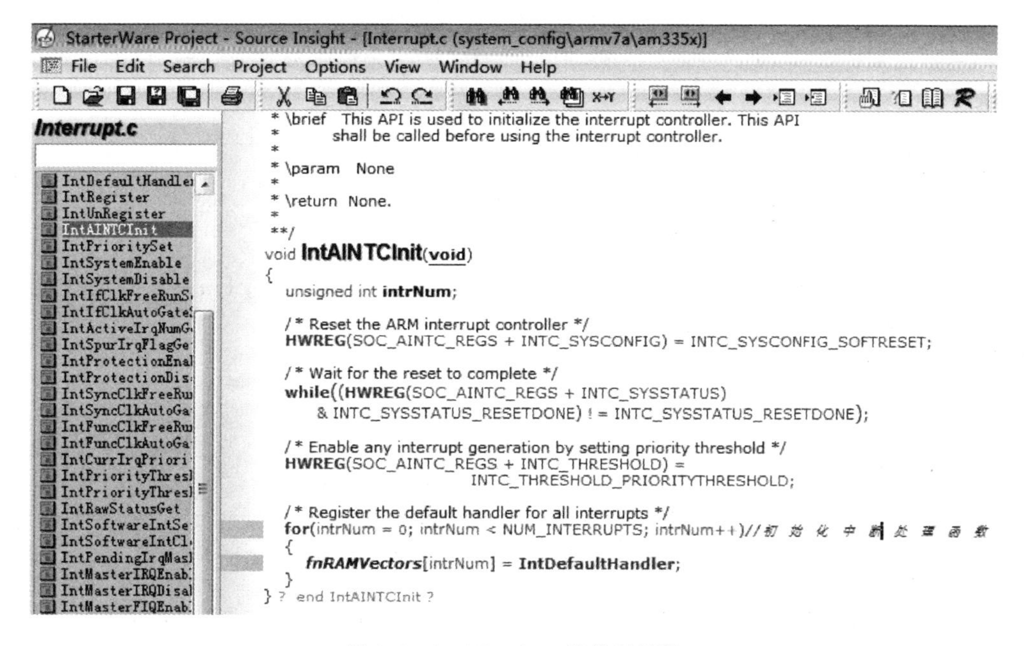

图7.2 IntAINTCInit()处理函数

IntAINTCInit()函数中INTC_SYSCONFIG_SOFTRESET有如下定义：

```
#define INTC_SYSCONFIG_SOFTRESET      (0x00000002u)
```

具体的寄存器功能说明参考AM335x数据手册，或如图7.3所示。

INTC_SYSSTATUS特殊功能寄存器说明如图7.4所示。

因此，由上述寄存器INTC_SYSCONFIG_SOFTRESET和INTC_SYSSTATUS可知，IntAINTCInit()函数的第一条语句进行了软件复位，而接着的第二条语句为等

Table 6-5. INTC_SYSCONFIG Register Field Descriptions

Bit	Field	Type	Reset	Description
31-5	RESERVED	R/W	0h	
4-3	RESERVED	R	0h	Write 0's for future compatibility. Reads returns 0
2	RESERVED	R/W	0h	
1	SoftReset	R/W	0h	Software reset. Set this bit to trigger a module reset. The bit is automatically reset by the hardware. During reads, it always returns 0. 0h(Read) = always_Always returns 0 1h(Read) = never_never happens
0	Autoidle	R/W	0h	Internal OCP clock gating strategy 0h = clkfree : OCP clock is free running 1h = autoClkGate : Automatic OCP clock gating strategy is applied, bnased on the OCP interface activity

图 7.3 INTC_SYSCONFIG 特殊功能寄存器(手册截图)

Table 6-6. INTC_SYSSTATUS Register Field Descriptions

Bit	Field	Type	Reset	Description
31-8	RESERVED	R	0h	
7-1	RESERVED	R	0h	Reserved for OCP socket status information Read returns 0
0	ResetDone	R	X	Internal reset monitoring 0h = rstOngoing : Internal module reset is on-going 1h = rstComp : Reset completed

图 7.4 INTC_SYSSTATUS 特殊功能寄存器(手册截图)

待软件复位完成。

如图 7.5 所示的 INTC_THRESHOLD 寄存器中的 PriorityThreshold 为优先级设置。

Table 6-14. INTC_THRESHOLD Register Field Descriptions

Bit	Field	Type	Reset	Description
31-8	RESERVED	R	0h	Reads returns 0
7-0	PriorityThreshold	R/W	FFh	Priority threshold. Values used are 00h to 3Fh. Value FFh disables the threshold.

图 7.5 INTC_THRESHOLD 特殊功能寄存器(手册截图)

所以,IntAINTCInit()函数的第三条语句为禁止中断优先级,而最后一个 for 循环是初始化各个中断处理函数,即将其设置为进去就出来的空函数。

如图 7.6 所示为 system_config\armv7a\am335x 目录下的 Interrupt.c 文件。

其中,fnRAMVectors[]被定义为函数型数组,相当于中断向量表,专门用于保存处理器各个中断源的中断处理函数。

rastterDisplay.c 文件 main()函数中的 LCDAINTCConfigure()函数为 LCD 中断配置的函数,如图 7.7 所示。

IntRegister(SYS_INT_LCDCINT,LCDIsr)函数的调用是将 LCDIsr()函数作

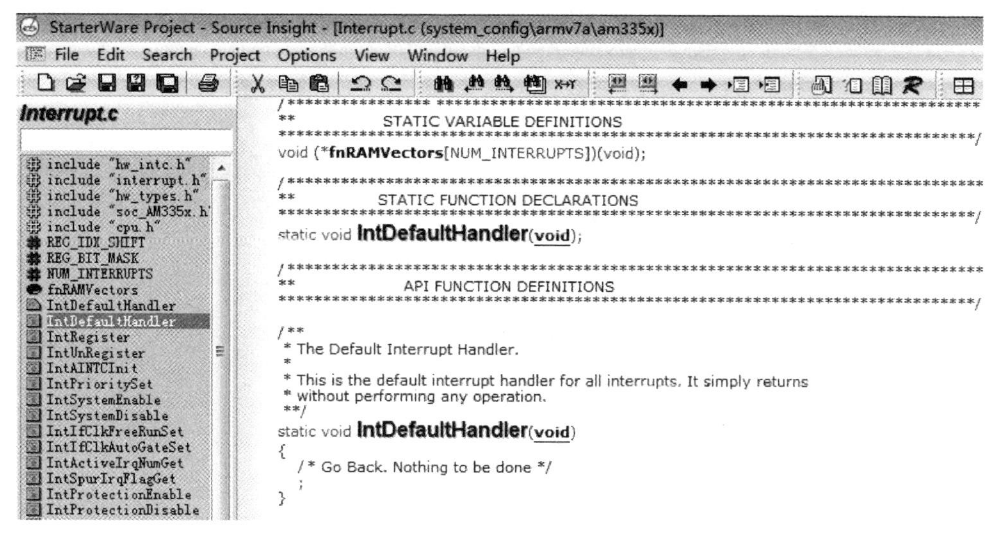

图 7.6　fnRAMVectors[]中断向量表

```
rasterDisplay.c
268 */
269 static void LCDAINTCConfigure(void)
270 {
271     /* Register the ISR in the Interrupt Vector Table.*/
272     IntRegister(SYS_INT_LCDCINT, LCDIsr);
273
274     IntPrioritySet(SYS_INT_LCDCINT, 0, AINTC_HOSTINT_ROUTE_IRQ );
275
276     /* Enable the System Interrupts for AINTC.*/
277     IntSystemEnable(SYS_INT_LCDCINT);
278 }
```

图 7.7　LCDAINTCConfigure()函数

为 LCD 中断处理函数存入中断向量表 fnRAMVectors[]数组中，如图 7.8 所示。

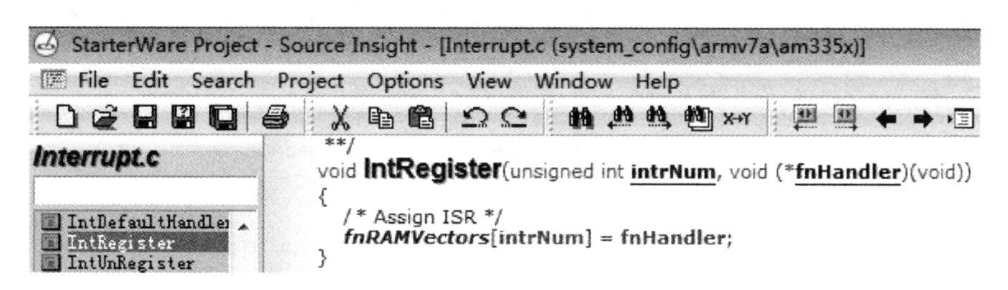

图 7.8　IntRegister()函数

如图 7.9 所示为 LCD 中断处理函数 LCDIsr()。

在 LCDIsr()中断处理函数中，调用 RasterIntStatus()函数读取中断状态标志位，而 RasterClearGetIntStatus()函数则清除中断状态标志位，同时返回中断状态标志位。

```
/*
** For each end of frame interrupt base and ceiling is reconfigured
*/
static void LCDIsr(void)
{
    unsigned int  status;

    status = RasterIntStatus(LCDC_INSTANCE,RASTER_END_OF_FRAME0_INT_STAT |
                                   RASTER_END_OF_FRAME1_INT_STAT );//读 中 断 状 态 标 志 位

    status = RasterClearGetIntStatus(LCDC_INSTANCE, status);

    if (status & RASTER_END_OF_FRAME0_INT_STAT)
    {
        RasterDMAFBConfig(LCDC_INSTANCE,
                          (unsigned int)image1,
                          (unsigned int)image1 + sizeof(image1) - 2,
                          0);
    }

    if(status & RASTER_END_OF_FRAME1_INT_STAT)
    {
        RasterDMAFBConfig(LCDC_INSTANCE,
                          (unsigned int)image1,
                          (unsigned int)image1 + sizeof(image1) - 2,
                          1);
    }
} ? end LCDIsr ?
```

图 7.9　LCDIsr()函数

如图7.10所示为IRQSTATUS中断状态寄存器。

13.5.1.23　IRQSTATUS Register (offset = 5Ch) [reset = 0h]

IRQSTATUS is shown in Figure 20-10 and described in Table 20-15.

Figure 13-41. IRQSTATUS Register

31	30	29	28	27	26	25	24
RESERVED							
R/W-0h							
23	**22**	**21**	**20**	**19**	**18**	**17**	**16**
RESERVED							
R/W-0h							
15	**14**	**13**	**12**	**11**	**10**	**9**	**8**
RESERVED						eof1_en_clr	eof0_en_clr
R/W-0h						R/W-0h	R/W-0h
7	**6**	**5**	**4**	**3**	**2**	**1**	**0**
RESERVED	pl_en_clr	fuf_en_clr	RESERVED	acb_en_clr	sync_en_clr	recurrent_raster_done_en_clr	done_en_clr
R/W-0h	R/W-0h	R/W-0h	R/W-0h	R/W-0h	R/W-0h	R/W-0h	R/W-0h

LEGEND: R/W = Read/Write; R = Read only; W1toCl = Write 1 to clear bit; -*n* = value after reset

Table 13-36. IRQSTATUS Register Field Descriptions

Bit	Field	Type	Reset	Description
31-10	RESERVED	R/W	0h	
9	eof1_en_clr	R/W	0h	DMA End-of-Frame 1 Enabled Interrupt and Clear Read indicates enabled (masked) status. 0 = inactive. 1 = active. Writing 1 will clear interrupt enable. Writing 0 has no effect.
8	eof0_en_clr	R/W	0h	DMA End-of-Frame 0 Enabled Interrupt and Clear Read indicates enabled (masked) status. 0 = inactive. 1 = active. Writing 1 will clear interrupt enable. Writing 0 has no effect.

图 7.10　IRQSTATUS中断状态寄存器(手册截图)

如图7.11所示为RasterIntStatus()函数，用于读取中断状态标志位。

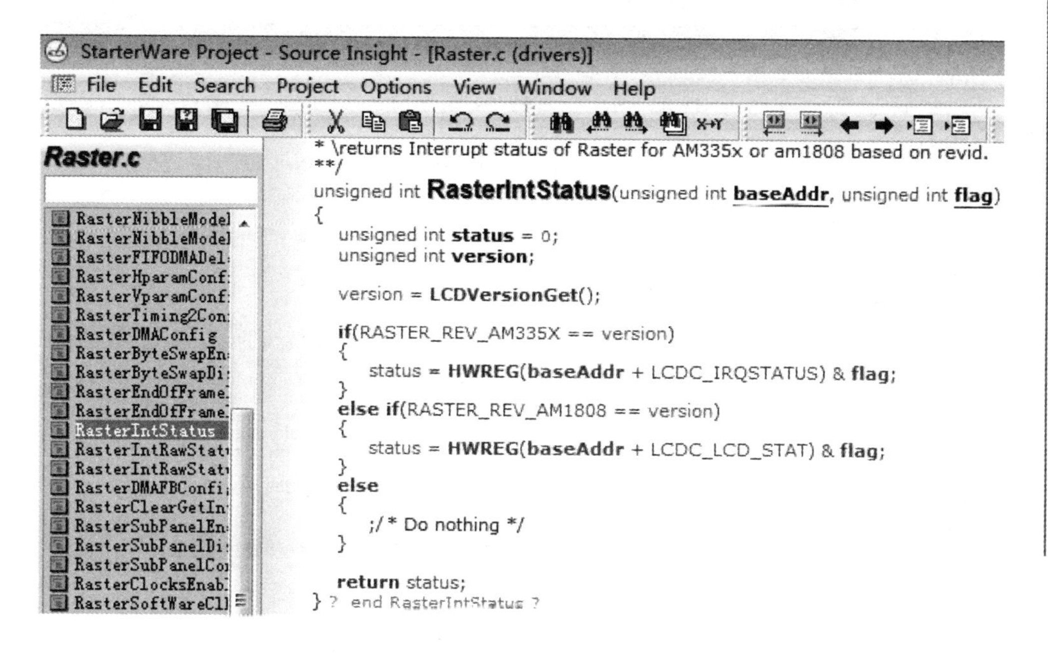

图7.11　RasterIntStatus()函数

在LCDIsr()的LCD中断处理函数中，处理完中断状态标志后，接着是两个if语句，分别用于判断是否为LCD的缓冲区0和1的DMA写结束产生的中断，如果是则重新配置对应缓冲区的基地址和结束地址，如图7.12所示的RasterClearGetIntStatus()函数。

在LCDAINTCConfigure()中断配置函数中，IntPrioritySet(SYS_INT_LCDCINT,0,AINTC_HOSTINT_ROUTE_IRQ)函数调用设定LCD中断的优先级，如图7.13所示。

其中，有两个定义如下所示：

```
#define    SOC_AINTC_REGS      (0x48200000)
#define    INTC_ILR(n)         (0x100 + ((n) * 0x04))
```

INTC_ILR寄存器如图7.14所示。

因此，IntPrioritySet()函数中的语句就是将LCD控制器设置成IRQ中断，且将优先级设置为最高的0。

在LCDAINTCConfigure()中断配置函数中，调用IntSystemEnable(SYS_INT_LCDCINT)使能LCD中断。如图7.15所示为IntSystemEnable()函数的定义。

```
}
/**
* \brief This function clear status interrupt status register.
*
* \param baseAddr is the Memory Address of the LCD Module.
*
* \param flag     is the value which detemines status of which
*                 interrupt to be cleared.\n
*
*        flag can take following values.\n
*
*        RASTER_ACBIAS_COUNT_INT_STAT   -  ACBias Count interrupt status.\n
*        RASTER_PALETTE_LOADED_INT_STAT -  Palette loaded interrupt status.\n
*        RASTER_SYNC_LOST_INT_STAT      -  Sync lost interrupt status.\n
*        RASTER_FIFO_UNDRFLOW_INT_STAT  -  FIFO underflow interrupt status.\n
*        RASTER_FRAME_DONE_INT_STAT     -  frame done interrupt status.\n
*        RASTER_END_OF_FRAME0_INT_STAT  -  end of frame 0 interrupt status.\n
*        RASTER_END_OF_FRAME1_INT_STAT  -  end of frame 1 interrupt status.\n
*        RASTER_DONE_INT                -  Raster Done interrupt.\n
*
* \return Interrupt status of Raster for AM335x or am1808 based on rev id.
*
**/
unsigned int RasterClearGetIntStatus(unsigned int baseAddr,
                                     unsigned int flag)//清除中断标志位
{
    unsigned int saveStatus = 0;
    unsigned int version;

    version = LCDVersionGet();

    if(RASTER_REV_AM335X == version)
    {
        saveStatus = HWREG(baseAddr + LCDC_IRQSTATUS) & flag;

        HWREG(baseAddr + LCDC_IRQSTATUS) = saveStatus;
    }
    else if(RASTER_REV_AM1808 == version)
    {
        saveStatus = HWREG(baseAddr + LCDC_LCD_STAT) & flag;

        HWREG(baseAddr + LCDC_LCD_STAT) = saveStatus;
    }
    else
    {
        ;/*Do nothing */
    }

    return (saveStatus);
} ? end RasterClearGetIntStatus ?
/**
```

图 7.12 RasterClearGetIntStatus()函数

如图 7.16 所示为 INTC_MIR_CLEAR0 特殊功能寄存器定义。

同时，有如下定义：

```
#define INTC_MIR_CLEAR(n)      (0x88 + ((n) * 0x20))
#define REG_IDX_SHIFT          (0x05)
```

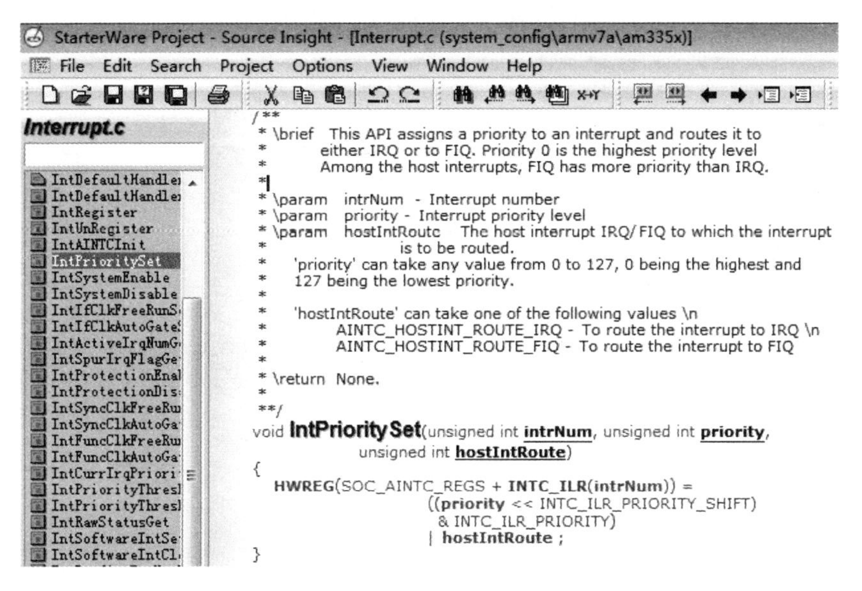

图 7.13 IntPrioritySet()函数

6.5.1.44 INTC_ILR_0 to INTC_ILR_127 Register (offset = 100h to 2FCh) [reset = 0h] 0x4820_0000

Register mask: FFFFFFFFh

INTC_ILR_0 to INTC_ILR_127 is shown in Figure 6-47 and described in Table 6-47.

The INTC_ILRx registers contain the priority for the interrupts and the FIQ/IRQ steering.

Figure 6-47. INTC_ILR_0 to INTC_ILR_127 Register

31	30	29	28	27	26	25	24
RESERVED							
R-0h							
23	22	21	20	19	18	17	16
RESERVED							
R-0h							
15	14	13	12	11	10	9	8
RESERVED							
R-0h							
7	6	5	4	3	2	1	0
Priority						RESERVED	FIQnIRQ
R/W-0h						R/W-0h	R/W-0h

LEGEND: R/W = Read/Write; R = Read only; W1toCl = Write 1 to clear bit; *-n* = value after reset

Table 6-47. INTC_ILR_0 to INTC_ILR_127 Register Field Descriptions

Bit	Field	Type	Reset	Description
31-8	RESERVED	R	0h	Write 0's for future compatibility. Reads returns 0
7-2	Priority	R/W	0h	Interrupt priority
1	RESERVED	R/W	0h	
0	FIQnIRQ	R/W	0h	Interrupt IRQ FiQ mapping 0h = IntIRQ : Interrupt is routed to IRQ. 1h = IntFIQ : Interrupt is routed to FIQ (this selection is reserved on GP devices).

图 7.14 INTC_ILR 特殊功能寄存器(手册截图)

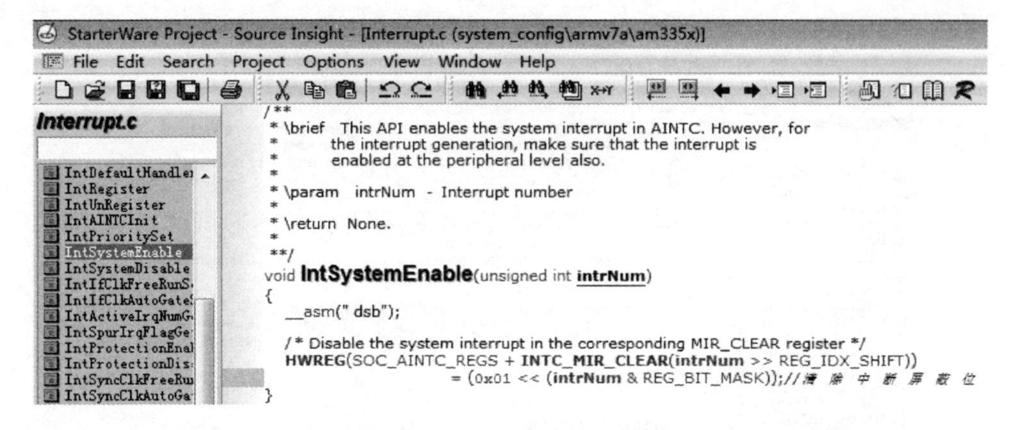

图 7.15　IntSystemEnable( )函数

6.5.1.14　INTC_MIR_CLEAR0 Register (offset = 88h) [reset = 0h]

Register mask: 0h

INTC_MIR_CLEAR0 is shown in Figure 6-17 and described in Table 6-17.

This register is used to clear the interrupt mask bits.

Figure 6-17. INTC_MIR_CLEAR0 Register

31–0
MirClear
W-X

LEGEND: R/W = Read/Write; R = Read only; W1toCl = Write 1 to clear bit; -*n* = value after reset

Table 6-17. INTC_MIR_CLEAR0 Register Field Descriptions

Bit	Field	Type	Reset	Description
31-0	MirClear	W	X	Write 1 clears the mask bit to 0, reads return 0

图 7.16　INTC_MIR_CLEAR0 特殊功能寄存器(手册截图)

这里为何要“(n) * 0x20”呢？这是因为 INTC_MIR_CLEAR0、INTC_MIR_CLEAR1 和 INTC_MIR_CLEAR2 分别相差 0x20。而 2^5 等于 32，所以 REG_IDX_SHIFT 被定义为 0x05。

总的来说就是中断向量号对应的 mask 位按顺序分别从小到大排列在 INTC_MIR_CLEAR0、INTC_MIR_CLEAR1 和 INTC_MIR_CLEAR2 的每位中。

如图 7.17 所示为 INTC 寄存器。

88h	INTC_MIR_CLEAR0	Section 6.5.1.14
8Ch	INTC_MIR_SET0	Section 6.5.1.15
90h	INTC_ISR_SET0	Section 6.5.1.16
94h	INTC_ISR_CLEAR0	Section 6.5.1.17
98h	INTC_PENDING_IRQ0	Section 6.5.1.18
9Ch	INTC_PENDING_FIQ0	Section 6.5.1.19
A0h	INTC_ITR1	Section 6.5.1.20
A4h	INTC_MIR1	Section 6.5.1.21
A8h	INTC_MIR_CLEAR1	Section 6.5.1.22
ACh	INTC_MIR_SET1	Section 6.5.1.23
B0h	INTC_ISR_SET1	Section 6.5.1.24
B4h	INTC_ISR_CLEAR1	Section 6.5.1.25
B8h	INTC_PENDING_IRQ1	Section 6.5.1.26
BCh	INTC_PENDING_FIQ1	Section 6.5.1.27
C0h	INTC_ITR2	Section 6.5.1.28
C4h	INTC_MIR2	Section 6.5.1.29
C8h	INTC_MIR_CLEAR2	Section 6.5.1.30
CCh	INTC_MIR_SET2	Section 6.5.1.31
D0h	INTC_ISR_SET2	Section 6.5.1.32
D4h	INTC_ISR_CLEAR2	Section 6.5.1.33
D8h	INTC_PENDING_IRQ2	Section 6.5.1.34
DCh	INTC_PENDING_FIQ2	Section 6.5.1.35
E0h	INTC_ITR3	Section 6.5.1.36
E4h	INTC_MIR3	Section 6.5.1.37
E8h	INTC_MIR_CLEAR3	Section 6.5.1.38
ECh	INTC_MIR_SET3	Section 6.5.1.39

图 7.17 INTC寄存器(手册截图)

7.2.3 LCD背光设置

LCDBackLightEnable()函数调用使能LCD背光。

如图7.18所示为处理器AM335x的GPIO口驱动输出GPIO3_17。

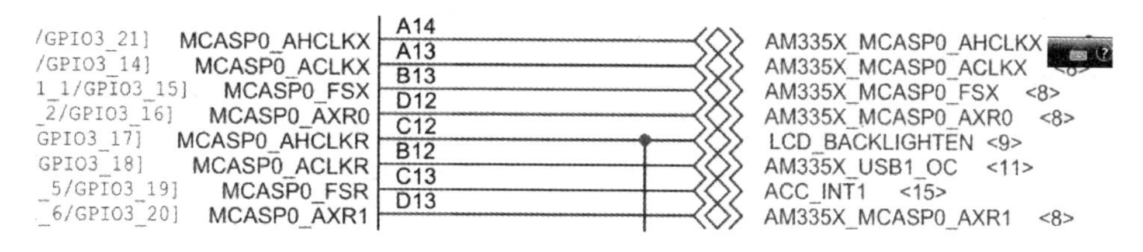

图 7.18 LCD背光驱动接口GPIO3_17

如图7.19所示为LCD背光驱动调节电路，由AM335x的GPIO3_17接LCD_BACKLIGHTEN驱动。

LCD背光使能由GPIO3_17控制，低电平使能，高电平禁止。

如图7.20所示为LCDBackLightEnable()函数定义。

调用GPIO3ModuleClkConfig()函数，配置GPIO3的时钟模块。

调用GPIO_PMUX_OFFADDR_VALUE(3,17,PAD_FS_RXE_PD_PUPDE(7))函

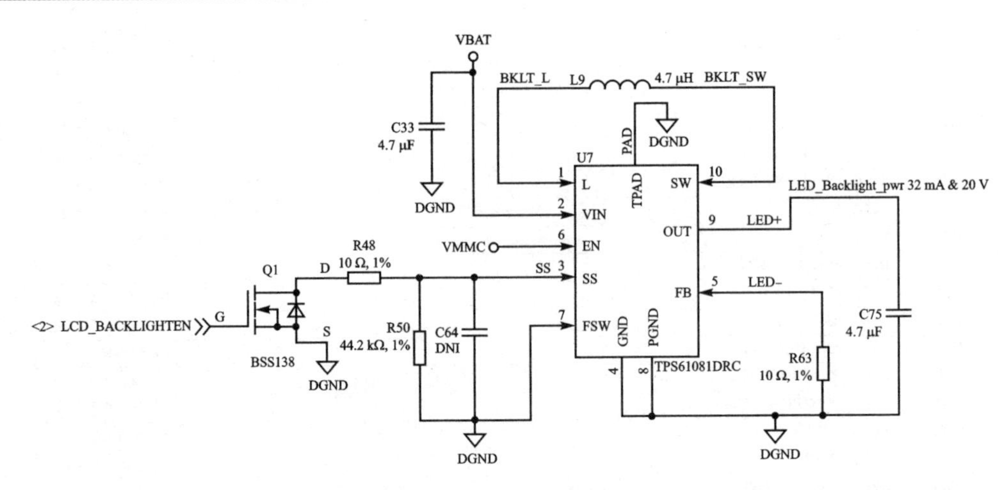

图 7.19 LCD 背光驱动电路

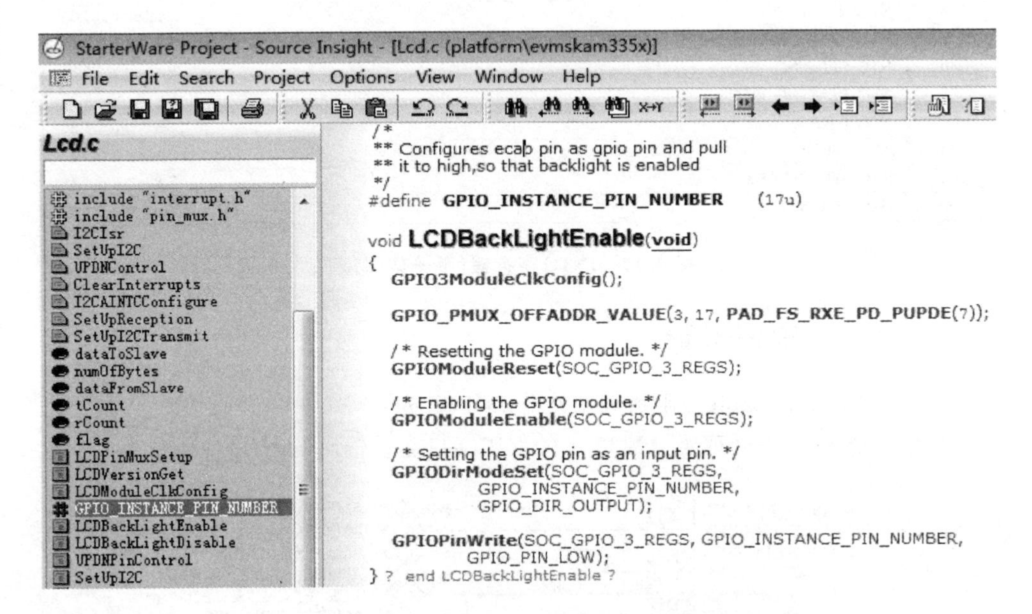

图 7.20 LCDBackLightEnable()函数

数，配置 GPIO3_17 的引脚复用功能为 mode 7，CONTROL_CONF_RXACTIVE 使能 GPIO3_17 的输入功能和选择内部电阻下拉，如图 7.21 和图 7.22 所示。

```
/* Fast Slew Rate - Receiver Enabled - Pulldown - PU/PD feature Enabled. */
#define PAD_FS_RXE_PD_PUPDE(n)        (CONTROL_CONF_RXACTIVE | \
                                       CONTROL_CONF_MUXMODE(n))
```

图 7.21 PAD_FS_RXE_PD_PUPDE(n) 定义

GPIO_PMUX_OFFADDR_VALUE(3,17,PAD_FS_RXE_PD_PUPDE(7))函

```
#define CONTROL_CONF_PULLUDDISABLE    0x00000008
#define CONTROL_CONF_PULLUPSEL        0x00000010
#define CONTROL_CONF_RXACTIVE         0x00000020
#define CONTROL_CONF_SLOWSLEW         0x00000040
#define CONTROL_CONF_MUXMODE(n)       (n)
```

图 7.22　CONTROL_CONF_RXACTIVE 定义

数定义如图 7.23 所示。

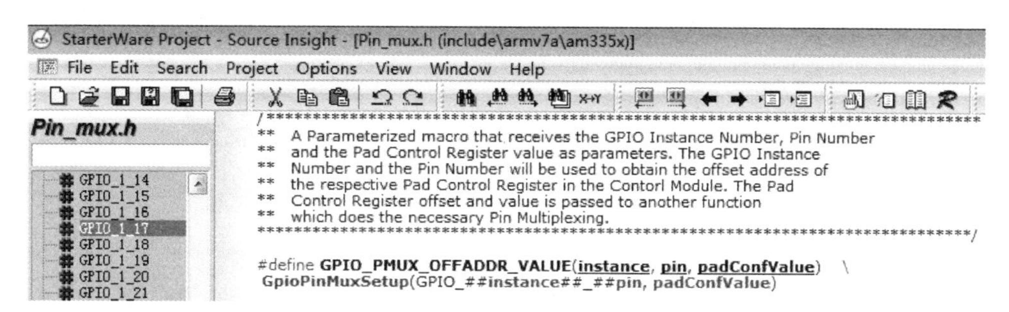

图 7.23　GPIO_PMUX_OFFADDR_VALUE 定义

“GPIO_##instance##_##pin”语句等效于 GPIO_1_17。

GPIO_1_17 定义如图 7.24 所示。

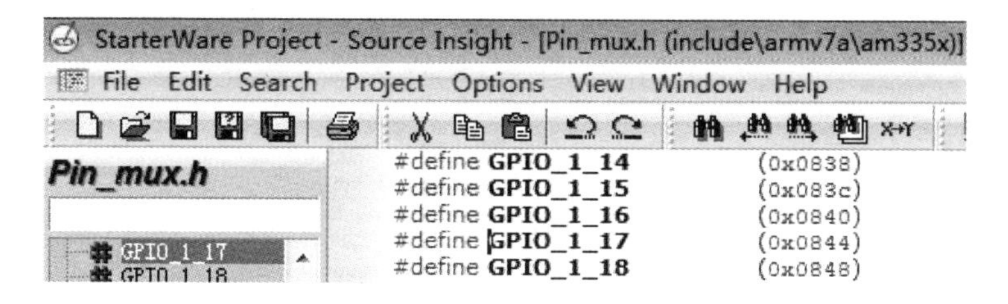

图 7.24　GPIO_1_17 定义

参考 AM335x 数据手册可查看到如图 7.25 和 7.26 所示的定义。

840h	conf_gpmc_a0	Section 9.3.1.49
844h	conf_gpmc_a1	Section 9.3.1.49
848h	conf_gpmc_a2	Section 9.3.1.49

图 7.25　conf_gpmc_a1 定义(手册截图)

如图 7.27 所示为模式选择。

GPIOModuleReset(SOC_GPIO_3_REGS)函数，为软件复位 GPIO3 模块。

GPIOModuleEnable(SOC_GPIO_3_REGS)函数，为使能 GPIO3 模块。

GPIODirModeSet(SOC_GPIO_3_REGS，GPIO_INSTANCE_PIN_NUMBER，GPIO_DIR_OUTPUT)函数，设置 GPIO3_17 为输出口，宏定义如下：

0x44e1_0000

9.3.1.49 conf_<module>_<pin> Register (offset = 800h–A34h)

See the device datasheet for information on default pin mux configurations. Note that the device ROM may change the default pin mux for certain pins based on the SYSBOOT mode settings.

See Table 9-10, *Control Module Registers Table*, for the full list of offsets for each module/pin configuration.

conf_<module>_<pin> is shown in Figure 9-51 and described in Table 9-60.

Figure 9-51. conf_<module>_<pin> Register

31	30	29	28	27	26	25	24
Reserved							
R-0h							

23	22	21	20	19	18	17	16
Reserved				Reserved			
R-0h				R-0h			

15	14	13	12	11	10	9	8
Reserved							
R-0h							

7	6	5	4	3	2	1	0
Reserved	conf_<module>_<pin>_slewctrl	conf_<module>_<pin>_rxactive	conf_<module>_<pin>_putypesel	conf_<module>_<pin>_puden	conf_<module>_<pin>_mmode		
R-0h	R/W-0h	R/W-1h	R/W-0h	R/W-0h	R/W-0h		

LEGEND: R/W = Read/Write; R = Read only; W1toCl = Write 1 to clear bit; -*n* = value after reset

Table 9-60. conf_<module>_<pin> Register Field Descriptions

Bit	Field	Type	Reset	Description
31-20	Reserved	R	0h	
19-7	Reserved	R	0h	
6	conf_<module>_<pin>_slewctrl	R/W	X	Select between faster or slower slew rate 0: Fast 1: Slow Reset value is pad-dependent.
5	conf_<module>_<pin>_rxactive	R/W	1h	Input enable value for the PAD 0: Receiver disabled 1: Receiver enabled
4	conf_<module>_<pin>_putypesel	R/W	X	Pad pullup/pulldown type selection 0: Pulldown selected 1: Pullup selected Reset value is pad-dependent.
3	conf_<module>_<pin>_puden	R/W	X	Pad pullup/pulldown enable 0: Pullup/pulldown enabled 1: Pullup/pulldown disabled Reset value is pad-dependent.
2-0	conf_<module>_<pin>_mmode	R/W	X	Pad functional signal mux select. Reset value is pad-dependent.

图 7.26　conf_module_pin 寄存器定义(手册截图)

Table 9-2. Mode Selection

MUXMODE	Selected Mode
000b	Primary Mode = Mode 0
001b	Mode 1
010b	Mode 2
011b	Mode 3
100b	Mode 4
101b	Mode 5
110b	Mode 6
111b	Mode 7

图 7.27　Mode Selection(手册截图)

```
#define   GPIO_INSTANCE_PIN_NUMBER          (17u)
#define   GPIO_DIR_OUTPUT                   0
```

GPIODirModeSet()函数定义，如图 7.28 所示。

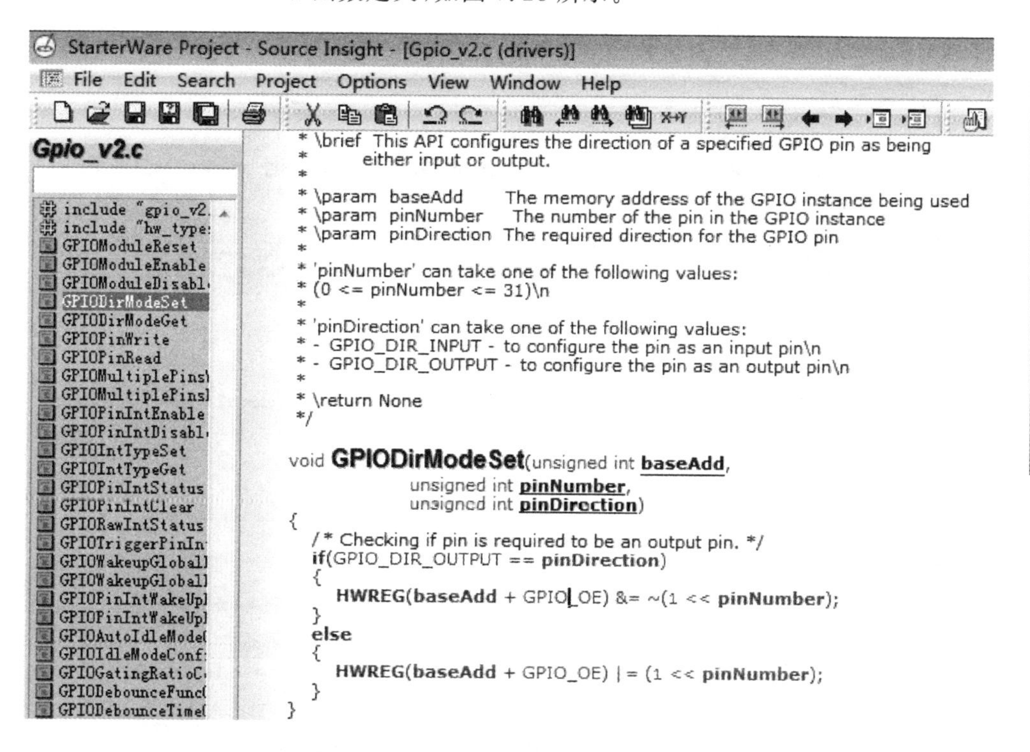

图 7.28　GPIODirModeSet()函数定义

如图 7.29 所示为 GPIO_OE 特殊功能寄存器定义。

25.4.1.16　GPIO_OE Register (offset = 134h) [reset = FFFFFFFFh]　　0x481A_E

GPIO_OE is shown in Figure 25-22 and described in Table 25-21.

The GPIO_OE register is used to enable the pins output capabilities. At reset, all the GPIO related pins are configured as input and output capabilities are disabled. This register is not used within the module, its only function is to carry the pads configuration. When the application is using a pin as an output and does not want interrupt generation from this pin, the application can/has to properly configure the Interrupt Enable registers.

Figure 25-22. GPIO_OE Register

31 – 0
OUTPUTEN[n]
R/W-FFFFFFFFh

LEGEND: R/W = Read/Write; R = Read only; W1toCl = Write 1 to clear bit; -n = value after reset

Table 25-21. GPIO_OE Register Field Descriptions

Bit	Field	Type	Reset	Description
31-0	OUTPUTEN[n]	R/W	FFFFFFFFh	Output Data Enable 0h = The corresponding GPIO port is configured as an output. 1h = The corresponding GPIO port is configured as an input.

图 7.29　GPIO_OE 特殊功能寄存器定义(手册截图)

GPIOPinWrite(SOC_GPIO_3_REGS,PIO_INSTANCE_PIN_NUMBER,GPIO_PIN_LOW)函数,设置GPIO3_17输出低电平,GPIO_PIN_LOW宏定义如下:

```
#define GPIO_PIN_LOW                    0
```

GPIOPinWrite()函数定义,如图7.30所示。

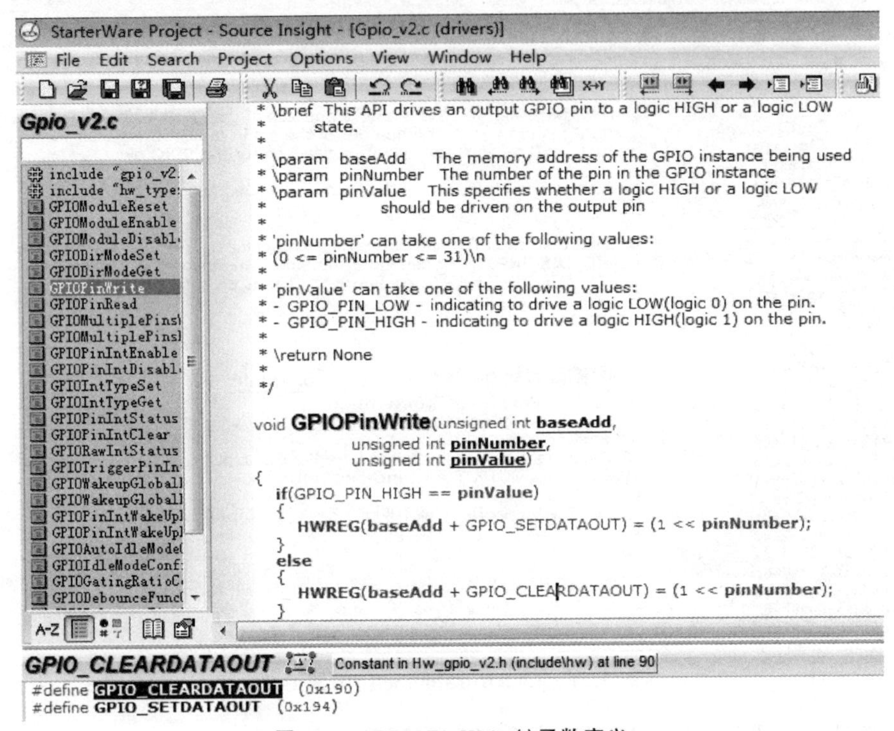

图7.30　GPIOPinWrite()函数定义

GPIO_CLEARDATAOUT特殊功能寄存器定义如图7.31所示。

25.4.1.25 GPIO_CLEARDATAOUT Register (offset = 190h) [reset = 0h]　　0x481A_E

GPIO_CLEARDATAOUT is shown in Figure 25-31 and described in Table 25-30.

Writing a 1 to a bit in the GPIO_CLEARDATAOUT register clears to 0 the corresponding bit in the GPIO_DATAOUT register; writing a 0 has no effect. A read of the GPIO_CLEARDATAOUT register returns the value of the data output register (GPIO_DATAOUT).

Figure 25-31. GPIO_CLEARDATAOUT Register

31 30 29 28 27 26 25 24 23 22 21 20 19 18 17 16 15 14 13 12 11 10 9 8 7 6 5 4 3 2 1 0
INTLINE[n]
R/W-0h

LEGEND: R/W = Read/Write; R = Read only; W1toCl = Write 1 to clear bit; -n = value after reset

Table 25-30. GPIO_CLEARDATAOUT Register Field Descriptions

Bit	Field	Type	Reset	Description
31-0	INTLINE[n]	R/W	0h	Clear Data Output Register 0h = No effect 1h = Clear the corresponding bit in the GPIO_DATAOUT register.

图7.31　GPIO_CLEARDATAOUT特殊功能寄存器定义(手册截图)

7.2.4 LCD显示模块配置

在 rasterDisplay. c 的 main()函数中，SetUpLCD()函数的调用用于配置 LCD 控制器 Raster 显示图像。如图 7.32 所示为 SetUpLCD()函数定义。

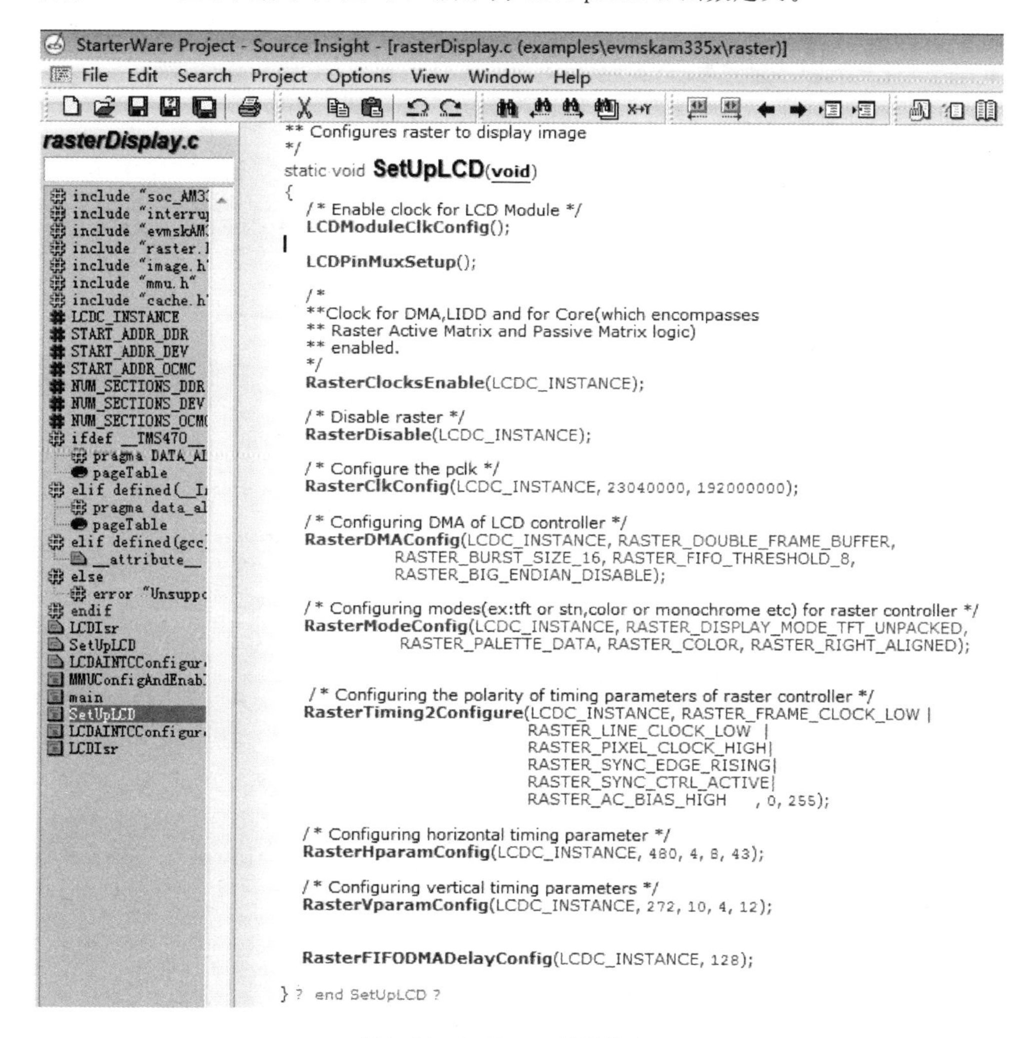

图 7.32 SetUpLCD()函数定义

LCDModuleClkConfig()使能 LCD 模块的时钟。

LCDPinMuxSetup()配置 LCD 所使用的引脚的复用功能。

RasterClocksEnable(LCDC_INSTANCE)使能 LCD 控制器 Raster 模块时钟。

RasterDisable(LCDC_INSTANCE)禁止 Raster 模块。

RasterClkConfig(LCDC_INSTANCE,23040000,192000000)配置 LCD 控制器 Raster 从 PLL 输入的时钟 PCLK 的分频数。

RasterDMAConfig(LCDC_INSTANCE,RASTER_DOUBLE_FRAME_BUFFER,RASTER_BURST_SIZE_16,RASTER_FIFO_THRESHOLD_8,RASTER_BIG_ENDIAN_DISABLE)配置LCD控制器Raster的DMA模块方式。

RasterModeConfig(LCDC_INSTANCE,RASTER_DISPLAY_MODE_TFT_UNPACKED,RASTER_PALETTE_DATA,RASTER_COLOR,RASTER_RIGHT_ALIGNED)配置LCD控制器Raster支持的模式、类型和颜色等。

如图7.33所示为RasterModeConfig()函数定义说明。

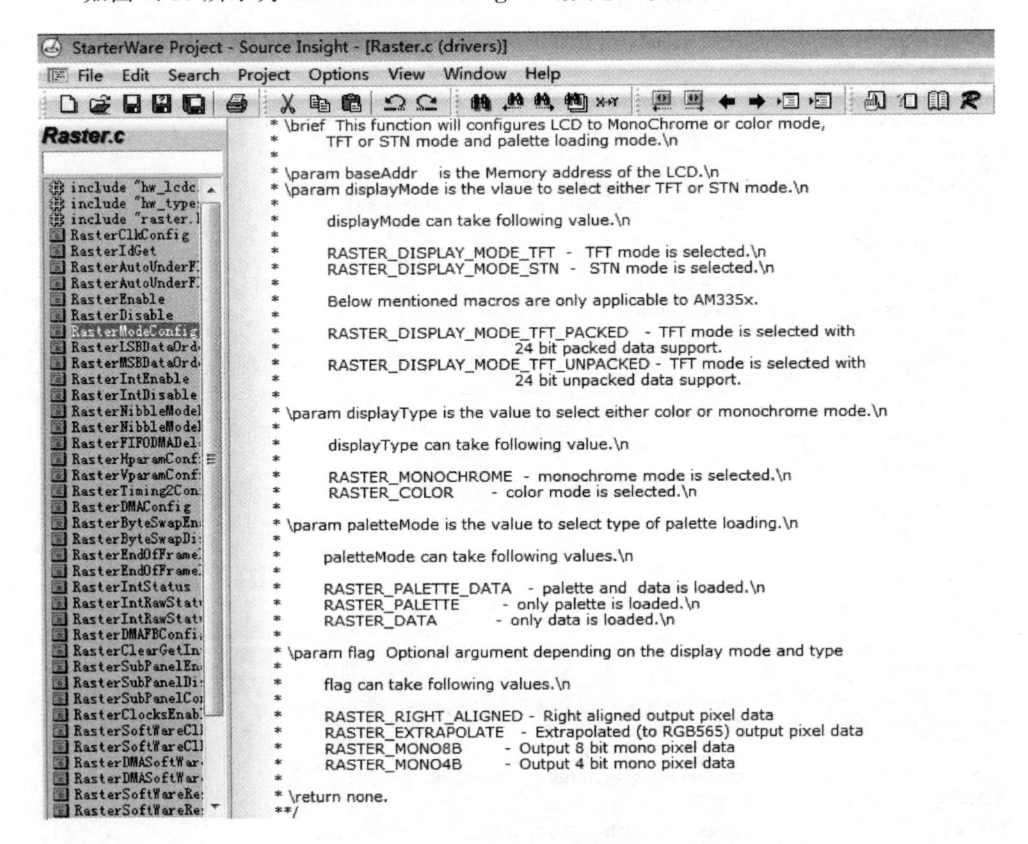

图7.33 RasterModeConfig()函数定义说明

如图7.34所示为RasterModeConfig()函数定义,配置LCD控制器Raster为TFT或STN模式。

SetUpLCD()函数中,RasterTiming2Configure(LCDC_INSTANCE,RASTER_FRAME_CLOCK_LOW|RASTER_LINE_CLOCK_LOW|RASTER_PIXEL_CLOCK_HIGH|RASTER_SYNC_EDGE_RISING|RASTER_SYNC_CTRL_ACTIVE|RASTER_AC_BIAS_HIGH,0,255)配置LCD控制器Raster的驱动时序,需要注意的是不同的屏需要做相应的调整,如图7.35所示。

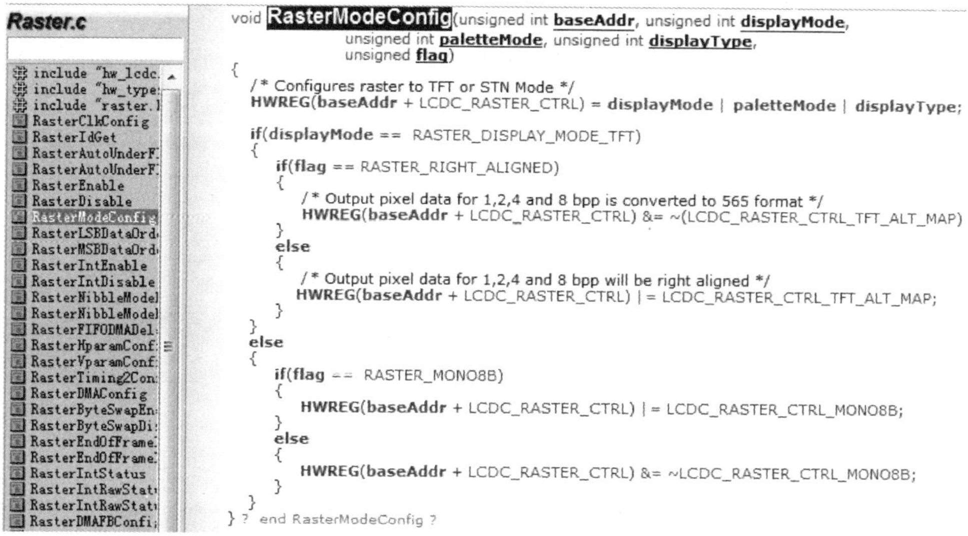

```
Raster.c

include "hw_lcdc.
include "hw_type:
include "raster.l
RasterClkConfig
RasterIdGet
RasterAutoUnderF
RasterAutoUnderF
RasterEnable
RasterDisable
RasterModeConfig
RasterLSBDataOrd
RasterMSBDataOrd
RasterIntEnable
RasterIntDisable
RasterNibbleMode
RasterNibbleMode
RasterFIFODMADel
RasterHparamConf
RasterVparamConf
RasterTiming2Con
RasterDMAConfig
RasterByteSwapEn
RasterByteSwapDi
RasterEndOfFrame
RasterEndOfFrame
RasterIntStatus
RasterIntRawStat
RasterIntRawStat
RasterDMAFBConfi
```

```
void RasterModeConfig(unsigned int baseAddr, unsigned int displayMode,
                      unsigned int paletteMode, unsigned int displayType,
                      unsigned flag)
{
   /* Configures raster to TFT or STN Mode */
   HWREG(baseAddr + LCDC_RASTER_CTRL) = displayMode | paletteMode | displayType;

   if(displayMode ==  RASTER_DISPLAY_MODE_TFT)
   {
       if(flag == RASTER_RIGHT_ALIGNED)
       {
           /* Output pixel data for 1,2,4 and 8 bpp is converted to 565 format */
           HWREG(baseAddr + LCDC_RASTER_CTRL) &= ~(LCDC_RASTER_CTRL_TFT_ALT_MAP)
       }
       else
       {
           /* Output pixel data for 1,2,4 and 8 bpp will be right aligned */
          HWREG(baseAddr + LCDC_RASTER_CTRL) |= LCDC_RASTER_CTRL_TFT_ALT_MAP;
       }
   }
   else
   {
       if(flag == RASTER_MONO8B)
       {
           HWREG(baseAddr + LCDC_RASTER_CTRL) |= LCDC_RASTER_CTRL_MONO8B;
       }
       else
       {
           HWREG(baseAddr + LCDC_RASTER_CTRL) &= ~LCDC_RASTER_CTRL_MONO8B;
       }
   }
} ? end RasterModeConfig ?
```

图 7.34 RasterModeConfig( )函数定义

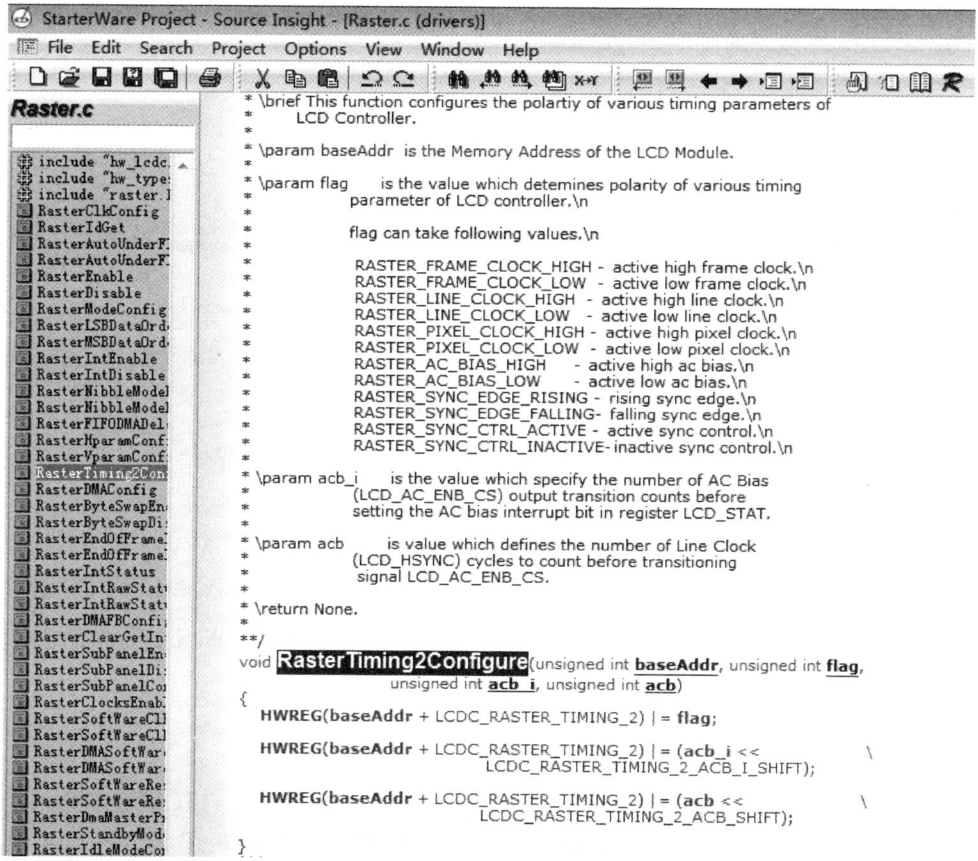

StarterWare Project - Source Insight - [Raster.c (drivers)]

File Edit Search Project Options View Window Help

```
Raster.c

include "hw_lcdc.
include "hw_type:
include "raster.l
RasterClkConfig
RasterIdGet
RasterAutoUnderF
RasterAutoUnderF
RasterEnable
RasterDisable
RasterModeConfig
RasterLSBDataOrd
RasterMSBDataOrd
RasterIntEnable
RasterIntDisable
RasterNibbleMode
RasterNibbleMode
RasterFIFODMADel
RasterHparamConf
RasterVparamConf
RasterTiming2Con
RasterDMAConfig
RasterByteSwapEn
RasterByteSwapDi
RasterEndOfFrame
RasterEndOfFrame
RasterIntStatus
RasterIntRawStat
RasterIntRawStat
RasterDMAFBConfi
RasterClearGetIn
RasterSubPanelEn
RasterSubPanelDi
RasterSubPanelCo
RasterClocksEnab
RasterSoftWareCl
RasterSoftWareCl
RasterDMASoftWar
RasterDMASoftWar
RasterSoftWareRe
RasterSoftWareRe
RasterDmaMasterP
RasterStandbyMod
RasterIdleModeCo
```

```
 * \brief This function configures the polartiy of various timing parameters of
 *        LCD Controller.
 *
 * \param baseAddr  is the Memory Address of the LCD Module.
 *
 * \param flag      is the value which detemines polarity of various timing
 *                  parameter of LCD controller.\n
 *
 *                  flag can take following values.\n
 *
 *                   RASTER_FRAME_CLOCK_HIGH - active high frame clock.\n
 *                   RASTER_FRAME_CLOCK_LOW  - active low frame clock.\n
 *                   RASTER_LINE_CLOCK_HIGH  - active high line clock.\n
 *                   RASTER_LINE_CLOCK_LOW   - active low line clock.\n
 *                   RASTER_PIXEL_CLOCK_HIGH - active high pixel clock.\n
 *                   RASTER_PIXEL_CLOCK_LOW  - active low pixel clock.\n
 *                   RASTER_AC_BIAS_HIGH     - active high ac bias.\n
 *                   RASTER_AC_BIAS_LOW      - active low ac bias.\n
 *                   RASTER_SYNC_EDGE_RISING - rising sync edge.\n
 *                   RASTER_SYNC_EDGE_FALLING- falling sync edge.\n
 *                   RASTER_SYNC_CTRL_ACTIVE - active sync control.\n
 *                   RASTER_SYNC_CTRL_INACTIVE- inactive sync control.\n
 *
 * \param acb_i     is the value which specify the number of AC Bias
 *                  (LCD_AC_ENB_CS) output transition counts before
 *                  setting the AC bias interrupt bit in register LCD_STAT.
 *
 * \param acb       is value which defines the number of Line Clock
 *                  (LCD_HSYNC) cycles to count before transitioning
 *                   signal LCD_AC_ENB_CS.
 *
 * \return None.
 *
 **/
void RasterTiming2Configure(unsigned int baseAddr, unsigned int flag,
                            unsigned int acb_i, unsigned int acb)
{
   HWREG(baseAddr + LCDC_RASTER_TIMING_2) |= flag;

   HWREG(baseAddr + LCDC_RASTER_TIMING_2) |= (acb_i <<                    \
                                LCDC_RASTER_TIMING_2_ACB_I_SHIFT);

   HWREG(baseAddr + LCDC_RASTER_TIMING_2) |= (acb <<                      \
                               LCDC_RASTER_TIMING_2_ACB_SHIFT);
}
```

图 7.35 RasterTiming2Configure( )函数定义

SetUpLCD()函数中，RasterHparamConfig(LCDC_INSTANCE,480,4,8,43)配置 LCD 屏的水平像素时序，即常说的行同步信号(HS)。

RasterVparamConfig(LCDC_INSTANCE,272,10,4,12)配置 LCD 屏的垂直像素时序，即常说的场同步信号(VS)。

RasterFIFODMADelayConfig(LCDC_INSTANCE,128)配置 LCD 控制器 Raster 先进先出 FIFO 队列的 DMA 延时时间。

如图 7.36 所示为 RasterDMAFBConfig()函数定义。

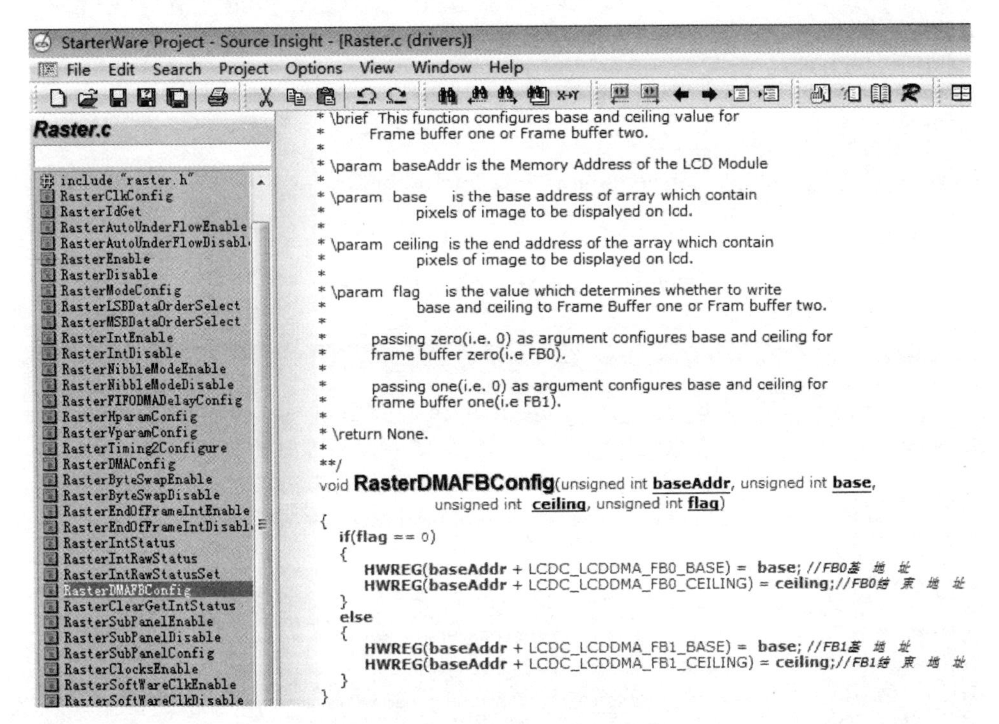

图 7.36 RasterDMAFBConfig()函数定义

RasterDMAFBConfig()函数将要显示的图像 image1 数组的首地址传给 LCD 控制器 Raster 的 DMA 缓冲区的基地址寄存器(Frame Buffer 0 Base Address pointer)，且将图像 image1 数组的末地址传给 LCD 控制器 Raster 的 DMA 缓冲区的结束地址寄存器(Frame Buffer 0 Ceiling Address pointer)。另外，AM335x 共有两块 LCD 控制器 Raster 的 DMA 缓冲区，通过 flag 标志确定图像 image1 传给的是哪一块。

7.2.5 LCD 控制器 Raster 及中断使能

回归到 rasterDisplay.c 的 main()函数的最后几条语句的函数调用，“RasterIntEnable(LCDC_INSTANCE,RASTER_END_OF_FRAME0_INT | RASTER_

END_OF_FRAME1_INT);"，使能了 LCD 控制器 Raster 的 DMA END_OF_FRAME 中断，即当 LCD 缓冲区 FB0 和 FB1 的 DMA 写完成后，将发出各自的中断信号，其定义如下：

```
#define RASTER_END_OF_FRAME0_INT          LCDC_IRQENABLE_SET_EOF0
#define RASTER_END_OF_FRAME1_INT          LCDC_IRQENABLE_SET_EOF1
#define LCDC_IRQENABLE_SET_EOF1           (0x00000200u)
#define LCDC_IRQENABLE_SET_EOF0           (0x00000100u)
```

AM335x 数据手册中，对应的中断使能设置寄存器如图 7.37 所示。

Table 13-37. IRQENABLE_SET Register Field Descriptions

Bit	Field	Type	Reset	Description
31-10	RESERVED	R/W	0h	
9	eof1_en_set	R/W	0h	DMA End-of-Frame 1 Interrupt Enable Set Read indicates enabled (mask) status. 0 = disabled. 1 = enabled. Writing 1 will set interrupt enable. Writing 0 has no effect.
8	eof0_en_set	R/W	0h	DMA End-of-Frame 0 Interrupt Enable Set Read indicates enabled (mask) status. 0 = disabled. 1 = enabled. Writing 1 will set interrupt enable. Writing 0 has no effect.

图 7.37 IRQENABLE_SET 寄存器描述(手册截图)

main()主函数中 RasterEnable()函数调用使能了 LCD 控制器 Raster，如图 7.38 所示。

```
209        RasterEnable(LCDC_INSTANCE);
```

图 7.38 main()函数中 RasterEnable()函数的调用

如图 7.39 所示为 RasterEnable()函数定义。

```
void RasterEnable(unsigned int baseAddr)
{
    HWREG(baseAddr + LCDC_RASTER_CTRL) |= LCDC_RASTER_CTRL_RASTER_EN;
}
```

图 7.39 RasterEnable()函数定义

如图 7.40 所示为 LCD 控制器 Raster 使能/禁止寄存器描述。

main()函数的最后是 while(1)死循环。当其运行后将在 AM335x Starter Kit 开发板的 LCD 屏上显示如图 7.41 所示的图像。

注：因手机拍摄角度原因，显示的画面有点倾斜，但显示是正常的。

0	lcden	R/W	0h	LCD Controller Enable. 0 = LCD controller disabled. 1 = LCD controller enabled.

图 7.40 LCD 控制器 Raster 使能/禁止寄存器(手册截图)

图 7.41 LCD 显示画面

7.3 LCD 显示修改实验

当我们分析完 LCD 部分的源代码后，为了进一步巩固、消化和应用，最好还是亲自去修改其中的部分代码。

7.3.1 demo 工程中关于 LCD 显示代码的对比分析

我们先参考 demo 例程中关于 LCD 显示部分的源代码。

demoMain.c 的主函数 main()中，关于 LCD 的语句如图 7.42 所示(图中已经将与 LCD 无关的代码语句去除，只摘取 LCD 部分的语句及函数)。

Raster0IntRegister()函数调用，用于注册 LCD 控制器 Raster0 的中断及中断处理程序，函数定义如图 7.43 所示。

Raster0IntRegister()函数定义中注册函数中的 Raster0Isr 为中断处理函数，其定义如图 7.43 所示。

这里 LCD 的中断处理函数 Raster0Isr()与 rasterDisplay 里的 LCD 中断处理函数 LCDIsr()相比，不同点只是 LCD 控制器 Raster 的 DMA 缓冲存储器的起始地址和结束地址不同。

```
811    /* Register the ISRs */
812    Raster0IntRegister();

823    /* Enable system interrupts */
824    IntPrioritySet(SYS_INT_LCDCINT, 0, AINTC_HOSTINT_ROUTE_IRQ);
825    IntSystemEnable(SYS_INT_LCDCINT);

884    Raster0Init();

922    Raster0EOFIntEnable();

929    imageCount = 0;
930
931    frameBufIdx = 0;
932
933    /* Extract banner image to Frame buffer */
934    ImageArrExtract(bannerImage,
935                    (unsigned int*)(g_pucBuffer[!frameBufIdx]+PALETTE_OFFSET));
936
937    CacheDataCleanBuff((unsigned int) &g_pucBuffer[0]+PALETTE_OFFSET,
938          GrOffScreen24BPPSize(LCD_WIDTH, LCD_HEIGHT, PIXEL_24_BPP_UNPACKED));
939    CacheDataCleanBuff((unsigned int) &g_pucBuffer[1]+PALETTE_OFFSET,
940          GrOffScreen24BPPSize(LCD_WIDTH, LCD_HEIGHT, PIXEL_24_BPP_UNPACKED));
941
942    Raster0Start();

952    /* Extract base image to uncomp buffer */
953    ImageArrExtract(baseImage, (unsigned int*)baseUnCompImage);
954
955    /* Copy base image to FB */
956    memcpy((void *)((g_pucBuffer[frameBufIdx]+PALETTE_OFFSET)),
957           (const void *)baseUnCompImage, (LCD_SIZE+PALETTE_SIZE));
958
959    CacheDataCleanBuff((unsigned int) &g_pucBuffer[0]+PALETTE_OFFSET,
960          GrOffScreen24BPPSize(LCD_WIDTH, LCD_HEIGHT, PIXEL_24_BPP_UNPACKED));
961    CacheDataCleanBuff((unsigned int) &g_pucBuffer[1]+PALETTE_OFFSET,
962          GrOffScreen24BPPSize(LCD_WIDTH, LCD_HEIGHT, PIXEL_24_BPP_UNPACKED));

982    /*  Create menu page */
983    imageCount = 1;
984    updatePage(imageCount);
```

图7.42 demo项目工程中demoMain.c的main()函数中关于LCD显示的语句

LCD的显示缓存定义如图7.44所示。

LCD参数定义如图7.45所示。

g_pucBuffer[]中GrOffScreen24BPPSize()定义如图7.46所示。

g_pucBuffer为二维数组，每维包含480×272×4+8×4+4 =(482×272+8)×4+4，而LCD控制器Raster的DMA缓存的起始地址等于g_pucBuffer+4，结束地址就等于g_pucBuffer+4+(482×272+8)×4 −1。另外，g_pucBuffer的最前面4字节为调色板的数据。

“IntPrioritySet(SYS_INT_LCDCINT, 0, AINTC_HOSTINT_ROUTE_IRQ);”和“IntSystemEnable(SYS_INT_LCDCINT);”，这两个函数和rasterDisplay中的函数是一样的。

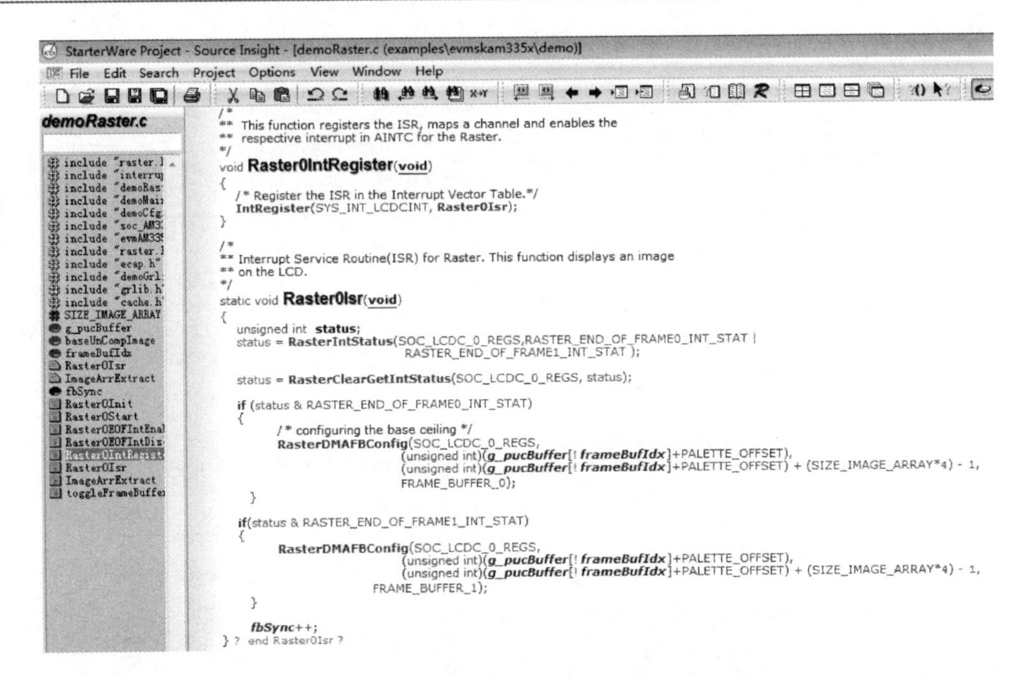

图 7.43　Raster0IntRegister()和 Raster0Isr()函数定义

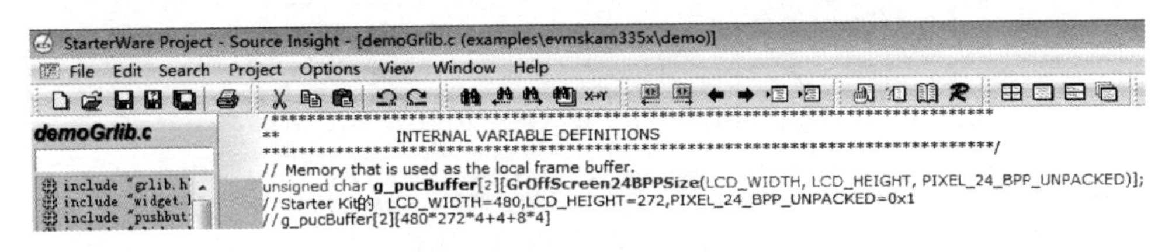

图 7.44　g_pucBuffer[]缓存定义

```
#define LCD_SIZE (480*272*4)
#define PALETTE_SIZE 32
#define LCD_WIDTH 480
#define LCD_HEIGHT 272
#define PALETTE_OFFSET  4
#define PIXEL_24_BPP_PACKED      (0x0)
#define PIXEL_24_BPP_UNPACKED    (0x1)
#define FRAME_BUFFER_0  0
#define FRAME_BUFFER_1  1
```

图 7.45　LCD 参数定义

Raster0Init()函数和 rasterDisplay 里的 SetUpLCD()函数也是一样的。

继续 demoMain.c 的主函数 main()，Raster0EOFIntEnable()函数定义如图 7.47 所示，用于使能 Raster0 的帧 frame0 和 frame1 读/写结束的中断。

而在 rasterDisplay 里，RasterIntEnable()是直接在 main()函数中定义的，如图 7.48 所示。

```
//*****************************************************************************
//
//! Determines the size of the buffer for a 24 BPP off-screen image.
//!
//! \param lWidth is the width of the image in pixels.
//! \param lHeight is the height of the image in pixels.
//!
//! This function determines the size of the memory buffer required to hold a
//! 16 BPP off-screen image of the specified geometry.
//!
//! \return Returns the number of bytes required by the image.
//
//*****************************************************************************
#define GrOffScreen24BPPSize(lWidth, lHeight, pack)          \
    (4 + (8*4) + (lWidth * lHeight * (3+pack)))
```

图 7.46 GrOffScreen24BPPSize()定义

```
/*
** A wrapper function which enables the End-Of-Frame interrupt of Raster.
*/
void Raster0EOFIntEnable(void)
{
   /* Enable End of frame0/frame1 interrupt */
   RasterIntEnable(SOC_LCDC_0_REGS, RASTER_END_OF_FRAME0_INT |
                                    RASTER_END_OF_FRAME1_INT);

}
```

图 7.47 Raster0EOFIntEnable()函数定义

```
205     /* Enable End of frame0/frame1 interrupt */
206     RasterIntEnable(LCDC_INSTANCE, RASTER_END_OF_FRAME0_INT |
207                                    RASTER_END_OF_FRAME1_INT);
```

图 7.48 rasterDisplay 中 RasterIntEnable()函数调用

继续 demoMain.c 的主函数 main()，ImageArrExtract(bannerImage,(unsigned int *)(g_pucBuffer[! frameBufIdx]+PALETTE_OFFSET))函数将 bannerImage 存入 g_pucBuffer[]缓冲区。

ImageArrExtract()函数定义，如图 7.49 所示。

ImageArrExtract()函数需要和 bannerImage 数据结构对比进行分析，如 bannerImage 中的一个数据 0x02f1f1fe，它表示 2 个相同的数据 0x00f1f1fe，所以在执行完该函数时，将 0x00f1f1fe、0x00f1f1fe 存入目标地址的连续两个存储单元中。

继续 demoMain.c 的主函数 main()，“CacheDataCleanBuff((unsigned int) &g_pucBuffer[0]+PALETTE_OFFSET,GrOffScreen24BPPSize(LCD_WIDTH,LCD_HEIGHT,PIXEL_24_BPP_UNPACKED));”和“CacheDataCleanBuff((unsigned int) &g_pucBuffer[1]+PALETTE_OFFSET,GrOffScreen24BPPSize(LCD_WIDTH,LCD_HEIGHT,PIXEL_24_BPP_UNPACKED));”，这两个函数调用清除 g_pucBuffer[]对应的 D-Cache 缓冲区段，如图 7.50 所示。

CP15DCacheCleanBuff()函数定义为汇编实现的代码，一般不会去修改和深入

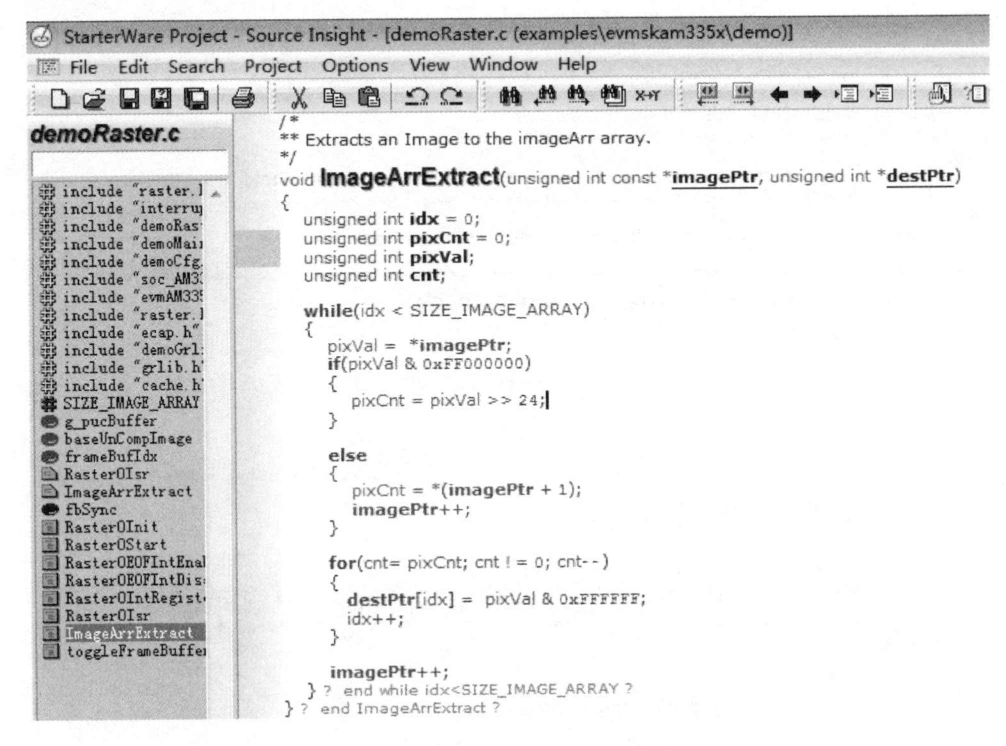

图 7.49 ImageArrExtract()函数定义

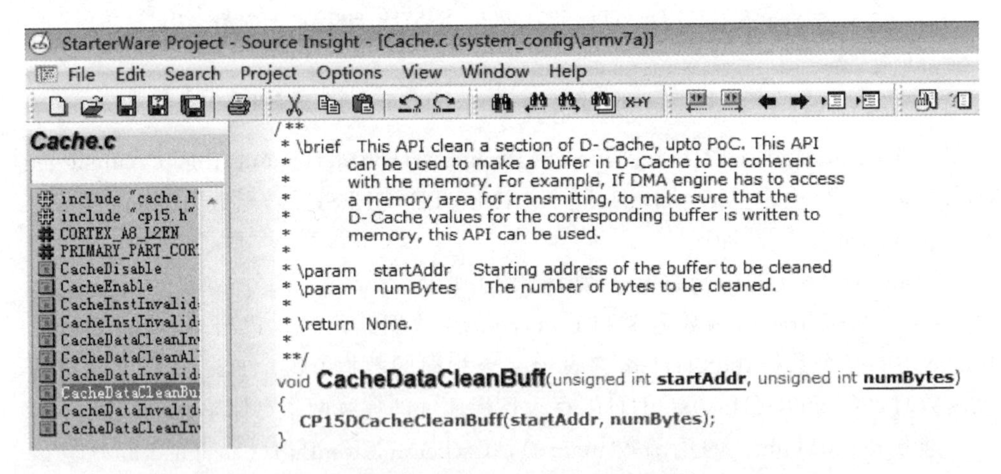

图 7.50 CacheDataCleanBuff()函数定义

研究，感兴趣的读者可以自己去参考分析。

继续 demoMain.c 的主函数 main()，Raster0Start()开始显示图像，其函数定义如图 7.51 所示。

如图 7.52 所示为 DMA 配置寄存器描述。

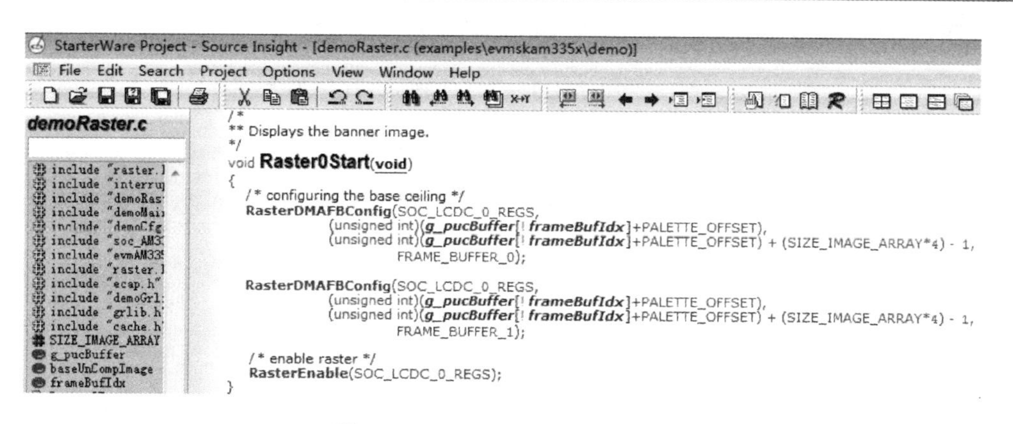

图 7.51　Raster0Start()函数定义

Table 13-5. Register Configuration for DMA Engine Programming

Register	Configuration
LCDDMA_CTRL	Configure DMA data format
LCDDMA_FB0_BASE LCDDMA_FB0_CEILING	Configure frame buffer 0
LCDDMA_FB1_BASE LCDDMA_FB1_CEILING	Configure frame buffer 1. (If only one frame buffer is used, these two registers will not be used.)

图 7.52　DMA 配置寄存器描述(手册截图)

7.3.2　LCD 显示实验调试曾出现的问题及解决方法

问题 1：在 C:\ti\AM335X_StarterWare_02_00_01_01\examples\evmsk-AM335x\raster 下将 image.h 复制成 image2.h，并将文件里的 0x00ffffff 替换成 0x00777777，将 image1 修改成 image2 后保存，然后在 rasterDisplay.c 中将 image1 修改成image2 后重新编译生成目标文件，再将目标文件放到 TF 卡上运行，但图像的背景颜色没有修改。

原因：只修改了主函数里的 image1，而没有修改中断处理函数里的 image1。

问题 2：CCS6＋TMDSSK3358 仿真调试 StarterWare 中的 rasterDisplay 时，单步进入外部函数时提示错误。

如图 7.53 所示为 evmskAM335x.h 中定义的外部函数。

在程序中会调用很多项目目录之外的函数，而此时仿真调试时如果想单步进入这些函数就会出错，如图 7.54 所示。

TI 技术工程师在 TI 官方论坛上给出的解答回复如图 7.55 所示。

解决方法：选择如图 7.56 所示的 Locate File；装载对应的驱动文件，如图 7.57 所示；此时就能正确进入 raster.c 的 RasterDMAFaConfig()函数执行单步运行了，如图 7.58 所示。

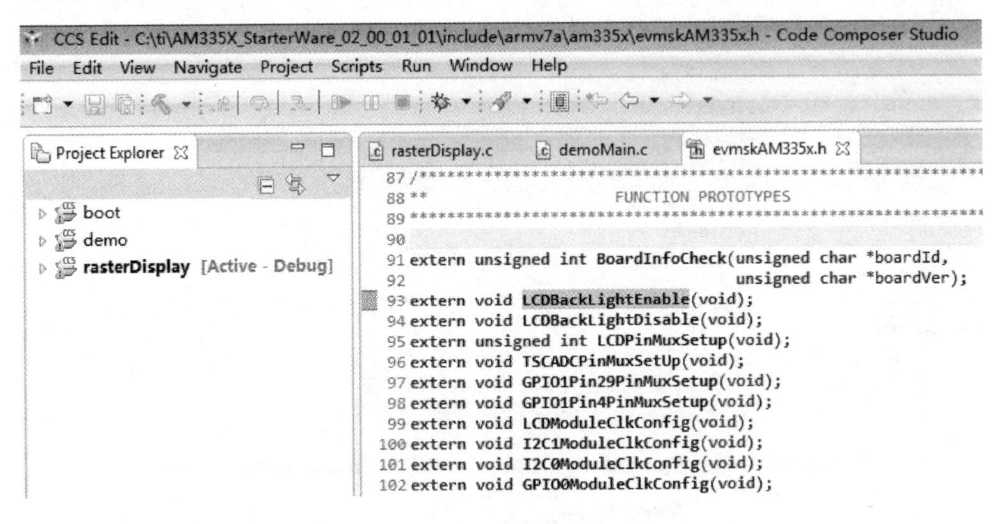

图 7.53　外部函数定义

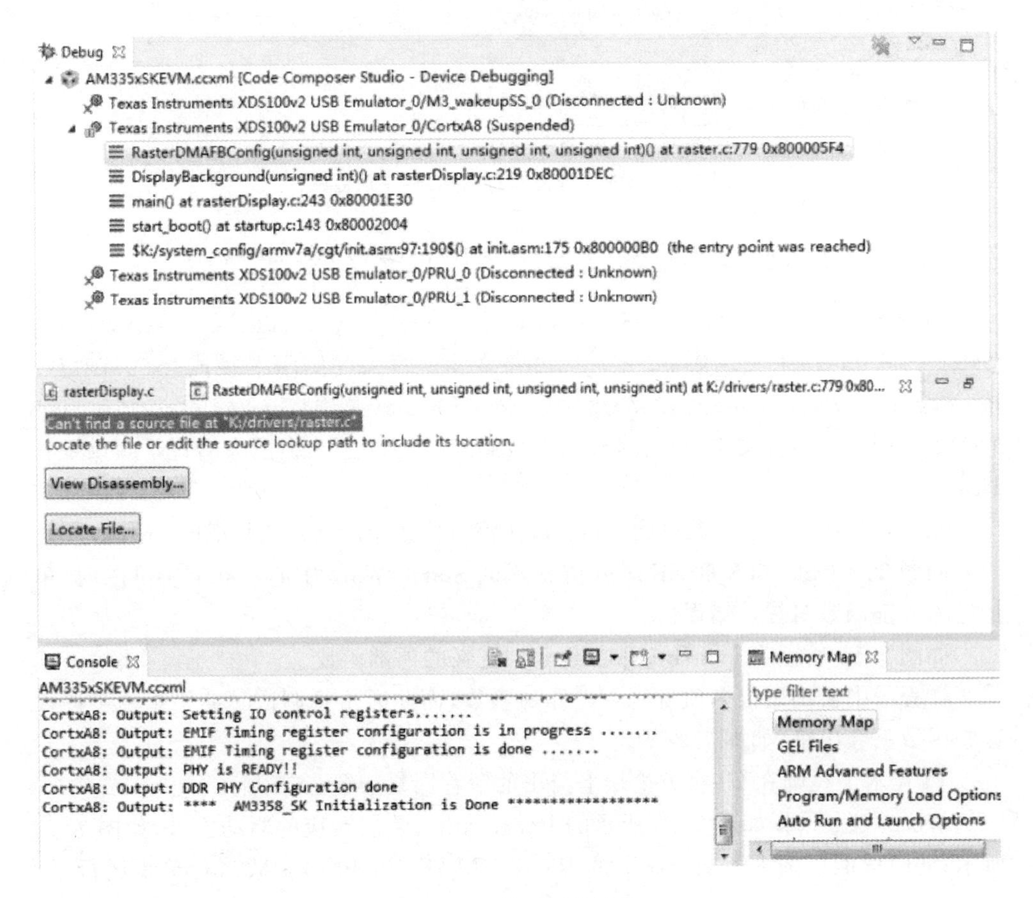

图 7.54　单步调试错误

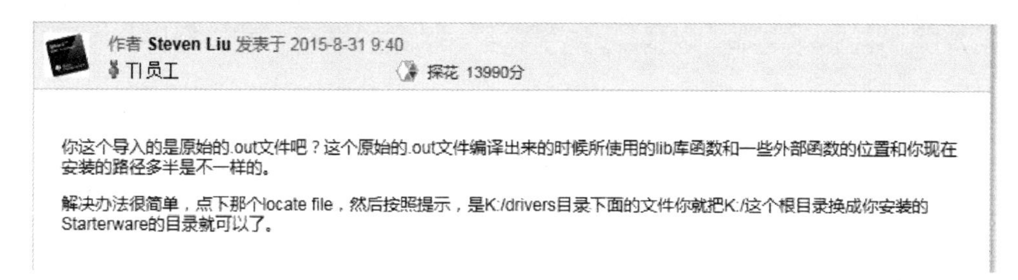

图 7.55　TI 技术工程师的回复

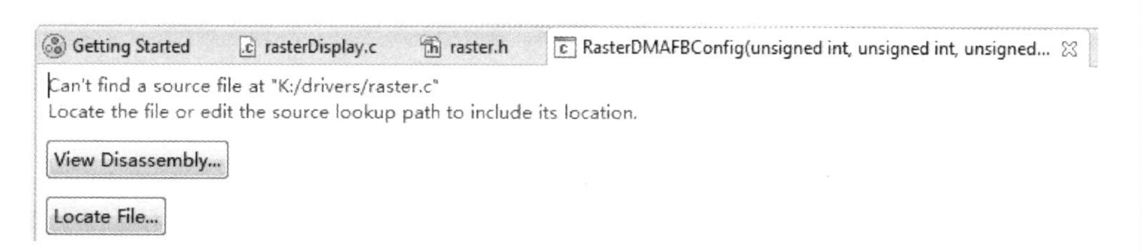

图 7.56　选择 Locate File

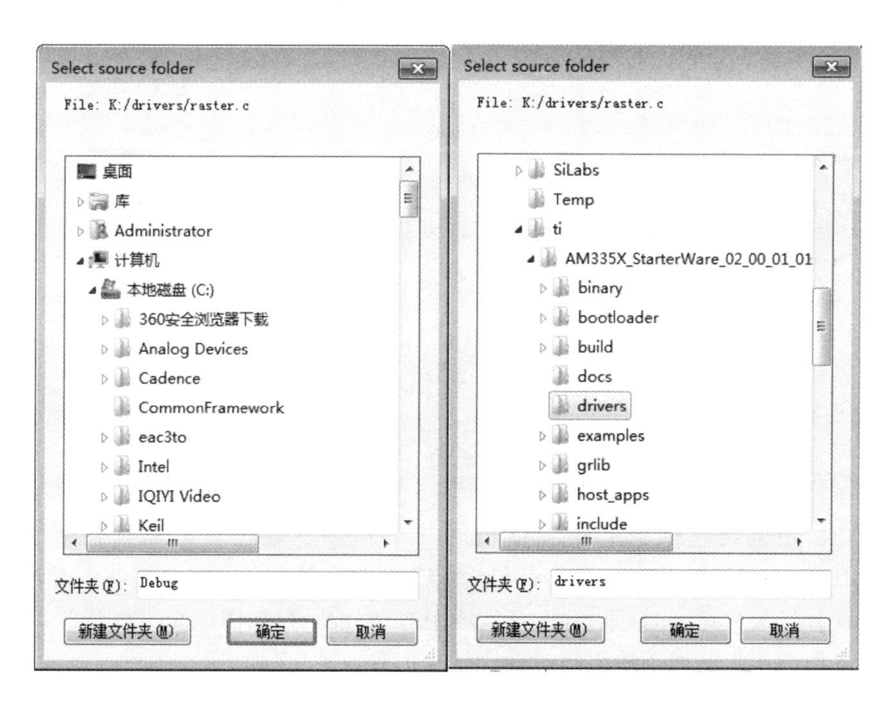

图 7.57　装载驱动文件

问题 3： LCD 显示问题。因为我们要自己掌握 LCD 的显示规律及相关的知识，所以需要自己修改代码来调试一下。用 AM335x Starter Kit 开发板的实验，在原来的 rasterDisplay.c 中定义了一个数据缓冲区：

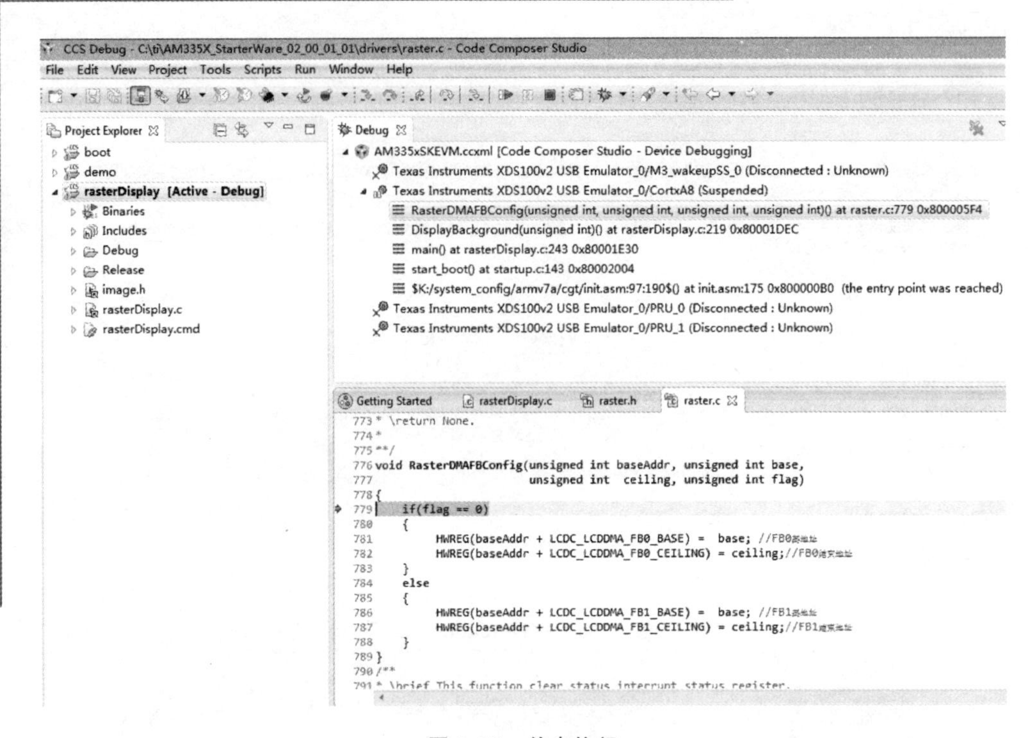

图 7.58 单步执行

```
#define LCD_WIDTH 480
#define LCD_HEIGHT 272
unsigned int LCDBuffer[LCD_WIDTH * LCD_HEIGHT + 8];
```

然后在程序中将原来的显示代码去掉，修改成自己加的代码，如下：

```
memcpy(LCDBuffer,image1,sizeof(LCDBuffer));
RasterDMAFBConfig(LCDC_INSTANCE,(unsigned int)LCDBuffer,(unsigned
int)LCDBuffer + sizeof(LCDBuffer) - 2,0);
RasterDMAFBConfig(LCDC_INSTANCE,(unsigned int)LCDBuffer,(unsigned
int)LCDBuffer + sizeof(LCDBuffer) - 2,1);
```

代码的用意就是将原来显示的 image1 数组中的数据全部复制到新定义的 LCDBuffer 中，再将 LCDBuffer 指针传给 LCD DMA 的起始和终止缓冲区寄存器。

但在开发板中运行时，显示的 LCD 屏中有部分乱码，如图 7.59 所示。

将上面的语句换回下述语句时就会显示正常：

```
RasterDMAFBConfig(LCDC_INSTANCE,(unsigned int)image1,(unsigned
int)image1 + sizeof(LCDBuffer) - 2,0);
RasterDMAFBConfig(LCDC_INSTANCE,(unsigned int)image1,(unsigned
int)image1 + sizeof(LCDBuffer) - 2,1);
```

图 7.59　LCD 显示乱码

LCD 显示正常图像，如图 7.41 所示。

问题 4：如图 7.60 所示。image1 数组中，前面 8 个数据是什么？为何会有该 8 个数据，而且也是直接传给 DMA 寄存器的？

```
_init.asm   bl_main.c   bl_platform.c   image.h
40
41 #ifdef __IAR_SYSTEMS_ICC__
42 #pragma data_alignment=4
43 unsigned int const image1[] = {
44
45 #elif defined(__TMS470__)
46 #pragma DATA_ALIGN(image1, 4);
47 unsigned int const image1[] = {
48
49 #else
50 unsigned int const image1[] __attribute__((aligned(4)))= {
51 #endif
52 0x4000u, 0x0000u, 0x0000u, 0x0000u, 0x0000u, 0x0000u, 0x0000u, 0x0000u,
53 0x00ffffff, 0x00ffffff, 0x00ffffff, 0x00ffffff, 0x00ffffff, 0x00ffffff, 0x00ffffff,
54 0x00ffffff, 0x00ffffff, 0x00ffffff, 0x00ffffff, 0x00ffffff, 0x00ffffff, 0x00ffffff,
55 0x00ffffff, 0x00ffffff, 0x00ffffff, 0x00ffffff, 0x00ffffff, 0x00ffffff, 0x00ffffff,
56 0x00ffffff, 0x00ffffff, 0x00ffffff, 0x00ffffff, 0x00ffffff, 0x00ffffff, 0x00ffffff,
57 0x00ffffff, 0x00ffffff, 0x00ffffff, 0x00ffffff, 0x00ffffff, 0x00ffffff, 0x00ffffff,
58 0x00ffffff, 0x00ffffff, 0x00ffffff, 0x00ffffff, 0x00ffffff, 0x00ffffff, 0x00ffffff,
```

图 7.60　image.h 定义

但是当在主函数中再次引用 LCDBuffer 时，显示会正常些，也就是乱码没有那么多，但还是有一点乱码。

另外，如果加了如图 7.61 所示的 delay(1000)函数，显示也会出现问题，我觉得 delay.h 中的延时调用有些问题。于是，自己写了一个 delay()函数，如图 7.62 所示。

调用 delay()函数，如图 7.63 所示。

```
    while(1)
    {
//        delay(1000);
        unsigned int i;
        for (i =0; i < sizeof(LCDBuffer)/4; i++){
            if (LCDBuffer[i] != image1[i]) break;
        }
```

图 7.61　delay()函数调用

```
228 void delay(unsigned int time)
229 {
230     unsigned int j;
231     for (j=0; j<time; j++);
232 }
```

图 7.62　delay()函数定义


```
259     while(1)
260     {
261         delay(1000);
262         unsigned int i;
263
264         for (i =0; i < sizeof(LCDBuffer)/4; i++){
265             if (LCDBuffer[i] != image1[i]) break;
266         }
```

图 7.63　调用自己写的 delay()函数

我们还尝试了 CCS 的优化设置，如图 7.64 所示，但仍然没有解决问题。

因此，我们还是需要回到 demo 工程中，因为 demo 工程的显示是正确无误的。在 demo 中反复测试，我们发现当颜色数据写入缓冲区后，需要调用一个清除缓冲区 Cache 的函数，源代码如下：

```
/* Extract banner image to Frame buffer */
ImageArrExtract(bannerImage,
        (unsigned int *)(g_pucBuffer[!frameBufIdx] + PALETTE_OFFSET));
CacheDataCleanBuff((unsigned int) &g_pucBuffer[0] + PALETTE_OFFSET,
        GrOffScreen24BPPSize(LCD_WIDTH,LCD_HEIGHT,PIXEL_24_BPP_UNPACKED));
CacheDataCleanBuff((unsigned int) &g_pucBuffer[1] + PALETTE_OFFSET,
        GrOffScreen24BPPSize(LCD_WIDTH,LCD_HEIGHT,PIXEL_24_BPP_UNPACKED));
```

所以，我们也需要在 raster 工程中增加 CacheDataCleanBuff()函数的调用，如图 7.65 所示。

我们定义自己的颜色值，代码如下：

```
//FRM8:8:8(红:绿:蓝)
#define clWhite         0x00FFFFFF                          //白色
```

```
#define clBlack       0x00000000            //黑色
#define clDRed        0x00800000            //浅红色,为全色的一半
#define clLRed        0x00FF0000            //深红色
#define clDMagenta    0x00800080            //暗紫色
#define clLMagenta    0x00FF00FF            //亮紫色
#define clGreen       0x0000FF00            //绿色
#define clDBlue       0x00000080            //暗蓝色
#define clLBlue       0x000000FF            //亮蓝色
#define clDCyan       0x00008800            //暗青色
#define clLCyan       0x0000FFFF            //亮青色
#define clDYellow     0x00808000            //暗黄色
#define clLYellow     0x00FFFF00            //亮黄色
#define clDGray       0x00808080            //暗灰色
#define clLGray       0x00FCFCFC            //亮灰色
#define clLArgent     0x00E3E3E3            //亮银色
```

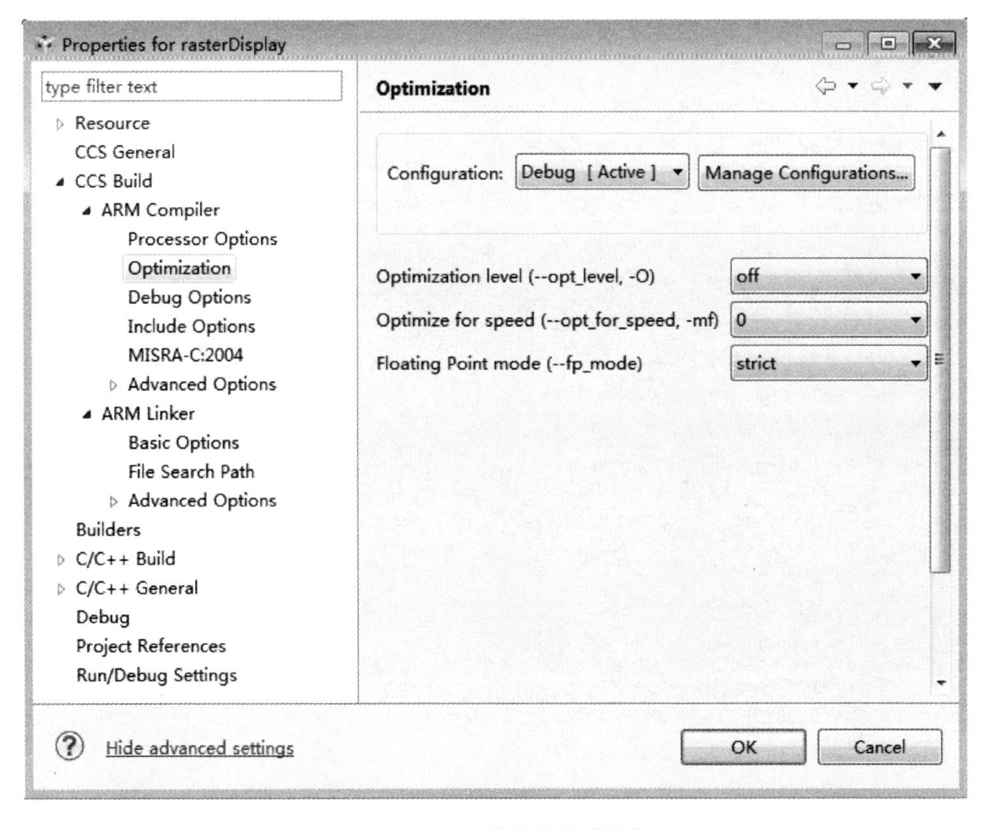

图 7.64　CCS 优化设置对话框

增加一个显示背景色的函数,代码如下:

```
void DisplayBackground(unsigned int * destPtr,unsigned long color){
    unsigned int i;
    for (i = 0; i < 130560; i++){
        destPtr[i] = color;
    }
    CacheDataCleanBuff((unsigned int)destPtr,LCD_WIDTH * LCD_HEIGHT * 4);
}
```

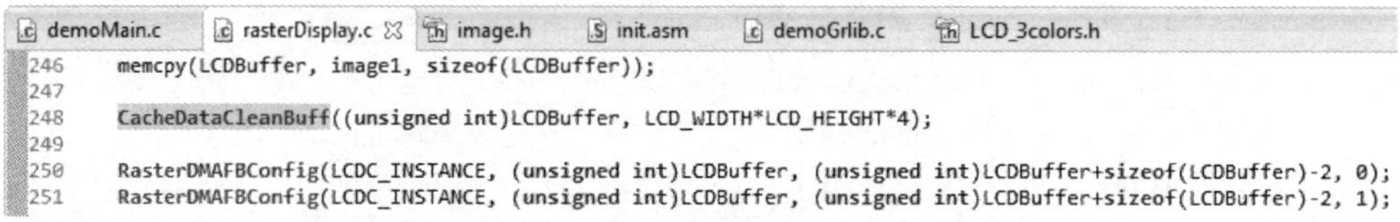

```
demoMain.c   rasterDisplay.c   image.h   init.asm   demoGrlib.c   LCD_3colors.h
246    memcpy(LCDBuffer, image1, sizeof(LCDBuffer));
247
248    CacheDataCleanBuff((unsigned int)LCDBuffer, LCD_WIDTH*LCD_HEIGHT*4);
249
250    RasterDMAFBConfig(LCDC_INSTANCE, (unsigned int)LCDBuffer, (unsigned int)LCDBuffer+sizeof(LCDBuffer)-2, 0);
251    RasterDMAFBConfig(LCDC_INSTANCE, (unsigned int)LCDBuffer, (unsigned int)LCDBuffer+sizeof(LCDBuffer)-2, 1);
```

图 7.65　CacheDataCleanBuff()函数调用

再在 main()函数的 while(1)中增加各种颜色的测试代码如下：

```
while(1){
    delay(50000000);
    DisplayBackground(clWhite);
    delay(50000000);
    DisplayBackground(clBlack);
    delay(50000000);
    DisplayBackground(clDRed);
    delay(500000000);
    DisplayBackground(clLRed);
    delay(50000000);
    DisplayBackground(clDMagenta);
    delay(500000000);
    DisplayBackground(clLMagenta);
    delay(500000000);
    DisplayBackground(clGreen);
    delay(50000000);
    DisplayBackground(clDBlue);
    delay(500000000);
    DisplayBackground(clLBlue);
    delay(50000000);
    DisplayBackground(clDCyan);
    delay(500000000);
    DisplayBackground(clLCyan);
    delay(500000000);
    DisplayBackground(clDYellow);
    delay(500000000);
```

```
        DisplayBackground(clLYellow);
        delay(500000000);
        DisplayBackground(clDGray);
        delay(500000000);
        DisplayBackground(clLGray);
        delay(500000000);
        DisplayBackground(clLArgent);

        unsigned int i;
        for (i = 0; i < sizeof(LCDBuffer)/4; i++){
        if (LCDBuffer[i] != image1[i]) break;
        }
    }
```

还是需要回到 demo 工程中。

在 demo 工程中增加显示测试代码，如图 7.66 所示。

```
demoMain.c    rasterDisplay.c    demoRaster.c    image.h
988      /* Extract banner image to Frame buffer */
989 //     ImageArrExtract(bannerImage,
990 //                     (unsigned int*)(g_pucBuffer[!frameBufIdx]+PALETTE_OFFSET));
991      memcpy(g_pucBuffer[0], image1, GrOffScreen24BPPSize(LCD_WIDTH, LCD_HEIGHT, PIXEL_24_BPP_UNPACKED));
992      memcpy(g_pucBuffer[1], image1, GrOffScreen24BPPSize(LCD_WIDTH, LCD_HEIGHT, PIXEL_24_BPP_UNPACKED));
993
994      CacheDataCleanBuff((unsigned int) &g_pucBuffer[0]+PALETTE_OFFSET,
995            GrOffScreen24BPPSize(LCD_WIDTH, LCD_HEIGHT, PIXEL_24_BPP_UNPACKED));
996      CacheDataCleanBuff((unsigned int) &g_pucBuffer[1]+PALETTE_OFFSET,
997            GrOffScreen24BPPSize(LCD_WIDTH, LCD_HEIGHT, PIXEL_24_BPP_UNPACKED));
998 //     DisplayBackground(clWhite);
999      Raster0Start();
1000     EcapBkLightEnable();
1001
1002     ConsoleUtilsPrintf("\n test 3");
1003
1004     Timer4Start();
1005     while(FALSE == tmr4Flag);
1006     tmr4Flag = FALSE;
1007     Timer4Stop();
1008
1009     ConsoleUtilsPrintf("\n test 4");
1010     while(1);
```

图 7.66　demoMain.c 中增加显示测试代码

将原来的 logo 修改成 rasterDisplay 中的 logo，显示正常。但增加各种颜色测试时，有些色没有反映，都和灰色一样。

rasterDisplay 乱码真正的解决方法是，在显示缓冲区写后和配置显示 DMA 缓冲区前，清除数据 cache，如图 7.67 所示。

显示图像如图 7.68 所示。但右边的图像有个错位。

红色的右下角也少了几个颜色值，如图 7.69 所示。

解决方法是在数组的前面多定义 8 字节数据，如图 7.70 所示。

```
demoMain.c   rasterDisplay.c   demoRaster.c   image.h   demoGrlib.c   init.asm
257     memcpy(LCDBuffer, image1, sizeof(LCDBuffer));
258
259     CacheDataCleanBuff((unsigned int)LCDBuffer, LCD_WIDTH*LCD_HEIGHT*4);
260
261     RasterDMAFBConfig(LCDC_INSTANCE, (unsigned int)LCDBuffer, (unsigned int)LCDBuffer+sizeof(LCDBuffer)-2, 0);
262     RasterDMAFBConfig(LCDC_INSTANCE, (unsigned int)LCDBuffer, (unsigned int)LCDBuffer+sizeof(LCDBuffer)-2, 1);
263
264 //    DisplayBackground(clWhite);
265     /* Enable End of frame0/frame1 interrupt */
266     RasterIntEnable(LCDC_INSTANCE, RASTER_END_OF_FRAME0_INT |
267                                    RASTER_END_OF_FRAME1_INT);
268     /* Enable raster */
269     RasterEnable(LCDC_INSTANCE);
```

图7.67 LCD显示乱码的解决方法

图7.68 LCD显示图像错位

图7.69 LCD红色显示的图像错位

```
47 const unsigned long LCD_3colors[130568] = {
48 0x4000u, 0x0000u, 0x0000u, 0x0000u, 0x0000u, 0x0000u, 0x0000u, 0x0000u,
49 0x00FFFFFF,0x00FFFFFF,0x00FFFFFF,0x00FFFFFF,0x00FFFFFF,0x00FFFFFF,0x00FFFFFF,0x00FFFFFF,0x00FFFFFF,0x00FFFFFF,
50 0x00FFFFFF,0x00FFFFFF,0x00FFFFFF,0x00FFFFFF,0x00FFFFFF,0x00FFFFFF,0x00FFFFFF,0x00FFFFFF,0x00FFFFFF,0x00FFFFFF,
51 0x00FFFFFF,0x00FFFFFF,0x00FFFFFF,0x00FFFFFF,0x00FFFFFF,0x00FFFFFF,0x00FFFFFF,0x00FFFFFF,0x00FFFFFF,0x00FFFFFF,
```

图7.70 LCD_3colors[]定义

第8章

触摸屏例程源代码分析

8.1 触摸屏例程源代码目录结构

在 CCS 左边的 Project Explorer 中，右击选择 Import→CCS Projects 打开 Import CCS Eclipse Projects 对话框，单击 Browse 按钮选择目录 C:\ti\AM335X_StarterWare_02_00_01_01\build\armv7a\cgt_ccs\am335x\evmskAM335x\touchScreen，导入 touchScreen 触摸屏项目例程，如图 8.1 所示。

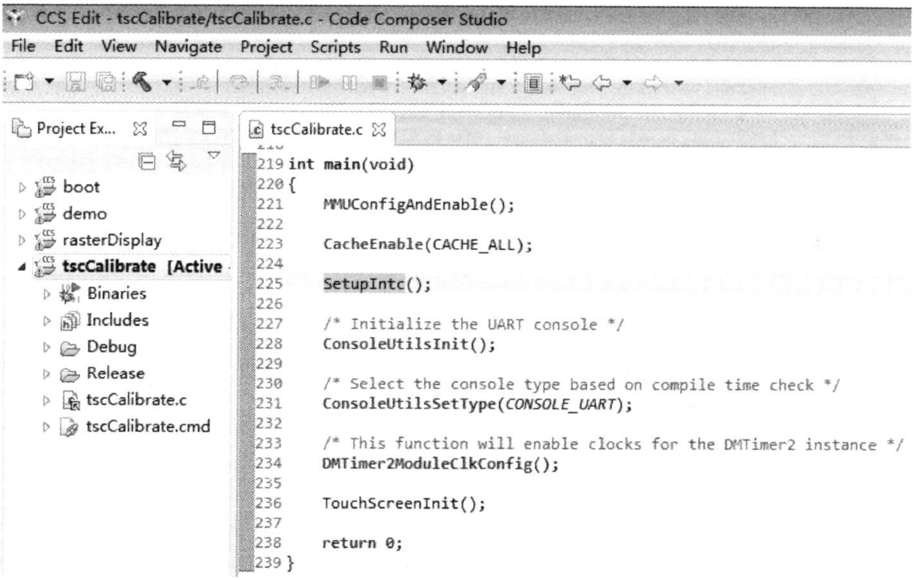

图 8.1 touchScreen 触摸屏项目例程

和 rasterDisplay 项目工程一样，虽然 touchScreen 项目工程中只有 tscCalibrate.c 源文件，但仍会调用如 system_config、platform 等目录下的一些文件。

8.2　tscCalibrate.c 文件分析

在图 8.1 所示的 tscCalibrate.c 的主函数 main()里，基本也都是初始化调用函数。

8.2.1　内存管理和高速缓存的配置

MMUConfigAndEnable()函数为配置使能 MMU 内存管理单元。

CacheEnable()函数使能 Cache 高速缓冲区。

8.2.2　中断使能和注册

SetupIntc()函数配置使能中断，并注册触摸屏中断函数，如图 8.2 所示。


```
241 static void SetupIntc(void)
242 {
243     /* Enable IRQ in CPSR.*/
244     IntMasterIRQEnable();
245
246     /* Initialize the ARM Interrupt Controller.*/
247     IntAINTCInit();
248
249     IntRegister(SYS_INT_ADC_TSC_GENINT, TouchScreenIsr);//注册触摸屏中断
250
251     IntPrioritySet(SYS_INT_ADC_TSC_GENINT, 0, AINTC_HOSTINT_ROUTE_IRQ);
252
253     IntSystemEnable(SYS_INT_ADC_TSC_GENINT);
254 }
```

图 8.2　SetupInt()中断设置函数定义

IntMasterIRQEnable()函数使能 ARM 处理器的 IRQ 中断。

IntAINTCInit()函数为初始化 AM335x ARM 处理器的中断控制器，而且通常在使用中断控制器之前完成调用。

IntRegister(SYS_INT_ADC_TSC_GENINT，TouchScreenIsr)函数调用是将 TouchScreenIsr ()函数作为触摸屏中断处理函数存入中断向量表 fnRAMVectors[]数组中，用于触摸屏中断处理服务。TouchScreenIsr()触摸屏中断处理函数主要是读取触摸屏 X、Y 轴的 A/D 采样值。

IntPrioritySet(SYS_INT_ADC_TSC_GENINT，0，AINTC_HOSTINT_ROUTE_IRQ)函数调用设定触摸屏中断的优先级，即将触摸屏中断设置成 IRQ 中断，且将优先级设置为最高的 0。

IntSystemEnable(SYS_INT_ADC_TSC_GENINT)使能触摸屏中断。

8.2.3　调试串口初始化设置

在主函数 main()中，ConsoleUtilsInit()函数初始化调试串口，而 ConsoleUtilsSetType()函数用于设置调试串口类型。

8.2.4　定时器初始化

在主函数 main()中，DMTimer2ModuleClkConfig()函数的调用用于初始化定时器 2 的时钟模块。

8.2.5　触摸屏函数分析

在主函数 main()的最后调用 TouchScreenInit()函数，用于触摸屏相关模块的初始化配置、触摸屏校准和触摸屏值的读取。

TouchScreenInit()函数定义如图 8.3 所示。

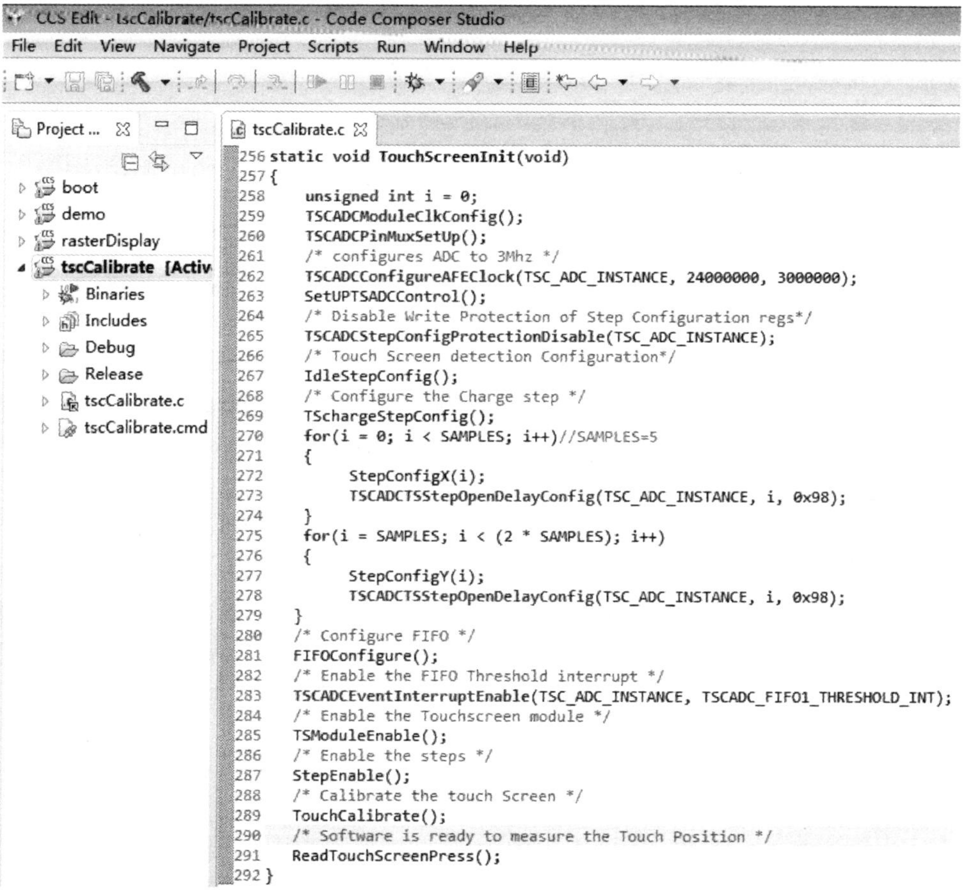

CCS Edit - tscCalibrate/tscCalibrate.c - Code Composer Studio

File　Edit　View　Navigate　Project　Scripts　Run　Window　Help

Project ...　　tscCalibrate.c

- boot
- demo
- rasterDisplay
- tscCalibrate [Activ
  - Binaries
  - Includes
  - Debug
  - Release
  - tscCalibrate.c
  - tscCalibrate.cmd

```
256 static void TouchScreenInit(void)
257 {
258     unsigned int i = 0;
259     TSCADCModuleClkConfig();
260     TSCADCPinMuxSetUp();
261     /* configures ADC to 3Mhz */
262     TSCADCConfigureAFEClock(TSC_ADC_INSTANCE, 24000000, 3000000);
263     SetUPTSADCControl();
264     /* Disable Write Protection of Step Configuration regs*/
265     TSCADCStepConfigProtectionDisable(TSC_ADC_INSTANCE);
266     /* Touch Screen detection Configuration*/
267     IdleStepConfig();
268     /* Configure the Charge step */
269     TSchargeStepConfig();
270     for(i = 0; i < SAMPLES; i++)//SAMPLES=5
271     {
272         StepConfigX(i);
273         TSCADCTSStepOpenDelayConfig(TSC_ADC_INSTANCE, i, 0x98);
274     }
275     for(i = SAMPLES; i < (2 * SAMPLES); i++)
276     {
277         StepConfigY(i);
278         TSCADCTSStepOpenDelayConfig(TSC_ADC_INSTANCE, i, 0x98);
279     }
280     /* Configure FIFO */
281     FIFOConfigure();
282     /* Enable the FIFO Threshold interrupt */
283     TSCADCEventInterruptEnable(TSC_ADC_INSTANCE, TSCADC_FIFO1_THRESHOLD_INT);
284     /* Enable the Touchscreen module */
285     TSModuleEnable();
286     /* Enable the steps */
287     StepEnable();
288     /* Calibrate the touch Screen */
289     TouchCalibrate();
290     /* Software is ready to measure the Touch Position */
291     ReadTouchScreenPress();
292 }
```

图 8.3　TouchScreenInit()函数定义

TouchScreenInit()函数中，前面几个函数的调用都是初始化配置等。这里比较重要的是两个 for 循环，其中的 StepConfigX()和 StepConfigY()实现了电阻触摸屏 4 个晶体管的切换控制，以满足模拟量的采集，并将模拟量值存入 FIFO 中。

StepConfigX()函数定义如图 8.4 所示。


```
361 static void StepConfigX(unsigned int stepSelc)
362 {
363     /* Configure ADC to Single ended operation mode */
364     TSCADCTSStepOperationModeControl(TSC_ADC_INSTANCE,
365                                      TSCADC_SINGLE_ENDED_OPER_MODE, stepSelc);
366     /* Configure reference volatage and input to charge step*/
367     TSCADCTSStepConfig(TSC_ADC_INSTANCE, stepSelc,TSCADC_NEGATIVE_REF_VSSA,
368                        TSCADC_POSITIVE_INP_CHANNEL3,TSCADC_NEGATIVE_INP_CHANNEL1,
369                        TSCADC_POSITIVE_REF_VDDA);
370     /* Configure the Analog Supply to Touch screen */
371     TSCADCTSStepAnalogSupplyConfig(TSC_ADC_INSTANCE, TSCADC_XPPSW_PIN_ON,
372                                    TSCADC_XNPSW_PIN_OFF, TSCADC_YPPSW_PIN_OFF,
373                                    stepSelc);
374     /* Configure the Analong Ground to Touch screen */
375     TSCADCTSStepAnalogGroundConfig(TSC_ADC_INSTANCE, TSCADC_XNNSW_PIN_ON,
376                                    TSCADC_YPNSW_PIN_OFF, TSCADC_YNNSW_PIN_OFF,
377                                    TSCADC_WPNSW_PIN_OFF, stepSelc);
378     /* select fifo 0 */
379     TSCADCTSStepFIFOSelConfig(TSC_ADC_INSTANCE, stepSelc, TSCADC_FIFO_0);
380     /* Configure in One short hardware sync mode */
381     TSCADCTSStepModeConfig(TSC_ADC_INSTANCE, stepSelc, TSCADC_ONE_SHOT_HARDWARE_SYNC);
382     TSCADCTSStepAverageConfig(TSC_ADC_INSTANCE, stepSelc, TSCADC_SIXTEEN_SAMPLES_AVG);
383 }
```

图 8.4　StepConfigX()函数定义

TouchScreenInit()函数中，TouchCalibrate()函数的调用实现了触摸屏的校准。

TouchCalibrate()函数定义如图 8.5 所示。

IsTSPress 在触摸屏中断处理函数中将其置位，以表示触摸屏被按下。从 for 循环中读取左上、右上、右下 3 点的 X、Y 对应的 A/D 值，setCalibrationMatrix()函数通过 3 点对应的坐标值实现校准，并将校准参数值存于 stMatrix 中，具体原理需要详细研究代码后总结出计算公式。

前面讲到的在中断设置函数中注册的 TouchScreenIsr()(触摸屏中断处理函数)主要是读取 X、Y 轴的 A/D 值。

TouchScreenInit()函数的最后，ReadTouchScreenPress()函数的调用用于读取触摸屏当前按下的坐标值。ReadTouchScreenPress()函数中最重要的函数调用是 getDisplayPoint()函数，它根据触摸屏的 A/D 值，转换成 LCD 屏的坐标值。

getDisplayPoint()函数定义如图 8.6 所示。

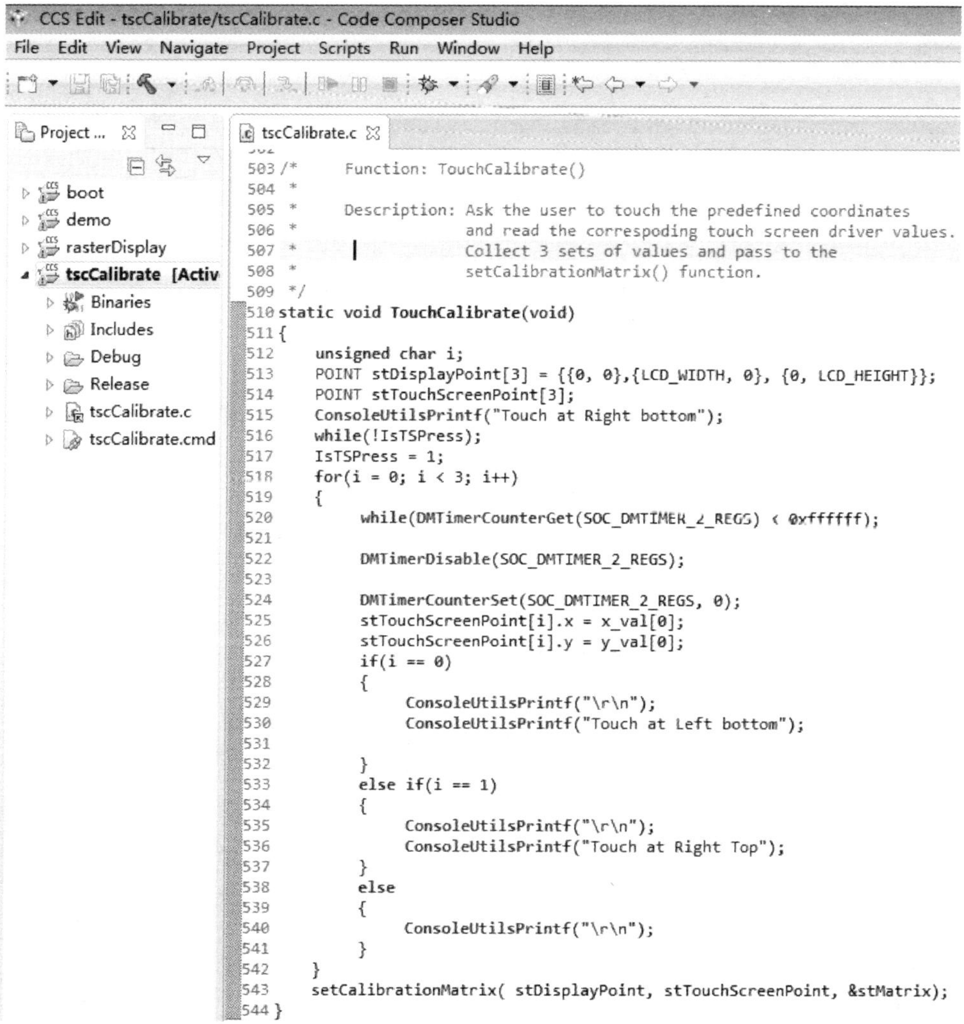

CCS Edit - tscCalibrate/tscCalibrate.c - Code Composer Studio

File Edit View Navigate Project Scripts Run Window Help

Project ...
- boot
- demo
- rasterDisplay
- tscCalibrate [Activ
  - Binaries
  - Includes
  - Debug
  - Release
  - tscCalibrate.c
  - tscCalibrate.cmd

tscCalibrate.c

```
503 /*     Function: TouchCalibrate()
504  *
505  *     Description: Ask the user to touch the predefined coordinates
506  *                  and read the correspoding touch screen driver values.
507  *                  Collect 3 sets of values and pass to the
508  *                  setCalibrationMatrix() function.
509  */
510 static void TouchCalibrate(void)
511 {
512     unsigned char i;
513     POINT stDisplayPoint[3] = {{0, 0},{LCD_WIDTH, 0}, {0, LCD_HEIGHT}};
514     POINT stTouchScreenPoint[3];
515     ConsoleUtilsPrintf("Touch at Right bottom");
516     while(!IsTSPress);
517     IsTSPress = 1;
518     for(i = 0; i < 3; i++)
519     {
520         while(DMTimerCounterGet(SOC_DMTIMER_2_REGS) < 0xffffff);
521 
522         DMTimerDisable(SOC_DMTIMER_2_REGS);
523 
524         DMTimerCounterSet(SOC_DMTIMER_2_REGS, 0);
525         stTouchScreenPoint[i].x = x_val[0];
526         stTouchScreenPoint[i].y = y_val[0];
527         if(i == 0)
528         {
529             ConsoleUtilsPrintf("\r\n");
530             ConsoleUtilsPrintf("Touch at Left bottom");
531 
532         }
533         else if(i == 1)
534         {
535             ConsoleUtilsPrintf("\r\n");
536             ConsoleUtilsPrintf("Touch at Right Top");
537         }
538         else
539         {
540             ConsoleUtilsPrintf("\r\n");
541         }
542     }
543     setCalibrationMatrix( stDisplayPoint, stTouchScreenPoint, &stMatrix);
544 }
```

图 8.5　TouchCalibrate()函数定义

```
682 /**********************************************************************
683  *
684  *          Function: getDisplayPoint()
685  *
686  *       Description: Given a valid set of calibration factors and a point
687  *                    value reported by the touch screen, this function
688  *                    calculates and returns the true (or closest to true)
689  *                    display point below the spot where the touch screen
690  *                    was touched.
691  *
692  *
693  *
694  *       Argument(s): displayPtr (output) - Pointer to the calculated
695  *                                          (true) display point.//返回的LCD坐标值
696  *                    ScreenPtr (input) -   Pointer to the reported touch
697  *                                          screen point.//触摸屏的A/D值
698  *                    matrixPtr (input) -   Pointer to calibration factors//校准后的计算参数值
699  *                                          matrix previously calculated
700  *                                          from a call to setCalibrationMatrix()
701  */
702 int getDisplayPoint(POINT *displayPtr,
703                     POINT *screenPtr,
704                     MATRIX *matrixPtr )//输入触摸屏的A/D值，再根据校准参数值，计算得到LCD的坐标值
705 {
706     int  retValue = 0 ;
707 
708     if( matrixPtr->Divider != 0 )
709     {
710         displayPtr->x = (((matrixPtr->An * screenPtr->x +
711                          matrixPtr->Bn * screenPtr->y) / 10000)
712                          + (matrixPtr->Cn));
713 
714      displayPtr->y = (((matrixPtr->Dn * screenPtr->x +
715               matrixPtr->En * screenPtr->y) / 10000)
716                          + (matrixPtr->Fn));
717     }
718     else
719     {
720         retValue = -1 ;
721     }
722 
723     return(retValue) ;
724 
725 } /* end of getDisplayPoint() */
```

图 8.6　getDisplayPoint()函数定义

第 9 章

StarterWare 对 BeagleBone Black 的支持

原本不打算在本书中加入 AM335x Starter Kit 开发板之外的内容，但仔细想想应该也有很多读者在使用 BeagleBone Black（简称 BBB），所以就有了本章节的内容。

9.1 补丁包 StarterWare_BBB_support.gz

StarterWare 的旧版本支持 BeagleBone（即白板），而不支持 BeagleBone Black（即黑板）。但最新版本 AM335X_StarterWare_02_00_01_01 可通过一个 Patch 包 StarterWare_BBB_support.gz 对 BeagleBone Black 实现支持。

从 http://software-dl.ti.com/dsps/dsps_public_sw/am_bu/starterware/02_00_01_01/index_FDS.html 页面单击如图 9.1 所示的补丁包下载项。

Beaglebone black patch	Beaglebone support files	84K

图 9.1 StarterWare 补丁包下载项

下载 StarterWare_BBB_support.gz 文件到 C:\ti\AM335X_StarterWare_02_00_01_01 目录，并解压覆盖原文件。

然后在 CCS 下编译 boot 和 demo 工程，分别将 boot_ti.bin 和 demo_ti.bin 重命名为 MLO 和 app，再拷贝到 TF 中，并插入到 BeagleBone Black 引导运行。此时终端将打印如图 9.2 所示信息。

图 9.2 StarterWare 在 BeagleBone Black 开发板上的启动信息

由 demo 工程中的 demoMain.c 的 main()函数(见图 9.3)可知,正常情况下应该还有更多的信息,所以此时应该是程序跑飞了。

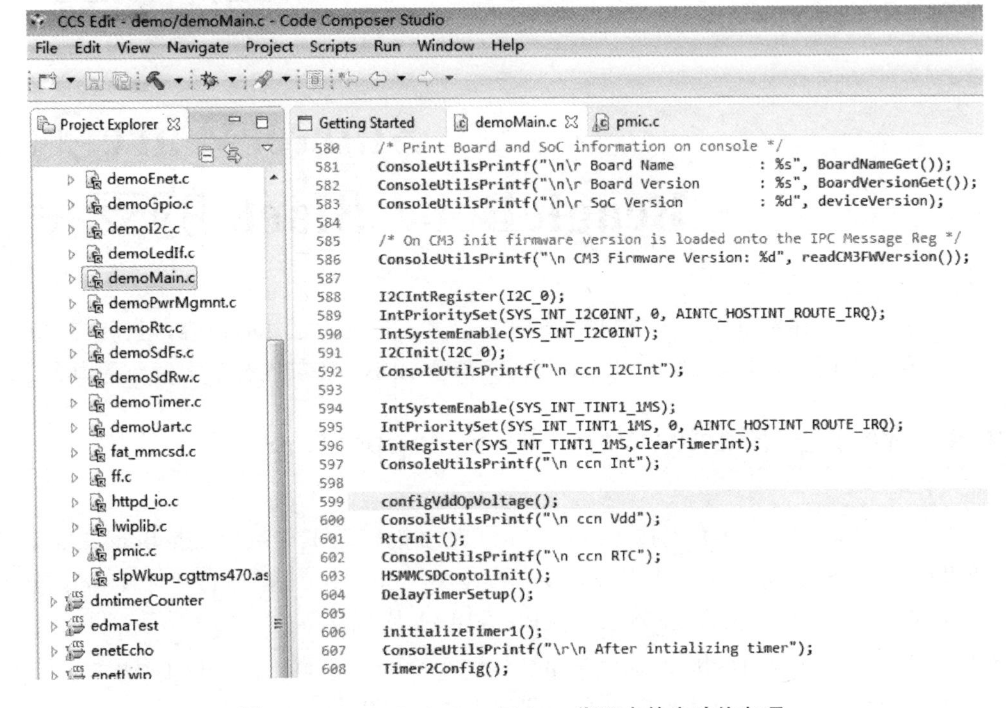

图 9.3　StarterWare demoMain.c 代码中的启动信息项

9.2　demo 在 BeagleBone Black 上死机现象的分析及追踪

在源代码中增加调试信息,以确定程序在哪个地方跑飞,如图 9.4 所示。

根据图 9.4 可以判断出,当设置 PMIC 的 LDO3 为 3.3 V 时,出现死机现象。查看 BeagleBone Black 原理图(见图 9.5)可知,LDO3 是给 HDMI 发送芯片 TDA19988 供电的,需为 1.8 V。

所以源码这里是错误的,用万用表实际测量了死机后,该电源确实为 3.3 V。但 BeagleBone Black 不从 TF 卡启动 Linux 系统,所测的该电源是 1.8 V,所以进一步判断该代码有问题。一方面,我们查看到图 9.4 所示程序中函数 configVddOpVoltage()定义在 pmic.c 源文件中,而该文件源码又位于 C:\ti\AM335X_StarterWare_02_00_01_01\examples\beaglebone\demo 目录;另一方面,我们查看到 Patch 解压包的目录结构及安装过程如图 9.6 所示,所以即使我们打包了 Patch,也没有更新 C:\ti\AM335X_StarterWare_02_00_01_01\examples 目录下的源代码,所以出现了如 demo 工程中的错误。

```
Getting Started | demoMain.c | pmic.c | hw_tps65217.h
595     IntPrioritySet(SYS_INT_TINT1_1MS, 0, AINTC_HOSTINT_ROUTE_IRQ);
596     IntRegister(SYS_INT_TINT1_1MS,clearTimerInt);
597     ConsoleUtilsPrintf("\n ccn Int");
598
599     configVddOpVoltage();
600     ConsoleUtilsPrintf("\n ccn Vdd");
601     RtcInit();
602     ConsoleUtilsPrintf("\n ccn RTC");
603     HSMMCSDContolInit();
604     DelayTimerSetup();
```

```
Getting Started | demoMain.c | pmic.c | hw_tps65217.h
155     TPS65217RegWrite(PROT_LEVEL_2, DEFSLEW, DCDC_GO, DCDC_GO);
156 }
157
158
159 void configVddOpVoltage(void)
160 {
161     /* Configure PMIC slave address */
162     I2CMasterSlaveAddrSet(SOC_I2C_0_REGS, PMIC_TPS65217_I2C_SLAVE_ADDR);
163     ConsoleUtilsPrintf("\n ccn I2CMasterSlaveAddrSet");
164
165     /* Increase USB current limit to 1300mA */
166     TPS65217RegWrite(PROT_LEVEL_NONE, POWER_PATH, USB_INPUT_CUR_LIMIT_1300MA,
167                      USB_INPUT_CUR_LIMIT_MASK);
168     ConsoleUtilsPrintf("\n ccn TPS65217RegWrite USB current ");
169
170     /* Set LDO3, LDO4 output voltage to 3.3V */
171     TPS65217RegWrite(PROT_LEVEL_2, DEFLS1, LDO_VOLTAGE_OUT_3_3, LDO_MASK);
172     ConsoleUtilsPrintf("\n ccn TPS65217RegWrite LDO3 voltage");
173
174     TPS65217RegWrite(PROT_LEVEL_2, DEFLS2, LDO_VOLTAGE_OUT_3_3, LDO_MASK);
175     ConsoleUtilsPrintf("\n ccn TPS65217RegWrite LDO4 voltage");
176 }
```

```
StarterWare
AM335x Boot Loader
Copying application image from MMC/SD card to RAM
Jumping to StarterWare Application...


 Board Name          : A335BNL
 Board Version       : 00A6
 SoC Version         : 1
 CM3 Firmware Version: 385
 ccn I2CInt
 ccn Int
 ccn I2CMasterSlaveAddrSet
 ccn TPS65217RegWrite USB current
```

图 9.4 StarterWare 在 BBB 跑飞代码的跟踪

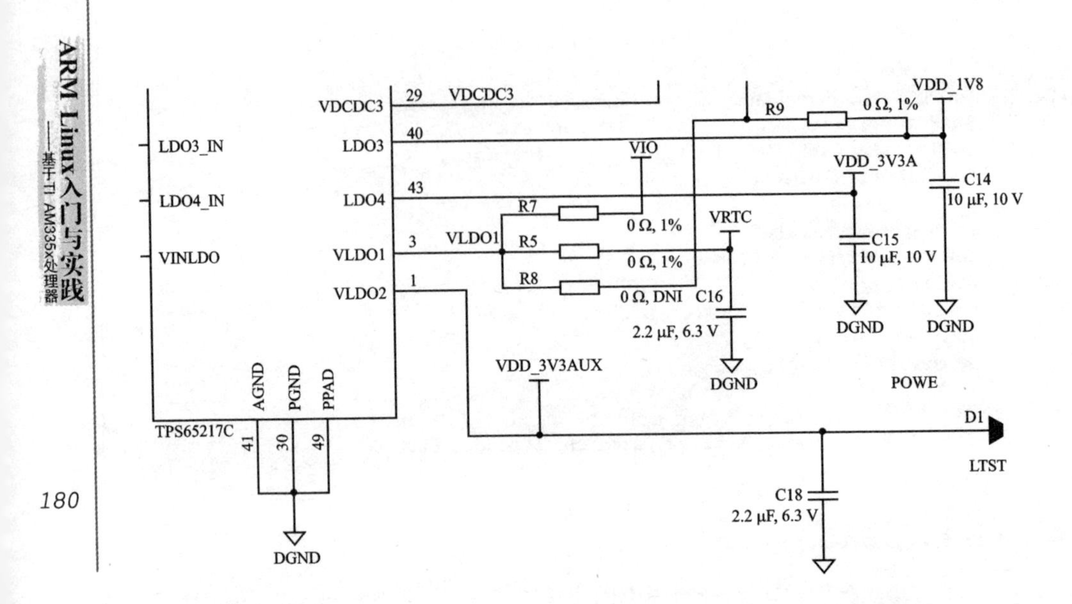

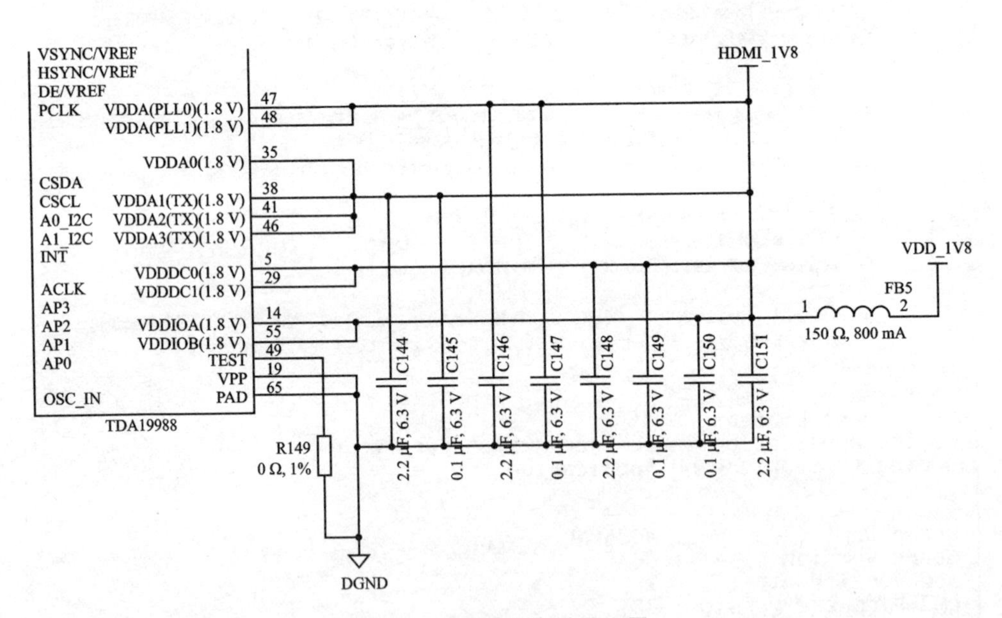

图 9.5　BBB 电源部分原理图

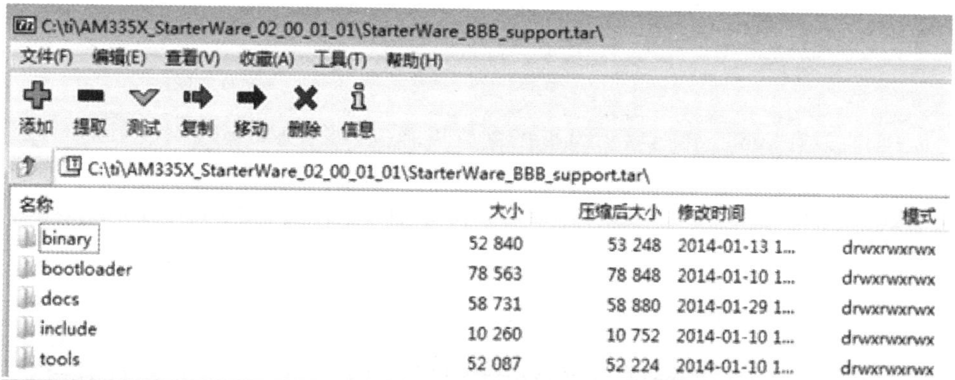

C:\ti\AM335X_StarterWare_02_00_01_01\StarterWare_BBB_support.tar\

文件(F)　编辑(E)　查看(V)　收藏(A)　工具(T)　帮助(H)

添加　提取　测试　复制　移动　删除　信息

C:\ti\AM335X_StarterWare_02_00_01_01\StarterWare_BBB_support.tar\

名称	大小	压缩后大小	修改时间	模式
binary	52 840	53 248	2014-01-13 1...	drwxrwxrwx
bootloader	78 563	78 848	2014-01-10 1...	drwxrwxrwx
docs	58 731	58 880	2014-01-29 1...	drwxrwxrwx
include	10 260	10 752	2014-01-10 1...	drwxrwxrwx
tools	52 087	52 224	2014-01-10 1...	drwxrwxrwx

```
ccn@ccn01:~/Downloads$ tar -xzvf StarterWare_BBB_support.tar.gz
binary/
binary/armv7a/
binary/armv7a/cgt_ccs/
binary/armv7a/cgt_ccs/am335x/
binary/armv7a/cgt_ccs/am335x/beaglebone/
binary/armv7a/cgt_ccs/am335x/beaglebone/bootloader/
binary/armv7a/cgt_ccs/am335x/beaglebone/bootloader/Release_MMCSD/
binary/armv7a/cgt_ccs/am335x/beaglebone/bootloader/Release_MMCSD/MLO
binary/armv7a/gcc/
binary/armv7a/gcc/am335x/
binary/armv7a/gcc/am335x/beaglebone/
binary/armv7a/gcc/am335x/beaglebone/bootloader/
binary/armv7a/gcc/am335x/beaglebone/bootloader/Release_MMCSD/
binary/armv7a/gcc/am335x/beaglebone/bootloader/Release_MMCSD/MLO
bootloader/
bootloader/include/
bootloader/include/armv7a/
bootloader/include/armv7a/am335x/
bootloader/include/armv7a/am335x/bl_platform.h
bootloader/src/
bootloader/src/armv7a/
bootloader/src/armv7a/am335x/
bootloader/src/armv7a/am335x/bl_platform.c
docs/
docs/BBB_support_manual.pdf
include/
include/armv7a/
include/armv7a/am335x/
include/armv7a/am335x/beaglebone.h
include/hw/
include/hw/hw_tps65217.h
tools/
tools/gel/
tools/gel/AM335X_beagleboneblack.gel
ccn@ccn01:~/Downloads$
```

图 9.6　BBB Patch 解压包目录结构和安装过程

9.3 StarterWare 在 BeagleBone Black 上死机现象的解决

我们可以在 pmic.c 中，将 LDO3 的配置屏蔽(因为从 boot 工程代码可知，已经将 LDO3 进行了配置)，或修改为 1.8 V，如图 9.7 所示。

```
    /* Set LDO3, LDO4 output voltage to 3.3V */
//     TPS65217RegWrite(PROT_LEVEL_2, DEFLS1, LDO_VOLTAGE_OUT_3_3, LDO_MASK);
    TPS65217RegWrite(PROT_LEVEL_2, DEFLS1, LDO_VOLTAGE_OUT_1_8, LDO_MASK);
    ConsoleUtilsPrintf("\n ccn TPS65217RegWrite LDO3 voltage");
```

图 9.7 pmic.c 源码中 LDO3 的配置语句

重新编译，再将目标文件复制到 μSD 上，并连接板子的网线，上电后正确的终端信息如图 9.8 所示。

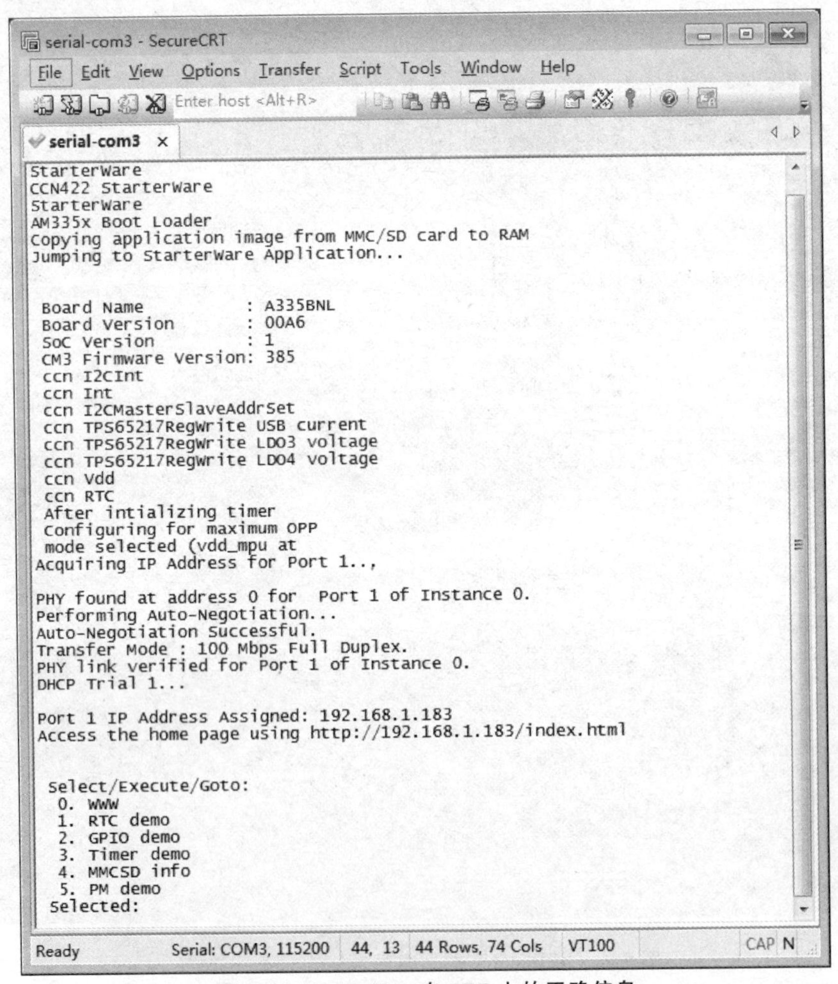

图 9.8 StarterWare 在 BBB 上的正确信息

注：需要连接网线且网络正常，demo 例程才能动态分配 IP 地址。

打开浏览器，直接输入动态分配的 IP 地址（此次为 192.168.1.183），如图 9.9 所示。

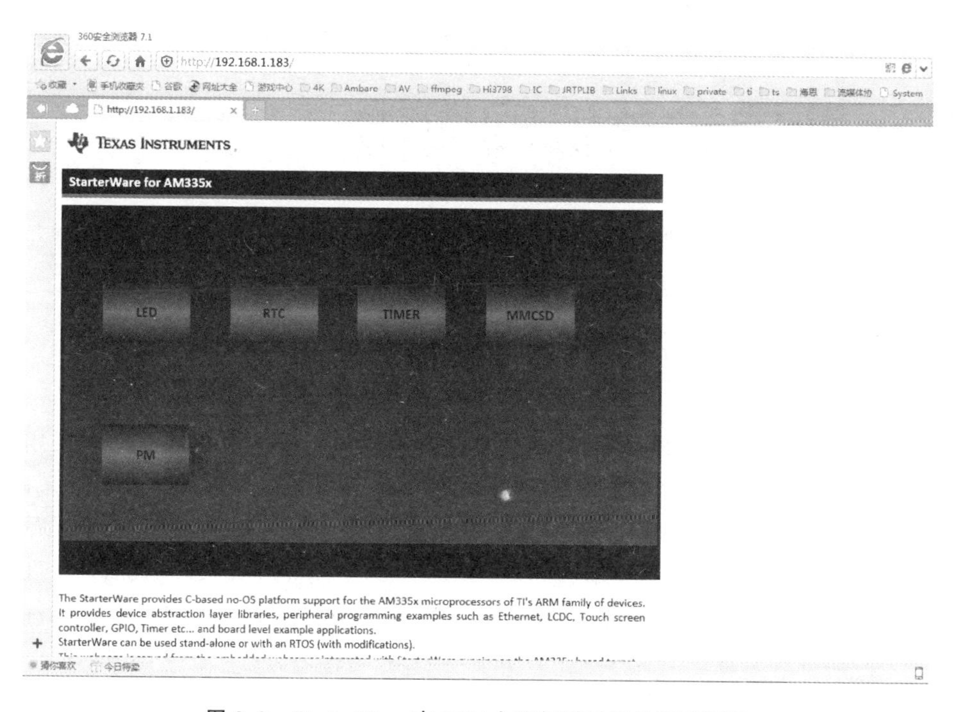

图 9.9　StarterWare 在 BBB 上运行时网页的测试页面

第 10 章

基于前后台系统的应用

其实，前面章节介绍的 StarterWare 库的内容都是基于前后台系统的，这章介绍实际应用过程还需要关注的一些知识点。

10.1 前后台系统概述

前后台系统是指不使用操作系统的嵌入式应用程序，是 8 位单片机广泛采用的方式。一般做法是，写一个无限循环的程序，在程序中查询每个事件是否发生，每个任务是否具备运行条件，如果是，则处理这个事件或执行这个任务。这可以看成是一个后台程序。而系统对中断的响应可以看成是前台程序。中断的打入表示某事件的发生，在中断服务子程序中，一般对中断做一些简单的处理，然后给出一个某事件发生了的标志，等待后台程序来处理。后台可以叫作任务级，前台叫作中断级。

前后台系统也被很多朋友称为“裸奔”！ARM9“裸奔”太浪费了吧？Cortex-A8 还“裸奔”，这不是更浪费？其实，这是很多朋友的疑问。这个疑问和 ARM 能否取代 8 位单片机是一个道理，它们俩也经常在论坛上被广泛争论。作为一名应用工程师，我们总是在考虑成本，物质成本、时间成本、后续维护成本，等等。总是希望用最熟悉、最简单的方法去完成项目。8 位 MCU 能完成的项目为何要用 ARM？“裸奔”就能满足、实现的，为何要引入复杂、庞大的操作系统？所以问题的关键是清楚地认识项目的需求，理解 MCU、ARM 处理器、操作系统的优点和缺点。比如有一个需要人机交互的终端产品，如果它只需要几个普通的按键和 128×64(或更高的分辨率)的单色 LCD 模块，那么我们自然会想着用 51 或 AVR 等 MCU 去实现。如果要把按键换成触摸屏，LCD 换成 640×480 STN 型(伪彩)彩色的，那么此时可能采用 ARM7 处理器 S3C44B0X 会更简单一些。如果 LCD 要换成 TFT 型(真彩)的 LCD 屏，那么可能需要选择 ARM9 处理器 S3C2410A，这不仅考虑 S3C44B0X 只能直接驱动640×480 的 STN 型 LCD，S3C2410A 可以直接驱动 TFT 屏，而且还要考虑刷屏的速度。当然最新的项目肯定会考虑像 Cortex-A8 内核的 AM335x 系列，因为 AM3352 的价格可以做得很低。如果在 LCD 屏上只是做些基本图片的显示和简单的交互，那么“裸奔”足以满足要求。如果一定要给它引入 Linux、WinCe 或 Android 等操作系统，那么不仅在技术上增加了难度和开发成本，而且系统还会存在不稳定性(因为不仅是应用程序，还有驱动等内核级的稳定），且系统响应、刷屏的速度等不一定有“裸奔”的快，因为前后台系统的任务如果不多，此时就专门检测用户的输入事件，然后立刻响应，而操作系统还需要花时间在内核调度上。如果需要在 LCD 屏上播放动画，实现 MP3、视频、网络传输等更复杂的功能，那么“裸奔”就不能满足了，需要使用操作系统来降低开发难度，以及进行各种任务的分时、并行处理。

10.2　Bootloader 的设计

基于嵌入式操作系统应用的 Boodloader，读者早有耳闻或已熟知了，有 vivi、redboot 等，最常用的当然还是 U-boot。它们用于下载、引导操作系统内核和文件系统等。那么前后台应用系统是否也需要 Bootloader 呢？早期的 S3C2410、S3C6410 等普遍使用 NAND Flash，当应用程序较大时，一般都需要 Bootloader。目前很多嵌入式处理器系统都使用 EMMC，有时还需要使用 SD 卡。AM335x Starter Kit 由于板子没有 NAND Flash 和 EMMC，而是直接使用 μSD 卡，所以它需要一个 Bootloader 来引导。但和之前的 S3C2410 一样，TI 的 StarterWare 也已经提供了一个 Bootloader，就是前面章节介绍的 Boot 工程，因此读者可以直接参考前面的章节来分析和应用，当然也可以对它进行修改和增强。

10.3　简易文件系统设计

文件系统对于读者来说一定不陌生，如 Windows 的 FAT16、FAT32，嵌入式 Linux 下的 JFFS、YAFFS，等等。对于前后台系统是否也需要这样的文件系统呢？比如 wav 声音文件，我们把 wav 文件直接与代码编译在一起，但是当需要的 wav 文件较多时，总的代码量就会很大，而且 wav 文件编译进去后不能灵活地更换。除了 wav 文件外，在应用中可能还有图片文件、记事本文件，甚至可执行的文件，等等，都需要一个文件系统来管理它们。另外，如果使用的是 NAND Flash，那么最严重的一个问题是 NAND Flash 存在坏块的现象，这就需要文件系统对坏块进行管理，防止由于坏块而破坏数据。目前，嵌入式系统中比较常用的是 FAT16、FAT32 等，特别是类似于 U 盘之类的应用，但对于一般的应用，它们还是过于复杂，读者也不易理解。这里我们实现一个非常简易的文件系统，或者说只是个链表，对于读者来说容易理解，并能满足大部分的前后台系统。

注：10.3 节提到的文件系统最初的设计是使用 S3C2410 的 NAND Flash，因此很多针对 NAND Flash 的定义都还保留着，如果只是单纯地使用在 AM335x Starter Kit 上，则可以做进一步的修改和优化。

10.3.1 文件系统结构

1. 文件表结构

```
struct fileTableType{                  //文件系统表结构
    int      vollable;                 //文件系统卷的序列号,格式化时被写入,当卷号不对时
                                       //说明文件系统不存在或错误,需要重新格式化
    int      totalBlock;               //总的可供文件使用的 NAND Flash 用户块 = endBlock -
                                       //startBlock
    int      startBlock;               //用于存放文件的开始块号,即第一个存放文件的块
    int      endBlock;                 //用于存放文件的最后块号,即最后一个存放文件的块
    int      maxFileNum;               //最多可以保存的文件数目
    char     blockFlag[TOTAL_BLOCK + START_BLOCK];
                              //用于记录文件损坏及占用信息,1 表示已被占用或损坏,0 为未用
    int      exitFileNum;              //文件表已存在的文件数目
    short    blockIndex[TOTAL_BLOCK];                 //块索引表,记录文件占用的块号
    struct   timeType creatTime; //文件表创建的时间
    struct DfileHeadType fileHeads[MAX_FILE_NUM];
                                              //所有文件头结构,可参考"2.文件夹结构"
    int      badblock;                 //总的坏块数
};
```

文件系统表在 NAND Flash 中位于整个文件系统的开始处,它管理着整个文件系统,以及记录着该文件的有用信息,在格式化文件系统时被创建初始化。

```
union fileTableUnion{ //定义文件系统表的联合体,方便下述两种方式共同操作文件结构表
    struct fileTableType fileTable;
    unsigned char U8s[sizeof(struct fileTableType)];
};
```

fileTableUnion 为定义的文件结构表联合体,比直接使用文件表结构定义更加方便操作,特别是从 NAND Flash 读取文件系统表,或是将文件结构表写入 NAND Flash。另外,还有一些文件表中使用到常数,它们可以根据实际的 NAND Flash 使用需求进行更改。

```
#define START_BLOCK     204          //User file start block,不包括文件表
#define END_BLOCK       2047         //User file end Block
#define TOTAL_BLOCK     (END_BLOCK - START_BLOCK)   //User total block
#define MAX_FILE_NUM    100          //记录可以最大容纳的文件个数
#define FILE_BLOCK      200          //file system talble star blcok
#define BLOCKSIZE       16384        //1 block  = 512 Byte * 32 Pages = 16 384 Byte = 16 KByte
#define VOLCONST        0x85641586   //file system serial lable
```

2. 文件头结构

```
struct fileHeadType{                    //定义文件在文件登记表中的记录结构
    char     fileName[16];              //文件名字,最长 20 个字符
    char     fileType[4];               //文件类型,bmp、exe、wav etc,共可以表示 256 种
    char     fileDescribe[20];          //文件的描述
    struct   timeType modifyTime;       //文件的修改日期
    int      fileSize;    //文件大小,单位为 NAND Flash 的块 BLOCK,最大 16 * 65536 KByte
    char     fileStatus;                //文件的状态,open、close、removed
    int      startFileMark;             //文件占用区块索引值在文件区块表中的起始位置
    int      endFileMark;               //文件占用区块索引值在文件区块表中的结束位置
    int      fileHeadCrc;               //文件头的校验和,保证文件头的完整性
    int      fileDataCrc;               //文件数据的校验和,保证文件数据的完整性
};
```

文件头是文件系统表中的一部分,每创建一个文件就会生成一个文件头,用于管理和记录该文件的信息。

```
struct timeType{//定义时间的结构
char     seconds;
char     minutes;
char     hours;
char     date;
char     month;
short    year;
};
```

10.3.2　文件系统功能函数

1. 文件系统初始化

```
union fileTableUnion sysFileTable;        //定义一个文件系统表
static unsigned char   fileTp[16384];
static unsigned char   * Tp;
/*************************************************************
函数原形: void FileSystemInit(unsigned char flag)
功    能: 文件系统初始化
参    数: flag——确定是否往终端打印各文件信息,调试用
*************************************************************/
void FileSystemInit(unsigned char flag){
    static int i,j,page,space;
    Tp = fileTp;
    i = sizeof(sysFileTable);
```

```
    for(j = 0; j < 32; j++){                    //从 NAND Flash 读取文件系统表
        NF_ReadPage(FILE_BLOCK,j,Tp);
        Tp += 512;
    }
    for(i = 0; i < BLOCKSIZE; i++){             //初始化文件结构表
        sysFileTable.U8s[i] = fileTp[i];
    }
    if(sysFileTable.fileTable.vollable != VOLCONST){    //判断文件系统是否正确存在
        Uart_Printf("NO file system format begain! \n");
        Format();                               //重新格式化文件系统
    }
    j = 0;
    for(i = sysFileTable.fileTable.startBlock; i<sysFileTable.fileTable.endBlock; i++)
        j += sysFileTable.fileTable.blockFlag[i];   //计算 NAND Flash 的块使用情况
    space = (sysFileTable.fileTable.totalBlock - j) * 16;
                                                //计算可用空间,单位为 KByte
    Uart_Printf("File system load complete\n");
    Uart_Printf("Volume serial number is   ");
    Uart_Printf(" %x        \n",sysFileTable.fileTable.vollable);
    if (flag)
        showFileTable();                        //往终端打印文件信息
    Uart_Printf("                    Total file is ");
    Uart_Printf(" %d \n",sysFileTable.fileTable.exitFileNum);
    Uart_Printf("                    Nand flash free space is ");
    Uart_Printf(" %dK Byte\n",space);
    Uart_Printf("                    Nand flash bad block is ");
    Uart_Printf(" %d \n",sysFileTable.fileTable.badblock);
}
```


该函数一般在系统上电复位启动时被调用,它从 NAND Flash 的文件系统结构表的块读取文件系统表;然后初始化 sysFileTable 全局变量;函数中根据卷的序列号(vollable)判断文件系统的存在,以确定是否需要重新格式化文件系统;最后打印文件系统的信息。

2. 文件系统格式化

```
/*****************************************************************
函数原形: void Format(void)
功    能: 格式化文件系统
参    数: 无
*****************************************************************/
void Format(){
    int i,j;
```

```
        for(i = 0;i < BLOCKSIZE ;i++){
            sysFileTable.U8s[i] = 0xff;                    //clean old ram data
        }
        j = 0;
        for(i = START_BLOCK; i < END_BLOCK; i++){
            if(!NF_EraseBlock(i)){
                                                    //NAND Flash 块擦除,填充坏块标志
                sysFileTable.fileTable.blockFlag[i] = 1;
                                               //在 blockFlag 相应位置把占用标志置 1
                j++;                                       //坏块值累加
            }
            else{
                sysFileTable.fileTable.blockFlag[i] = 0;
                                               //在 blockFlag 相应位置把占用标志清 0
            }
        }
    sysFileTable.fileTable.vollable = VOLCONST;            //填入磁盘卷序列号
    sysFileTable.fileTable.startBlock = START_BLOCK;       //文件开始区块
    sysFileTable.fileTable.endBlock = END_BLOCK;           //文件结束区块
    sysFileTable.fileTable.totalBlock= TOTAL_BLOCK;        //总块数
    sysFileTable.fileTable.exitFileNum = 0;                //文件表中现存的文件数目为 0
    sysFileTable.fileTable.maxFileNum = MAX_FILE_NUM;      //记录最大文件数目
    sysFileTable.fileTable.badblock = j;                   //记录 NAND Flash 失效的 BLOCK
    sysFileTable.fileTable.creatTime = ReadCurrentTime();  //记录系统当前时间
    UpdateFileTable();                                     //将文件结构表写入 NAND Flash
}
```

格式化就是重新创建且初始化文件系统结构表,以前的文件及数据信息都将被擦除。这里先是将文件系统表的内容都清除(即写入 0xff);再对 NAND Flash 进行块擦除操作,记录坏块信息和可用块的标志,同时也将系统的其他信息初始化到文件系统结构表;最后将它写入到 NAND Flash 的结构表占用的块。

```
/***********************************************************
函数原形: struct timeType ReadCurrentTime()
功    能: 读当前 RTC 时钟
返    回: 当前时间
***********************************************************/
struct timeType ReadCurrentTime(){
    struct timeType tempTime;
    tempTime.seconds = rBCDSEC;
    tempTime.minutes = rBCDMIN;
    tempTime.hours = rBCDHOUR;
```

```
    tempTime.date = rBCDDATE;
    tempTime.month = rBCDMON;
    tempTime.year = (short)(0x2000 + rBCDYEAR);
    return tempTime;
}
```

3. 更新 NAND Flash 中的文件系统表

```
/***********************************************************
函数原形：void UpdateFileTable(void)
功    能：更新文件系统表,即将表写入 NAND Flash
参    数：无
***********************************************************/
void UpdateFileTable(void){
    int i;
    if(! NF_EraseBlock(FILE_BLOCK)){          //NAND Flash 的块写入前都需要先擦除
        Uart_Printf("flash FILE_BLOCK fail! \n");
                                              //文件系统表占用的块损坏,打印便于调试
        return;
    }
    else{
        for(i = 0;i <32;i ++ ){               //文件系统结构表共占用 1 块
            NF_WritePage(FILE_BLOCK,i,&sysFileTable.U8s[i * 512]);
        }
    }
}
```

无论是格式化文件系统，还是在系统中增加文件，都要调用它以同步刷新 NAND Flash 的文件系统表。

4. 写文件

```
/***********************************************************
函数原形：void WriteFile(unsigned char * fileData,int fileSize)
功    能：写入一个文件到文件系统
参    数：fileData——指向文件的数据缓冲区
        fileSize——文件尺寸
        newFile——fileHeadType 结构的全局变量,写入文件系统的当前文件
***********************************************************/
void WriteFile(unsigned char * fileData,int fileSize){
    int needBlock;
    int currentBlockIndex = 0;
    int totalBlockUsed = 0;
    int i,j,k,checksum;
```

```
    checksum = 0x0;
    for(i = 0; i < fileSize; i++){        //累加文件校验和
        checksum += *(fileData + i);
    }
    if(fileSize%(512*32) != 0)           //计算文件所要占用的 NAND Flash 块的数量
        needBlock = fileSize/(512*32) + 1;
    else
        needBlock = fileSize/(512*32);
    j = sysFileTable.fileTable.endBlock;  //计算文件系统中未使用的块数量
    i = sysFileTable.fileTable.startBlock;
    for(; i < j; i++){
        totalBlockUsed += sysFileTable.fileTable.blockFlag[i];
    }
    if(needBlock > j - totalBlockUsed - sysFileTable.fileTable.startBlock){
        Uart_Printf("There is no enogh room for the file! \n");//空间不够
        return;
    }
    if(sysFileTable.fileTable.exitFileNum >= sysFileTable.fileTable.maxFileNum){
        Uart_Printf("Too many files,please remove some of them! \n");//文件目录不足
        return;
    }
    for(i=0; i<sysFileTable.fileTable.exitFileNum; i++){
                                          //计算现有文件占用的 Index 空间
        currentBlockIndex = currentBlockIndex +
sysFileTable.fileTable.fileHeads[i].endFileMark -
sysFileTable.fileTable.fileHeads[i].startFileMark + 1;
    }
sysFileTable.fileTable.fileHeads[sysFileTable.fileTable.exitFileNum] = newFile;
sysFileTable.fileTable.exitFileNum += 1;  //从文件系统表的尾部插入新文件
    j = sysFileTable.fileTable.startBlock;
    for(i = 0; i < needBlock; i++){                  //开始寻找存放文件的块
        for(;j<sysFileTable.fileTable.endBlock;j++){//寻找未占用的块
            if(sysFileTable.fileTable.blockFlag[j] == 0){
                                                     //找到未占用的块,可以使用
                sysFileTable.fileTable.blockFlag[j] = 1;    //设置占有标志
                sysFileTable.fileTable.blockIndex[currentBlockIndex] = j;
                                                     //在 index 中记录占用的区块号
                currentBlockIndex++;                 //占用区块索引值后移
                NF_EraseBlock(j);                    //块擦除,准备写入数据
                for(k = 0;k <32;k++){                //写 1 块数据
                    NF_WritePage(j,k,fileData);
                    fileData += 512;
```

```
                    }
                    break;
                }
            }
        }
    sysFileTable.fileTable.fileHeads[sysFileTable.fileTable.exitFileNum - 1].
    startFileMark = currentBlockIndex - needBlock;          //保存文件起始块索引值
    sysFileTable.fileTable.fileHeads[sysFileTable.fileTable.exitFileNum - 1].
    endFileMark = currentBlockIndex - 1;                    //保存文件结束块索引值
    sysFileTable.fileTable.fileHeads[sysFileTable.fileTable.exitFileNum - 1].
    fileStatus = 1;
    sysFileTable.fileTable.fileHeads[sysFileTable.fileTable.exitFileNum - 1].
    fileDataCrc = checksum;                                 //保存文件数据的校验和
    sysFileTable.fileTable.creatTime = ReadCurrentTime();
    UpdateFileTable();                                      //更新 NAND Flash 的文件系统表
    Uart_Printf("Write file complete! \n");
}
```

写函数刚开始是累加写入文件的数据的校验和；计算需要占用的 NAND Flash 块数，再和文件系统空间剩余块数比较；在文件系统表中查找未使用的块，然后写数据，以及记录所写数据的块索引值，以供后续查找；记录文件的校验和、当前时间，以及更新 NAND Flash 的整个文件系统表。

5. 读文件

```
/*********************************************************
函数原形：unsigned char * ReadFile(char * openFileName)
功    能：从文件系统中读出一个文件
参    数：openFileName——指向文件名缓冲区
返    回：指向保存文件数据的缓冲区
*********************************************************/
unsigned char * ReadFile(char * openFileName){
    int i,j,k,temp;
    static int fileSize;
    static int fileStatus;
    static int startFileMark;
    static int endFileMark;
    static unsigned char * dataPtr;
    static unsigned char * dataPtr1;
    static char * fileNamePtr;
    dataPtr1 = (unsigned char * )NULL;
    for(i = 0; i < sysFileTable.fileTable.exitFileNum; i++){
                                                        //根据文件名在文件表中查找
```

```
        fileNamePtr = sysFileTable.fileTable.fileHeads[i].fileName;
        if(strncmp(openFileName,fileNamePtr,20) == 0){//如果文件名完全一样
            newFile   = sysFileTable.fileTable.fileHeads[i];
            filetp = &sysFileTable.fileTable.fileHeads[i];
            fileSize = sysFileTable.fileTable.fileHeads[i].fileSize;
//如果文件找到,则取出 index 中存放的开始块地址和结束块地址、长度
            temp = sysFileTable.fileTable.fileHeads[i].fileDataCrc;
            fileStatus = sysFileTable.fileTable.fileHeads[i].fileStatus;
            startFileMark = sysFileTable.fileTable.fileHeads[i].startFileMark;
            endFileMark = sysFileTable.fileTable.fileHeads[i].endFileMark;
            dataPtr = (unsigned char *)malloc(fileSize);//申请临时的内存空间
            dataPtr1 = dataPtr;
            for(j = startFileMark; j < (endFileMark + 1); j++){//读文件数据
                for(k = 0;k < 32;k++){
                    NF_ReadPage(sysFileTable.fileTable.blockIndex[j],k,dataPtr);
                    dataPtr += 512;
                }
            }
        }
    }
    j = 0;
    for(k = 0; k < fileSize ; k++ ) //累加文件数据的校验和
        j += *(dataPtr1 + k);
    if(dataPtr1 != NULL)
        free(dataPtr1);//释放临时的内存空间
    else
        Uart_Printf("No file or file name invalid  \n");
    if(j != temp){
        dataPtr1 = (unsigned char *)NULL;
        Uart_Printf("File CRC check EER! \n");
    }
    return dataPtr1;
}
```

读函数主要是根据提供的文件名在文件系统表中查找,找到后读出数据,将数据缓冲区指针返回给调用的函数,否则返回空指针。

6. 删除文件

```
/**********************************************************
函数原形: void DelFile(char * fileName)
功    能: 从文件系统中删除一个文件
参    数: openFileName——指向文件名
```

```
**************************************************************/
void DelFile(char * fileName){
    struct fileHeadType fileHead;
    char * fileNamePtr;
    int i,j,k;
    for(i = 0;i < sysFileTable.fileTable.exitFileNum;i++){
                                                //根据文件名在文件表中查找
        fileNamePtr = sysFileTable.fileTable.fileHeads[i].fileName;
        if(! strncmp(fileName,fileNamePtr,20)){//如果文件名完全一样
            fileHead = sysFileTable.fileTable.fileHeads[i];//取出文件头
            for(;i < sysFileTable.fileTable.exitFileNum;i++){
                                                    //将后边的文件头往前移
            sysFileTable.fileTable.fileHeads[i] = sysFileTable.fileTable.fileHeads
            [i+1];
            }
            sysFileTable.fileTable.exitFileNum -= 1;    //现存文件数目减 1
            for(j = fileHead.startFileMark;j<(fileHead.endFileMark + 1);j++){
                                                            //清除占有标志
            /* 先用该文件记录的 blockindex start 和 end 找到在 blockIndex 中的区间,
            在此区间中记录了占用的 block 号,再到该 block 号处把占有标志去掉 */
              sysFileTable.fileTable.blockFlag[sysFileTable.fileTable.blockIndex
              [j]] = 0;
            }
            /* 整理 BlockIndex,把该文件 endFileMark 之后的内容提到 startFileMark 处,
            并到各个文件处修改它们各自的 startFileMark 和 endFileMark */
            for(j = fileHead.endFileMark + 1;j<TOTAL_BLOCK - fileHead.endFileMark;j++){
                sysFileTable.fileTable.blockIndex[fileHead.startFileMark + j -
                fileHead.endFileMark - 1] = sysFileTable.fileTable.blockIndex[j];
            }
            /* 整理各个文件的 endFileMark 和 startFileMark */
            for(j = 0; j < sysFileTable.fileTable.exitFileNum; j++){
            /* 如果在 index 中排在删除文件后,要把它们提到前面去 */
           if(sysFileTable.fileTable.fileHeads[j].startFileMark> fileHead.
           endFileMark){
                    k = fileHead.endFileMark - fileHead.startFileMark + 1;
                    sysFileTable.fileTable.fileHeads[j].startFileMark =
                    sysFileTable.fileTable.fileHeads[j].startFileMark - k;
                    sysFileTable.fileTable.fileHeads[j].endFileMark =
                    sysFileTable.fileTable.fileHeads[j].endFileMark - k;
                }
            }
            break;
```

```
        }
    }
    sysFileTable.fileTable.creatTime = ReadCurrentTime();
    UpdateFileTable();
}
```

删除函数也是根据提供的文件名查找文件系统表中的文件，然后从文件表中删除，再重新整理文件系统表，以及更新等。

7. 搜索文件

```
/*****************************************************************
函数原形：unsigned char SearchFile( char * fileName)
功    能：在文件系统中查找指定的文件是否存在
参    数：openFileName——指向文件名
返    回：TRUE 为找到，FALSE 为没有找到
*****************************************************************/
unsigned char SearchFile(char * fileName){
    char * fileNamePtr;
    int i,j,k;
    for(i = 0; i < sysFileTable.fileTable.exitFileNum; i++){
        fileNamePtr = sysFileTable.fileTable.fileHeads[i].fileName;
        if(! strncmp(fileName,fileNamePtr,20))//如果文件名完全一样
        return TRUE;//找到返回 1
    }
    return  FALSE;
}
```

该函数的目的是防止在写入文件时，文件系统已存在有相同名字的文件。

8. 打印文件系统中存在的文件信息

```
/*****************************************************************
函数原形：void showFileTable(void)
功    能：向终端打印文件系统中存在的文件信息
参    数：无
*****************************************************************/
void showFileTable(void){
    char * fileNamePtr;
    char * fileType;
    int i,j;
    j = 8;
    Uart_Printf("Directory of C:\\ \n\n");
    for(i = 0; i < sysFileTable.fileTable.exitFileNum; i++){
```

```
            fileNamePtr = sysFileTable.fileTable.fileHeads[i].fileName;
            fileType = sysFileTable.fileTable.fileHeads[i].fileType;
            Uart_Printf("%-15s",fileNamePtr);
            Uart_Printf("%-8s",fileType);
            Uart_Printf("%10d",sysFileTable.fileTable.fileHeads[i].fileSize);
            Uart_Printf("%10x-",sysFileTable.fileTable.fileHeads[i].modifyTime.
                    year);
            Uart_Printf("%02x-",sysFileTable.fileTable.fileHeads[i].modifyTime.
                    month);
            Uart_Printf("%02x",sysFileTable.fileTable.fileHeads[i].modifyTime.date);
            Uart_Printf("  %2x:",sysFileTable.fileTable.fileHeads[i].modifyTime.
                    hours);
            Uart_Printf("%02x:",sysFileTable.fileTable.fileHeads[i].modifyTime.
                    minutes);
            Uart_Printf("%02x \n",sysFileTable.fileTable.fileHeads[i].modifyTime.
                    seconds);
            j = 0x30;
        }
    }
```

10.3.3 文件系统的测试

上一节给出了可以满足大部分应用的文件系统基本功能函数，读者可直接在应用程序中调用，也可以通过本节介绍的通过串口终端与应用程序交互的方式对它进行测试、调试和应用。

```
char command[25];//保存串口接收的字符串
char *string, *string2;
/***********************************************************
函数原形：void DownLoad()
功    能：检测串口是否收到命令，以确定进入命令处理函数
参    数：无
***********************************************************/
void DownLoad(){
    int temp;
    if(GetUartCommand() == TRUE){
    temp = rINTMSK;//当检测到有串口命令时，屏蔽中断
        rINTMSK = BIT_ALLMSK;
        UartCommand();//命令处理
        string = string2 = command;
    rINTMSK = temp;//恢复原来的中断
    }
```

```
}
/***********************************************************
函数原形：unsigned char GetUartCommand(void)
功    能：检测串口是否收到命令，以确定进入命令处理函数
参    数：string——全局变量，指向的缓冲区用于保存接收的命令
返    回：TRUE 表示收到有效命令，NULL 表示没有收到有效命令
***********************************************************/
unsigned char GetUartCommand(void){
    static char key;
    key = Uart_GetKey();//从串口接收数据
    if(key!= 0x0){
        if(key!= '\r'){
            if(key == '\b'){
                if((int)string2 < (int)string){
                    Uart_Printf("\b \b");
                    string-- ;
                }
            }
            else{
                * string++  = key;
                Uart_SendByte(key);//回显
            }
        }
        else if(key ==  '\r'){
            return TRUE;
        }
    }
    return NULL;
}
/***********************************************************
函数原形：void UartCommand(void)
功    能：串口命令处理
***********************************************************/
void UartCommand(void){
    int i;
    Uart_Printf("\n");
    if(string == command){
        Uart_Printf("C:\\");
        return;
    }
    string = string2 = command;
    if(strncmp("dir",string,3) == 0){//显示文件系统目录下的文件信息
```

```
        FileSystemInit(1);
    }
    else if(strncmp("downfile",string,8) == 0){
        int downfilesize = 0;
        downfilesize = UartDownLoad();//串口下载文件
        if(downfilesize != NULL)
            filedialog(downfilesize);
        else
            Uart_Printf("\ndownfilesize is 0");
    }
    else if( strncmp("del",string,3) == 0){
        string += 3;
        for( i = 0; i < 20; i++ )
            while( * string == 0x20) string++;//去除空格
        DelFile(string);
        string = command;
    }
}
```

这里只写了 3 个命令，分别为查看文件系统下的文件、下载创建文件和删除文件。

```
/**********************************************************
函数原形：void filedialog(int filesize)
功    能：检测串口是否收到命令，以确定进入命令处理函数
参    数：string——全局变量，指向的缓冲区用于保存接收的命令
返    回：TRUE 表示收到有效命令，NULL 表示没有收到有效命令
**********************************************************/
void filedialog(int filesize){
    int i;
    unsigned char key;
    char tempFileName[20];
    newFile.fileSize = filesize;
    Uart_Printf("Are you want write file y or n\n");
    while(1){
        key = Uart_Getch();
        if(key == 'y' || key == 'n' || key == 'Y' || key == 'N') break;
    }
    if(key == 'y' || key == 'Y'){
        Uart_Printf("Enter a file name! less char 20 \n");
        Uart_GetString(tempFileName);
        while(SearchFile(tempFileName)){
            Uart_Printf("You enter name allready exist!!! \n");
```

```
            DelFile(tempFileName);
        }
        memcpy(newFile.fileName,tempFileName,20);
        Uart_Printf("Enter a file extend name! less char 4\n");
        Uart_GetString(tempFileName);
        memcpy(newFile.fileType,tempFileName,4);
        newFile.modifyTime = ReadCurrentTime();
        WriteFile((void * )downloadAddress,newFile.fileSize);//写文件
    }
}
```

10.4　简易图形用户界面(GUI)的设计

当产品带有LCD时，往往需要在LCD上设计图形界面、菜单等，如果界面、菜单复杂，如PDA、手机等往往需要采用专业的GUI软件包，且往往是基于某些操作系统之下。这里我们一起实现一些可以直接在前后台运行的简易的GUI基本函数，以满足基本的前后台LCD应用。

10.4.1　字符和汉字的显示

在LCD上显示字符和汉字之前需要先提取点阵码，通常根据LCD的显示方式有两种提取方法。第一种是固定字符(或汉字)形式的方式，就是指在设计界面时，某个界面上显示的字符是固定的，是我们事先已经知道的，设计完成后也永远不会被修改的，此时一般用字模软件提取要显示字符的点阵码或直接以图片的形式提取(就是指贴图)，再以C语言数组的形式保存，需要显示时再由函数调用，也就是说需要显示什么样的字符(或汉字)就制作什么样的点阵码。第二种要显示的字符(或汉字)非常多，或者它们是任意的，类似一个文本框，今天显示这些字符，明天可能就显示另外一些字符，此时必须有完整的ASCII字符和汉字字库。由于ASCII字符字库比较小，一般的字模软件也可生成；汉字字库很大，通常可以使用UCDOS软件下的字库文件，如16×16点阵的HZK16文件，读者可以从UCDOS安装包上找到，也可以从网上直接搜索汉字字库文件下载。

1. 字模软件的使用

目前网上有很多免费的字模软件，Zimo就是其中使用比较长久的一个。下面就介绍利用Zimo字模软件来提取字符的方法：打开Zimo字模软件，在文字输入区输入所有ASCII字符，然后按Ctrl+Enter键，如图10.1所示。可以在图10.1左边栏的参数设置项打开字体设置对话框，设置在LCD显示的合适大小的字体。在其他选项中可以设置横向取模或纵向取模，以及字节倒序等，如果读者对它们不理解，可以根据不同的设置查看提取的结果。

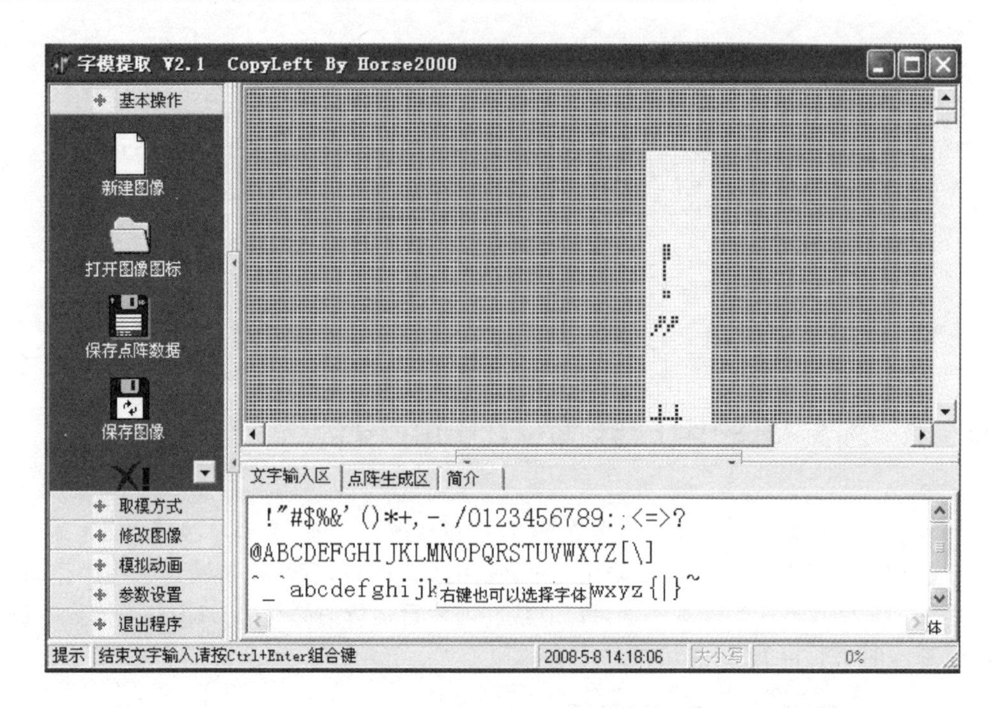

图 10.1　字模软件提供点阵码

最后在左边栏的取模方式中单击 C51 格式，将在点阵生成区中产生我们需要的字模点阵，可以将其复制到 C 语言程序中，且以数组形式保存，代码如下：

```
const char ascii_1620[] = {
/* --   文字：       -- */
/* --   宋体 15；  此字体下对应的点阵为：宽 × 高 = 11 × 20    -- */
/* --   宽度不是 8 的倍数，现调整为：宽度 × 高度 = 16 × 20    -- */
0x00,0x00,0x00,0x00,0x00,0x00,0x00,0x00,0x00,0x00,0x00,0x00,0x00,0x00,0x00,0x00,
0x00,0x00,0x00,0x00,0x00,0x00,0x00,0x00,0x00,0x00,0x00,0x00,0x00,0x00,0x00,0x00,
0x00,0x00,0x00,0x00,0x00,0x00,0x00,0x00,
/* --   文字：  !    -- */
/* --   宋体 15；  此字体下对应的点阵为：宽 × 高 = 11 × 20    -- */
/* --   宽度不是 8 的倍数，现调整为：宽度 × 高度 = 16 × 20    -- */
0x00,0x00,0x00,0x00,0x00,0x00,0x70,0x00,0x70,0x00,0x70,0x00,0x70,0x00,0x70,0x00,
0x60,0x00,0x60,0x00,0x60,0x00,0x60,0x00,0x60,0x00,0x00,0x00,0x00,0x00,0x70,0x00,
0x70,0x00,0x00,0x00,0x00,0x00,0x00,0x00,
……
/* --   文字：  ～   -- */
/* --   宋体 15；  此字体下对应的点阵为：宽 × 高 = 11 × 20    -- */
/* --   宽度不是 8 的倍数，现调整为：宽度 × 高度 = 16 × 20    -- */
0x3C,0x00,0x6E,0x06,0x66,0x07,0xC0,0x03,0x00,0x00,0x00,0x00,0x00,0x00,0x00,0x00,
```

```
0x00,0x00,0x00,0x00,0x00,0x00,0x00,0x00,0x00,0x00,0x00,0x00,0x00,0x00,0x00,0x00,
0x00,0x00,0x00,0x00,0x00,0x00,0x00,0x00
};
```

其他如单个或几个汉字,以及图像等也都可以使用 Zimo 字模软件提取不同大小的点阵码,读者可以自行尝试。

2. ASCII 字符的显示

当我们有了字模软件生成的字符点阵码之后,就可以在函数中获取它,输出到 LCD 屏上显示。

```
/************************************************************
函数原形: void print_ascii_1620(int line,int clue,char word,U16 fontcolor,U16 back-
        color)
功    能: 显示单个宽×高=11×20 的 ASCII 字符,格式为横向倒序,显示字库实际字体大
        小为 11×20
参    数: line,clue——显示字符的行和列,等同于字符起始左上角坐标(x,y)
        word——指向字符串
        fontcolor,backcolor——显示字符的前景色和背景色
************************************************************/
void print_ascii_1620(int line,int clue,char word,U16 fontcolor,U16 backcolor){
    int  x,y,va,cj,ci,bytedata;
    va = (word - 32) * 40;//用于查找点阵数组中字符对应的点阵码
    for (y = clue; y < clue + 20; y++){//ay
        bytedata = ascii_1620[va];//取一个像素点的点阵码
        va++;
        for (x = line; x < line + 8; x++){//ax
            if (bytedata & 0x01){//LCD 输出
                SetPixel(x,y,fontcolor);
            }
            else{
                SetPixel(x,y,backcolor);
            }
            bytedata >>= 1;
        }
        bytedata = ascii_1620[va];
        va++;
        for (; x < line + 11; x++){//去掉后面没用的 5 个空像素
            if (bytedata & 0x01){
                SetPixel(x,y,fontcolor);
            }
            else{
                SetPixel(x,y,backcolor);
```

```
                }
                bytedata >>= 1;
            }
        }
}
/**********************************************************
函数原形：lcdprinth(int x,int y,char * buf,U16 fontcolor,U16 backcolor)
功    能：在 LCD 上显示字符串
参    数：x,y——显示字符的起始左上角坐标(x,y)
        buf——指向字符串缓冲区
        fontcolor,backcolor——显示字符的前景色和背景色
**********************************************************/
void lcdprinth(int x,int y,char * buf,U16 fontcolor,U16 backcolor){
    char ex;
    ex = * buf;
    while(ex != 0){
        print_ascii_1620(x,y,ex,fontcolor,backcolor);
        x += 11;//字体的宽度为 11 个像素
        buf++;
        ex = * buf;
    }
}
```

我们可以在程序中使用下述的方式调用上述程序，代码如下：

```
……
lcdprinth(214,236,"System initialization...",clWhite,clBlack);//(640,480)中间显示
……
```

3. 汉字的显示

如果我们在 LCD 上显示具体的或数量不多的某个汉字或词组，则可以像显示 ASCII 码那种方法一样利用字模软件生成点阵码，再在程序中以数组的形式保存供函数获取，也就是指需要显示哪些汉字就制作哪些汉字的点阵码。具体实现请读者参考 ASCII 字符的显示，这里仅介绍基于字库文件的汉字显示。

首先我们先来了解一下汉字是如何在程序中存在的，以及编译之后生成的目标码是什么？大家通常都知道在程序中如果想直接引用某个字符，则使用单引号；如果直接引用字符串，则使用双引号。汉字也是一样的，使用双引号，当编译器编译后会生成汉字的机内码，一个汉字的机内码占两字节，为了和 ASCII 码区别，范围从 0xA1 开始(小于 0x80 的为 ASCII 码)，也就是说汉字两字节的机内码都会大于 0xA0。如图 10.2 所示为程序中汉字的定义和生成的机内码。

另外，我们还需要了解汉字的区位码，因为国家标准的汉字字符集(GB2312—

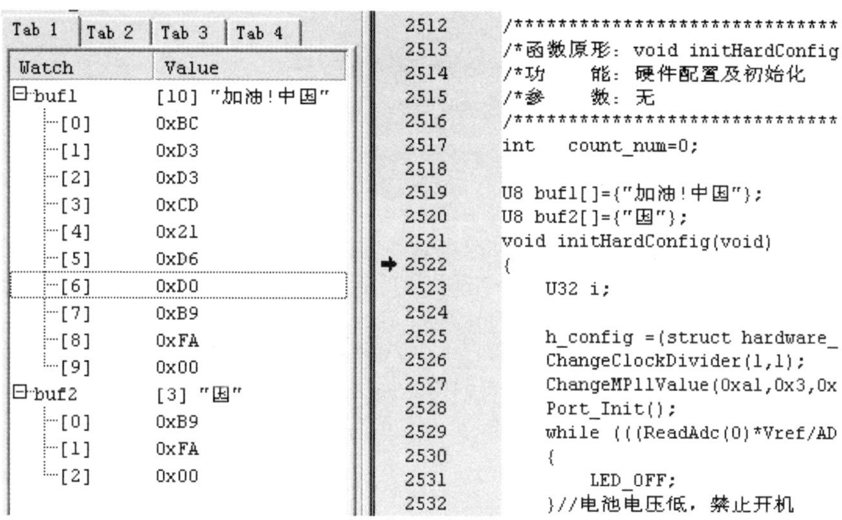

Tab 1 | Tab 2 | Tab 3 | Tab 4

Watch	Value
⊟buf1	[10] "加油!中国"
[0]	0xBC
[1]	0xD3
[2]	0xD3
[3]	0xCD
[4]	0x21
[5]	0xD6
[6]	0xD0
[7]	0xB9
[8]	0xFA
[9]	0x00
⊟buf2	[3] "国"
[0]	0xB9
[1]	0xFA
[2]	0x00

```
2512 /********************************
2513 /*函数原形: void initHardConfig
2514 /*功    能: 硬件配置及初始化
2515 /*参    数: 无
2516 /********************************
2517 int   count_num=0;
2518
2519 U8 buf1[]={"加油!中国"};
2520 U8 buf2[]={"国"};
2521 void initHardConfig(void)
2522 {
2523     U32 i;
2524
2525     h_config =(struct hardware_
2526     ChangeClockDivider(1,1);
2527     ChangeMPllValue(0xa1,0x3,0x
2528     Port_Init();
2529     while (((ReadAdc(0)*Vref/AD
2530     {
2531         LED_OFF;
2532     }//电池电压低，禁止开机
```

图 10.2 汉字定义和机内码

80)汉字库就是以区位码的形式存在的。汉字字库共分94个区，每个区共94个汉字，即每个汉字在字库中都有确定的区和位编号(区位码的高位字节为区号，低位字节为位号)，所以知道了汉字的区位码就可以知道该汉字在字库中的地址，而区位码就等于机内码的每字节各减去0xA0，如“国”字的机内码为0xB9FA，则每字节减去0xA0(即0xB9FA−0xA0A0)的区位码就等于0x195A。如图10.3所示为“国”字的机内码和区位码的对应关系。

类别	数值	高位字节								低位字节							
机内码	B9FAH	1	0	1	1	1	0	0	1	1	1	1	1	1	0	1	0
区位码	195AH	0	0	0	1	1	0	0	1	0	1	0	1	1	0	1	0

图 10.3 “国”字的机内码和区位码的对应关系

以UCDOS软件中HZK16文件为例，它为16×16点阵字库文件，每个汉字的点阵码占32字节，所以汉字的点阵码起始地址为

[(区号−1)×94+(位号−1)]×32 = [(机内码高位−0xA1)×94+(机内码低位−0xA1)]×32

“国”字的点阵码起始地址为

$$[(0xB9-0xA1)\times 94+(0xFA-0xA1)]\times 32 = 0x12520$$

我们可以以文件的形式将HZK16文件保存在Flash中，如果读者的系统不支持文件系统，也可以直接保存在某个Flash中或是装载到SDRAM中。下述为汉字的

显示程序：

```
/******************************************************************
函数原形：void print_hz(int x,int y,char hzjn[2],U16 fontcolor,U16 backcolor,U8
        * hzk16)
功    能：在 LCD 上显示单个汉字
参    数：x,y——显示汉字的起始左上角坐标(x,y)
        hzjn [2]——存放汉字机内码
        fontcolor,backcolor——显示字符的前景色和背景色
        hzk16——指向汉字库
******************************************************************/
void print_hz(int x,int y,char hzjn[2],U16 fontcolor,U16 backcolor,U8 * hzk16){
    int j,i;
    short bytedata,temp;
    U8 * zk;
    zk = hzku + ((hzjn[0] - 0xA1) * 94 + (hzjn[1] - 0xA1)) * 32;
                                                    //取汉字点阵起始地址
for (j = 0; j < 16; j++){ //ay
            temp = * zk++;
            bytedata = temp << 8 | * zk++;
            for (i = 0; i < 16; i++){//ax
                if (bytedata & 0x8000){
                    SetPixel(y + j,x + i,fontcolor);
                }
                else{
                    SetPixel(y + j,x + i,backcolor);
                }
                bytedata <<= 1;
            }
        }
}
/******************************************************************
函数原形：void print_HzString(int x,int y,char * string,U16 fontcolor,U16 backcolor,
        U8 * hzk16)
功    能：在 LCD 上显示字符(包括汉字)串
参    数：x,y——显示汉字的起始左上角坐标(x,y)
        string——指向的字符串
        fontcolor,backcolor——显示字符的前景色和背景色
        hzk16——指向汉字库
******************************************************************/
int print_HzString(int x, int y, char * string, U16 fontcolor, U16 backcolor, unsigned
char * hzku){
```

```
char hzjn[2];
    short i,j,k;
    while( * string != 0){
        if((( * string > 0xa0) && * (string + 1) > 0xa0)){
            hzjn[0] = * string++;
            hzjn[1] = * string++;
            print_hz(x,y,hzjn,fontcolor,backcolor,hzku);
        }
        else if( * string < 0xa0){
        print_ascii_1620(x,y, * string++ ,fontcolor,backcolor);
//这里可以更改为更合适大小的 ASCII 输出
        }
        x += 16;
    }
}
```

注：print_HzString 函数未处理回车换行等字符，只能在某一行显示，所以读者如果需要，请自己增加这些处理及功能。

10.4.2 基本图形和控件的绘制

1. 点

点是图形的最基本元素，描述一个点需要使用其坐标值(x,y)和颜色。对于 640×480 的 16 位 TFT 屏，一个点的颜色占用 2 字节(16 位)，最远坐标为(640,480)。如果希望点亮 LCD 屏的某个点，只需将该点的颜色值写入 LCD 屏映射的物理内存即可。在 LCD 显示部分，我们提到过有时会将 LCD 数据暂时写入 LCD 显示缓冲区以提高效率，两个版本的代码如下。

版本一：

```
/*****************************************************************
函数原形：void DrawPixel(int x,int y,U16 color)
功    能：在 LCD 物理内存中画点，即在 LCD 屏上显示点
参    数：x,y——显示点的坐标(x,y)
          color——显示点的颜色值
*****************************************************************/
void DrawPixel(int x,int y,U16 color){
        if (x < LCD_XSIZE_TFT_640480 && y < LCD_YSIZE_TFT_640480){
frameBuffer16BitTft640480A[y][x] = color;
}
}
```

版本二：

```
/*************************************************************
函数原形：void SetPixel (int x,int y,U16 color)
功    能：在LCD物理内存中画点,即在LCD屏上显示点
参    数：x,y——显示点的坐标(x,y)
          color——显示点的颜色值
*************************************************************/
void SetPixel (int x,int y,U16 color){
        if (x < LCD_XSIZE_TFT_640480 && y < LCD_YSIZE_TFT_640480){
frameBuffer16BitTft640480B[y][x] = color;
}
}
```

注：frameBuffer16BitTft640480A和frameBuffer16BitTft640480B的定义请参考LCD显示部分。x、y为坐标值(x,y),color为16位的颜色值。

2. 线

将多个点连接起来就成了线。线又有水平线(起点和终点的x坐标相同)、垂直线(起点和终点的y坐标相同)和一般直线(起点和终点的x、y坐标都不同)。前两者的绘制比较简单,只需保持某个坐标值不变,另外一个坐标值递增,而一般直线就比较复杂,通常需要采用Bresenham、DDA经典算法或其他算法完成。下述为Bresenham算法实现的画线程序,关于Bresenham原理请参考《计算机图形学的算法基础》一书,或上网搜索。

```
/*************************************************************
函数原形：void SetHLine (int x1,int y1,int x2,U16 color)
功    能：画水平线至LCD显示缓存
参    数：x1,y1——起始坐标(x1,y1)
          x2——终点坐标(x2,y1)
          color——颜色值
*************************************************************/
void SetHLine(int x1,int y1,int x2,U16 color){
    int temp;
    if (x1 > x2){
temp = x1; x1 = x2; x2 = temp;
}
for (temp = x1; temp <= x2; temp++){
SetPixel(temp,y1,color);
}
}
/*************************************************************
函数原形：void SetRLine(int x1,int y1,int y2,U16 color)
功    能：画垂直线至LCD显示缓存
```

```
参    数：x1,y1——起始坐标(x1,y1)
        y2——终点坐标(x2,y1)
        color——颜色值
*****************************************************************/
void SetRLine(int x1,int y1,int y2,U16 color){
    int temp;
    if (y1 > y2){
temp = y1; x1 = y2; y2 = temp;
}
for (temp = y1; temp <= y2; temp++){
SetPixel(x1,temp,color);
}
}
/*****************************************************************
函数原形：void SetLine(int x1,int y1,int x2,int y2,U16 color)
功    能：画直线至 LCD 显示缓存
参    数：x1,y1——起始坐标
        x2,y2——终点坐标
        color——颜色值
*****************************************************************/
void SetLine(int x1,int y1,int x2,int y2,U16 color){
    int dx = 0,dy = 0,e = 0;
    dx = x2 - x1;
    dy = y2 - y1;
    if(dx >= 0){
        if(dy >= 0){//dy>=0
            if(dx >= dy){//1/8 octant,第一个八分圆
                e = 2 * dy - dx;
                while (x1 <= x2){
                    SetPixel(x1,y1,color);
                    if(e > 0){
                        y1 += 1;
                        e -= 2 * dx;
                    }
                    x1 += 1;
                    e += 2 * dy;
                }
            }
            else{//2/8 octant
                e = 2 * dx - dy;
                while(y1 <= y2){
                    SetPixel(x1,y1,color);
```

```
                    if(e > 0){
                        x1 += 1;
                        e -= 2 * dy;
                    }
                    y1 += 1;
                    e += 2 * dx;
                }
            }
        }
        else{//dy<0
            dy = -dy;//dy = abs(dy)
            if(dx >= dy){//8/8 octant
                e = 2 * dy - dx;
                while(x1 <= x2){
                    SetPixel(x1,y1,color);
                    if(e > 0){
                        y1 -= 1;
                        e -= 2 * dx;
                    }
                    x1 += 1;
                    e += 2 * dy;
                }
            }
            else{//7/8 octant
                e = 2 * dx - dy;
                while(y1 >= y2){
                    SetPixel(x1,y1,color);
                    if(e > 0){
                        x1 += 1;
                        e -= 2 * dy;
                    }
                    y1 -= 1;
                    e += 2 * dx;
                }
            }
        }
    }
    else{//dx<0
        dx = -dx;//dx = abs(dx)
        if(dy >= 0){//dy>= 0
            if(dx >= dy){//4/8 octant
                e = 2 * dy - dx;
```

```
            while(x1 >= x2){
                SetPixel(x1,y1,color);
                if(e > 0){
                    y1 += 1;
                    e -= 2 * dx;
                }
                x1 -= 1;
                e += 2 * dy;
            }
        }
        else{//3/8 octant
            e = 2 * dx - dy;
            while(y1 <= y2){
                SetPixel(x1,y1,color);
                if(e > 0){
                    x1 -= 1;
                    e -= 2 * dy;
                }
                y1 += 1;
                e += 2 * dx;
            }
        }
    }
    else{//dy<0
        dy = -dy;//dy = abs(dy)
        if(dx >= dy){//5/8 octant
            e = 2 * dy - dx;
            while(x1 >= x2){
                SetPixel(x1,y1,color);
                if(e > 0){
                    y1 -= 1;
                    e -= 2 * dx;
                }
                x1 -= 1;
                e += 2 * dy;
            }
        }
        else{//6/8 octant
            e = 2 * dx - dy;
            while(y1 >= y2){
                SetPixel(x1,y1,color);
                if(e > 0){
```

```
                    x1 -= 1;
                    e -= 2 * dy;
                }
                y1 -= 1;
                e += 2 * dx;
            }
        }
    }
}
}
```

SetLine 为写入 LCD 的缓冲区内存，与之对应的写入 LCD 物理内存的函数为 DrawLine，其他不变，只需将函数中的 SetPixel 函数修改成 DrawPixel 即可。

3. 圆

圆是比较常见的图形，可以使用圆心和半径来描述，它的生成也可以使用 Bresenham、DDA 经典算法和其他等算法实现。Bresenham 算法是最简单有效的，其原理请参考《计算机图形学的算法基础》一书，或上网搜索。

```
/************************************************************
函数原形：void SetCircle(int x,int y,int r,U16 color)
功    能：画圆至 LCD 显示缓存
参    数：x,y——圆心坐标
          r——圆的半径
          color——颜色值
************************************************************/
void SetCircle(int x,int y,int r,U16 color){
    int x0,y0,di,e;
    x0 = 0;
    y0 = r;
    di = 2 * (1 - r);
    while (y0 >= 0){
        SetPixel(x + x0,y + y0,color);//1/4 quadrant,第一个四分圆
        SetPixel(x - x0,y + y0,color);//2/4 quadrant
        SetPixel(x - x0,y - y0,color);//3/4 quadrant
        SetPixel(x + x0,y - y0,color);//4/4 quadrant
        if(di < 0){//determine case 1 or 2
            e = 2 * (di + y0) - 1;
            if(e <= 0){
                x0 ++ ;
                di += 2 * x0 + 1;
            }
            else{
```

```
                x0++;
                y0--;
                di += 2*(x0 - y0 +1);
            }
        }
        else if(di > 0){//determine case 4 or 5
            e = 2*(di - x0) - 1;
            if(e <= 0){
                x0++;
                y0--;
                di += 2*(x0 - y0 +1);
            }
            else{
                y0--;
                di += (-2*y0 + 1);
            }
        }
        else{//case 3
            x0++;
            y0--;
            di += 2*(x0 - y0 +1);
        }
    }
}
```

程序在计算决策过程中，以(0,0)坐标为圆心，在绘制点时再加上实际圆心坐标的偏移量，计算和绘画第一个四分圆，然后根据对称关系画另外3个四分之一圆。如果将SetCircle程序中的SetPixel修改成DrawPixel，那么同样可以实现在LCD物理内存中画圆，而直接在LCD屏显示的画圆函数是DrawCircle。

4. 椭　圆

椭圆也是比较常见的图形，下面为采用中点算法绘制的程序，算法原理请参考《计算机图形学的算法基础》一书。

```
/**************************************************************
函数原形：void SetEllipse(int x1,int y1,int x2,int y2,U16 color)
功    能：画椭圆至LCD显示缓存
参    数：x1,x2——分别为最左和最右点的x坐标
          y1,y2——分别为最上和最下点的y坐标
          color——颜色值
**************************************************************/
void SetEllipse(int x1,int y1,int x2,int y2,U16 color){
```

```
int x0 = (x1 + x2)/2;//圆心坐标(x0,y0)
int y0 = (y1 + y2)/2;
int a = (x2 - x1)/2; //半长轴和半短轴
int b = (y2 - y1)/2;
int x = a + 1/2;//初始化(x,y)坐标变量
int y = 0;
int taa    = a * a;//定义临时变量,用于计算决策
int t2aa   = 2 * taa;
int t4aa   = 2 * t2aa;
int tbb    = b * b;
int t2bb   = 2 * tbb;
int t4bb   = 2 * t2bb;
int t2abb = a * t2bb;
int t2bbx = t2bb * x;
int tx     = x;
int di = t2bbx * (x - 1) + tbb/2 + t2aa * (1 - tbb);//决策,第1区域
while (t2bb * tx > t2aa * y){
    DrawPixel(x0 + x,y0 + y,color);//1/4 quadrant,第一个四分之一圆
    DrawPixel(x0 - x,y0 + y,color);//2/4 quadrant
    DrawPixel(x0 - x,y0 - y,color);//3/4 quadrant
    DrawPixel(x0 + x,y0 - y,color);//4/4 quadrant
    if (di < 0){
        y += 1;
        di += t4aa * y + t2aa;
        tx = x - 1;
    }
    else{
        x -= 1;
        y += 1;
        di = di - t4bb * x + t4aa * y + t2aa;
        tx = x;
    }
}
di = t2bb * (x * x + 1) - t4bb * x + t2aa * (y * y + y - tbb) + taa/2;
                                                        //决策,第2区域

while (x >= 0){
    DrawPixel(x0 + x,y0 + y,color);//1/4 quadrant,第一个四分之一圆
    DrawPixel(x0 - x,y0 + y,color);//2/4 quadrant
    DrawPixel(x0 - x,y0 - y,color);//3/4 quadrant
    DrawPixel(x0 + x,y0 - y,color);//4/4 quadrant
    if (di < 0){
        x -= 1;
```

```
            y += 1;
            di = di + t4aa * y - t4bb * x + t2bb;
        }
        else{
            x -= 1;
            di = di - t4bb * x + t2bb;
        }
    }
}
```

绘画椭圆和圆的程序一样，也是计算和绘画第一个四分之一圆，再根据对称关系绘画其他3个四分之一圆。由于圆是椭圆的特例，因此也可以使用该函数绘制圆。

5. 矩形和正方形

矩形的4条边分别为两条水平线和两条垂直线，所以在绘制时只需调用水平线和垂直线即可实现，而正方形为矩形的特例。

```
/************************************************************
函数原形：void SetRectangle(int x1,int y1,int x2,int y2,U16 color)
功    能：画矩形或正方形至LCD显示缓存
参    数：x1,y1——为左上角的坐标(x1,y1)
        x2,y2——为右下角的坐标(x2,y2)
        color——颜色值
************************************************************/
void SetRectangle(int x1,int y1,int x2,int y2,U16 color){
SetHLine(x1,y1,x2,color);
SetHLine(x1,y2,x2,color);
SetRLine(x1,y1,y2,color);
SetRLine(x2,y1,y2,color);
}
```

6. 填　充

有时经常需要在矩形、圆和椭圆内部填充，而且填充的颜色可能与边框的颜色不同。矩形的填充采用逐行地绘制水平线即可实现，如将SetRectangle函数修改如下：

```
/************************************************************
函数原形：void SetRect(int x1,int y1,int x2,int y2,U16 lineColor,U16 fillColor,U8
        frameWidth,U8 flag)
功    能：画矩形或正方形至LCD显示缓存
参    数：x1,y1——为左上角的坐标(x1,y1)
        x2,y2——为右下角的坐标(x2,y2)
        lineColor——边框线的颜色值
        fillColor——内部的填充颜色值
```

```
        frameWidth——边框宽度,单位为1个像素
        flag——为填充标志,0表示指定边框但内部不填充;1表示指定边框且内部填
              充;2表示没有边框只有填充
*****************************************************************/
void SetRect(int x1,int y1,int x2,int y2,U16 lineColor,U16 fillColor,U8 frameWidth,U8
        flag){
    U16 i = 0;
    if(flag == 0 || flag == 1){
        for(i = 0;i<frameWidth;i++){//画边框
            SetLine(x1 + i,y1 + i,x2 - i,y1 + i,lineColor);
            SetLine(x1 + i,y1 + i,x1 + i,y2 - i,lineColor);
            SetLine(x1 + i,y2 - i,x2 - i,y2 - i,lineColor);
            SetLine(x2 - i,y1 + i,x2 - i,y2 - i,lineColor);
        }
    }
    if(flag == 1){
        for(i = frameWidth;i<y2 - y1 - frameWidth + 1;i++){//边框内的填充
            SetLine(x1 + frameWidth,y1 + i,x2 - frameWidth,y1 + i,fillColor);
        }
    }
    if(flag == 2){
        for(i = 0;i<y2 - y1 + 1;i++){//填充
            SetLine(x1,y1 + i,x2,y1 + i,fillColor);
        }
    }
}
```

圆和椭圆则采用在绘制圆周上的点时再绘制以Y轴(即x为圆心的Y轴)为对称的两点间的水平线的方法,如将绘制圆的函数中画点的部分修改如下:

```
……
SetPixel(x + x0,y + y0,color);//1/4 quadrant,第一个四分之一圆
SetPixel(x - x0,y + y0,color);//2/4 quadrant
SetHLine(x - x0 + 1,y + y0,x + x0 - 1,fillColor)//假设边框只占用一个像素
SetPixel(x - x0,y - y0,color);//3/4 quadrant
SetPixel(x + x0,y - y0,color);//4/4 quadrant
SetHLine(x - x0 + 1,y - y0,x + x0 - 1,fillColor)
……
```

具体的实现这里就不一一举例,读者可以将填充功能直接加在圆和椭圆函数之中,也可另设函数。

7. 绘制图像

有时需要在LCD屏上直接显示某幅图像,如开机的LOGO,按钮上的图标,窗

口左上角的图标，对话框上的指示标志等。这些图像的大小不同，可能只是几个像素的方框，也可能是整个屏幕。通常我们通过字模软件将不同大小的图像转换成点阵数据，然后在程序中将点阵数据写入 LCD 内存。

```
/**************************************************************
函数原形：void DrawImage(int x1,int y1,int x2,int y2,U16 * addr)
功    能：画图像
参    数：x1,y1——为左上角的坐标(x1,y1)
        x2,y2——为右下角的坐标(x2,y2)
        addr——指向图像数据
**************************************************************/
void DrawImage(int x1,int y1,int x2,int y2,U16 * addr){
    int i,j;
    for(j = y1; j <= y2; j++){
        for(i = x1; i <= x2; i++){
            SetPixel(i,j, * addr++);
        }
    }
}
```

8. 按　钮

按钮与矩形或圆等图形的不同之处在于立体感，这种立体感的效果可以通过在边框周围增加一些淡淡的阴影部分实现，有时读者也称它为 2D，甚至更好的为 3D 效果；有时也会在按钮上增加字符或图标等。另外，当按钮被按下处于激活状态时，也应该有明显的表示，即与正常状态下的效果不同。如图 10.4 所示分别为按钮在正常和被按下处于激活状态的两种效果图。

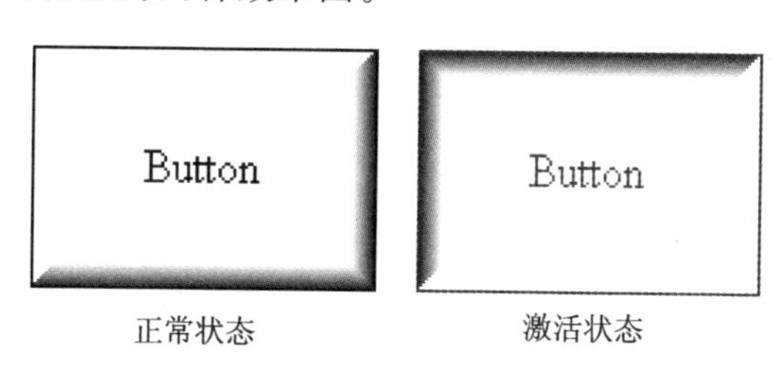

图 10.4　按钮效果图

```
/**************************************************************
函数原形：void SetBtn(int x1,int y1,int x2,int y2,U16 frameColor,U8 frameWidth)
功    能：画按钮
参    数：x1,y1——为左上角的坐标(x1,y1)
        x2,y2——为右下角的坐标(x2,y2)
        frameColor——边框颜色值
        frameWidth——阴影宽度
```

```
***********************************************************/
void SetBtn(int x1,int y1,int x2,int y2,U16 frameColor,U8 frameWidth){
        U16 i = 0;
DrawRect(x1,y1,x2,y2,frameColor,0,1,0);
        for(i = 0; i < frameWidth - 1; i++){
            SetLine(x1 + i + 1,y1 + i + 1,x2 - i - 1,y1 + i + 1,clWhite);//上
            SetLine(x1 + i + 1,y1 + i + 1,x1 + i + 1,y2 - i - 1,clWhite);//左
            SetLine(x1 + i + 2,y2 - i - 1,x2 - i - 1,y2 - i - 1,clDGray);//下
            SetLine(x2 - i - 1,y1 + i + 1,x2 - i - 1,y2 - i - 1,clDGray);//右
        }
}
/***********************************************************
函数原形：void SetPressedBtn(int x1,int y1,int x2,int y2,U16 frameColor,U8 frame-
          Width)
功    能：画按钮
参    数：x1,y1——为左上角的坐标(x1,y1)
          x2,y2——为右下角的坐标(x2,y2)
          frameColor——边框颜色值
          frameWidth——阴影宽度
***********************************************************/
void SetPressedBtn(int x1,int y1,int x2,int y2,U16 frameColor,U8 frameWidth){
    U16 i = 0;
DrawRect(x1,y1,x2,y2,frameColor,0,1,0);
    for(i = 0; i < frameWidth - 1; i++){
        SetLine(x1 + i + 1,y1 + i + 1,x2 - i - 1,y1 + i + 1,clDGray);
        SetLine(x1 + i + 1,y1 + i + 1,x1 + i + 1,y2 - i - 1,clDGray);
        SetLine(x1 + i + 2,y2 - i - 1,x2 - i - 1,y2 - i - 1,clWhite);
        SetLine(x2 - i - 1,y1 + i + 1,x2 - i - 1,y2 - i - 1,clWhite);
    }
}
```

上述绘制的都是简单的按钮，如果希望按钮更漂亮，则可以采用贴图。就是先在PC上用图像处理软件设计按钮图形，然后用字模块软件将图片转换成点阵，再在程序中绘制该图像。

9. 窗　口

有时系统界面可能不止一页，这时可以采用窗口的形式，即按某个按钮打开某个窗口，在窗口中单击右上角的关闭，关闭退出该窗口。

```
/***********************************************************
函数原形：void SetWindow(U16 left,U16 top,U16 right,U16 bottom)
功    能：与 DrawWindow 唯一的区别在于先画到显存里，而不是物理地址
参    数：
```

```
*************************************************************/
void SetWindow(U16 left,U16 top,U16 right,U16 bottom){
    SetLine(left,top,right - 1,top,clLGray);//画窗口的四条边线,单位两个像素,上
    SetLine(left,top,left,bottom - 1,clLGray);//左
    SetLine(left,bottom,right,bottom,clBlack); //下
    SetLine(right,top,right,bottom,clBlack);//右
    SetLine(left + 1,top + 1,right - 2,top + 1,clWhite);//第二个像素
    SetLine(left + 1,top + 1,left + 1,bottom - 2,clWhite);
    SetLine(left + 1,bottom - 1,right - 1,bottom - 1,clDGray);
    SetLine(right - 1,top + 1,right - 1,bottom - 1,clDGray);
    SetRect(left + 2,top + 2,right - 2,bottom - 2,clLGray,0,1,0);//画框,不填充
    SetRect(left + 3,top + 3,right - 3,bottom - 3,clLGray,0,1,0);
    SetRect(left + 4,top + 4,right - 4,top + 36,0,clDBlue,1,2);
    SetRect(left + 4,top + 36,right - 4,bottom - 4,0,clWhite,1,2);
    Set3DBtn(right - 34,top + 6,right - 6,top + 34,clBlack,3);//画右上角的关闭按钮
    SetRect(right - 34 + 2,top + 6 + 2,right - 6 - 2,top + 32,0,clLGray,1,2);
    SetLine(right - 34 + 2 + 2,top + 6 + 2 + 2,right - 6 - 2 - 2,top + 30,clBlack);
    SetLine(right - 6 - 2 - 2,top + 6 + 2 + 2,right - 34 + 2 + 2,top + 30,clBlack);
    Set_DrawImage(left + 6,top + 6,left + 35,top + 34,(P_U16)Image_window);
                                                        //窗口左上角的 LOGO
    lcdprinth(left + 35 + 6,top + 10,"Preferences Setup",clWhite,clDBlue);//窗口名称
}
```

这个程序实现的是一个非常简单的窗口,读者还可以根据实际情况添加更多丰富的内容。

10.4.3 触摸屏事件处理

上一小节我们一起学习了一些基本图形,将它们组合起来可以完成很多丰富的应用界面。可完成界面的绘制是一方面,此外还需要在界面上响应输入设备的操作、处理事件。嵌入式系统的输入设备通常是触摸屏,如果界面上的对象不多,那么最简单的方式是读取触摸屏的坐标值后与界面上每个对象的坐标范围进行比较,以确定用户是否按住某个对象及哪个对象。下面为示例程序:

```
......
while (1){
if(GetPos() == 1){
    if (doexit()){ //关闭窗口,函数内部是判断坐标是否在窗口右上角的关闭按钮范围内
        break;
    }
        if((X_POS> = 20)&&(X_POS< = 780)&&(Y_POS> = 20)&&(Y_POS< = 290)){
                                                        //第一个对象(或按钮)
```

```
        ……
        }
        if((x_pos>=20)&&(x_pos<=380)&&(y_pos>=310)&&(y_pos<=460)){
                                                          //第二个对象
        ……
    }
    ……
        if((x_pos>=417)&&(x_pos<=777)&&(y_pos>=310)&&(y_pos<=460)){
                                                          //第n个对象
        ……
        }
    }
    }
    ……
```


上述程序的结构非常简单，通常一个界面（或称窗口）需要一个死循环，在这个循环当中反复地扫描触摸屏输入，当扫描到坐标在具体对象时就做相应对象的处理。某些对象可能会进入另外一个界面，而另外一个界面也可能存在很多对象，那么系统也就在另外一个界面中做这样的循环扫描，同理可以实现很多个页面的嵌套。当在界面中扫描到关闭（或退出）按钮时，就关闭当前界面，而退回到上一个界面。

如果一个界面有很多对象（或称很多按钮），那么上述方法的程序将变得冗长和笨拙。如果界面中大部分对象（如按钮）是相同大小的，那么可以采用定义包含对象坐标等参数的公用结构体，这样可以在绘制对象和扫描对象时提供便利，如下述示例程序：

```
    typedef struct{//定义按钮结构
            U16 x;//按钮坐标
            U16 y;
            U8 state;//状态
            ……
    }sButton_T, * sButtonPtr_T;
    /**********************************************************
    函数原形：void DrawKeyButton(sButtonPtr_T pButton)
    功    能：画键盘的通用按钮
    参    数：pButton——参数结构
              size——x的偏移量
    **********************************************************/
    void DrawKeyButton(sButtonPtr_T pButton,U16 size){//按钮绘制函数
        ……
    SetRect(pButton->x+3,pButton->y+3,pButton->x+37+size,pButton->y+37,
BUTTON_COLOR,
```

```
BUTTON_COLOR,1,2);//画按钮的填充色
        pButton->state = 0;
    }
    SetHLine(pButton->x + 5,pButton->y,pButton->x + 35 + size,0);//4 条边线
……
}
/**********************************************************
函数原形：U8 FindButton_Key(sButtonPtr_T pButton,U8 count,U16 size)
功    能：查找键盘中的按钮
参    数：pButton——指向要查找的按钮结构
        count——总按钮数
**********************************************************/
U8 FindButton_Key(sButtonPtr_T pButton,U8 count,U16 size){
    U8 i;
    size += 40;
    for(i = 0;i<count;i++,pButton++){
        if(x_pos>pButton->x && x_pos<pButton->x + size && y_pos>pButton->y
&& y_pos<pButton->y + 40){
            pButton->state = 1;
            DrawKeyButton(pButton,size - 40);
            while(GetPos());
            pButton->state = 0;
            DrawKeyButton(pButton,size - 40);
            return i + 1;
        }
    }
    return 0;
}

/**********************************************************
函数原形：U8 KeyboardWin(char * buf,char * caption,U8 count,U8 IsPassword)
功    能：键盘输入
参    数：buf——指向键盘输入字符的缓冲区
        caption——标题
        count——输入字符的最大数
        IsPassword——1 为输入密码;0 为普通字符
返    回：1 为字符输入有效,0 为无效
**********************************************************/
U8 KeyboardWin(char * buf,char * caption,U8 count,U8 IsPassword){
    ……
    static sButtonPtr_T pButtonData = 0;
    sButtonPtr_T pButton;
```

```
    if(!pButtonData){
        pButtonData = (sButtonPtr_T)malloc(sizeof(sButton_T) * 56);//55
        pButton = pButtonData;
        for(i = 20;i<620;i += 40){//键盘初始化,第一行的 15 个按钮
                pButton->state = 2;
                pButton->x = i;
                pButton->y = 272;
                pButton++;
        }
    ……
}
        if(GetPos()){
            ret = FindButton_Key(pButtonData,48,0);//查找 48 个通用按钮
            if(ret){
……
                buf[i++] = KeyBoardChar[ret - 1];
            }
     ……
}
```

第三篇
基于 Linux 系统的应用

第 11 章

基于 PC 的 Linux 学习

如果读者以前没有接触过 Linux，那么就要先在 PC 上安装一个 Linux 的操作系统，然后找一本该版本的 Linux 的操作书籍来学习、熟悉。目前桌面的 Linux 操作系统很多，有 Ubuntu、Fedora、SUSE，还有 RedHat 等。阿南最初在学习开发 S3C2410 时使用的是 RedHat9. 0，之后一直就用 Ubuntu（Ubuntu 10. 04、Ubuntu 12. 04 和 Ubuntu 14. 04）。

11.1 RedHat Linux 系统下的常用操作

注：对于其他（如 Ubuntu 14. 04 等）的桌面 Linux 操作系统，下述操作基本上也是可用的。

11.1.1 RedHat Linux 9 下的常用操作问答

1. 在开机引导装载程序中，如何修改等待自动登录默认操作系统的时间？

答：如果引导装载程序是 GRUB，则修改/etc/grub. conf 文件中的 timeout＝秒数。

如果引导装载程序是 LILO，则修改/etc/lilo. conf 文件。可用 vi 等编辑器修改，下同。

2. 在字符（Text）模式下，如何关机、重启、注销？

答：关机，poweroff 或 shutdown -h now；重启，reboot 或 shutdown -r now；注销（即重新登入），logout。其中在 shutdown 指令中的 now 是指现在就执行，也可以指定多少时间后再执行此命令。

3. 如何使用 U 盘？

答：先创建/mnt/usb 目录，再执行 mount/dev/sda1/mnt/usb 挂载，此时/mnt/usb 就是 U 盘的目录，在拔出 U 盘时要执行 umount/mnt/usb 进行卸载。

4. 在字符模式（Text）下，如何进入 X Window 模式（Graphic）？在 X Window 模式下，如何返回字符模式？

答：执行 startx 命令启动 X Window 模式；单击 Main Menu（主菜单）→Log out（注销）打开对话框，选择“注销”进入字符模式；或按 CRTL＋ALT＋F1～F6 来进入不同的虚拟控制台(即文本模式下)。

5. 如何重新指定开机默认进入的执行模式(字符或X Window模式)?

答：修改/etc/inittab文件中的内容(id:5:initdefault:)。

其中,5表示以X Window模式(Graphic)登录,3为字符模式(Text)登录。

6. 在字符模式下,如何使用户登录时,系统不要求输入密码?如何恢复或更改用户密码?

答：取消输入密码,passwd -d 用户账号。如要取消root登录时的密码,则执行passwd -d root。也可以用vi打开/etc/shadow文件,以删除密码的方法取消。恢复或更改密码,则执行passwd用户账号(如果是取消自己,则不用)命令后会提示输入New password和Retype new password。

7. 字符模式下,如何新增用户账号?

答：使用"useradd 用户账号"命令增加,但在新增后还不能登录使用,还需要用passwd命令设置密码后才行。

8. 在X Window下,如何选择系统默认使用的语言?

答：单击"主菜单→系统设置→语言"打开选择语言对话框进行选择。

9. 用ls等命令查看的内容太多,超过一页时,如何分页显示?

答：可用ls | more或ls | less进行分页查看。其中,在用more浏览时,按空格键(Space)则会显示下一页的内容;按回车(Enter)键则会向下多显示一行;按q键则离开浏览模式。

在用less浏览时,按h键会出现在线使用说明;按q键离开浏览模式。

10. 如何获得命令的使用方法?

答：可利用在线手册——man(Manual),用法是输入man和待查的命令名称,如要查询ls命令的使用方法,则输入以下命令：man ls,也可以输入ls --help。

11. 搜索文件及目录,以及搜索包含特定字符串的文件?

答：搜索文件及目录可以用find命令,如要在根目录(/)上搜索apache文件则输入命令：find / -name apache -print,注意：如果没有指定目录,则系统会以当前的目录为搜索的范围;搜索包含特定字符串的文件可以用grep命令,如要在/etc目录下搜索包含字符串"password"的文件则输入：grep -n 'password' /etc/*.*,其中加入-n参数会标出符合指定的字符串的列数,另外不可指定在目录中搜索,否则会出现错误信息,如上述命令不能写成：grep -n 'password' /etc/。另外如果想停止搜索可以直接按Ctrl + C键结束该命令。现在笔者经常用：grep -ir password /etc。

12. 如何进行控制台间的切换?

答：在文本模式下,用Alt+F1～F6键来分别在6个虚拟控制台间切换,它们可分别用不同的用户名登录和执行不同的命令与程序,如果已经启动了X Window(如:在文本模式下用startx命令启动),则按Alt+F7切换到X Window图形模式。

在X Window图形模式下,用Crtl+Alt+F1～F6键分别切换到文本模式下的

6个虚拟控制台。Crtl + Alt + BackSpace结束图形模式。

因为Linux是多任务的系统，所以可以在不同的控制台下用不同(或同一)的用户登录来运行不同的程序。

我觉得这个功能很方便，因为有时在文本模式下，需要打开多个终端来处理显示多个任务，如：一个终端运行minicom作为目标板的控制，一个终端作为宿主机编译目标板要运行的文件，还有多个终端打开多个源文件浏览，等等。如果习惯在X Window模式下就例外，因为用右击就可以打开多个终端。

13. 如何查看PDF文档和浏览网页？

答：在X Window下打开shell终端，输入“xpdf filename.pdf”和“mizzo filename.html”命令分别查看。注：必须在X Window下才能运行这两个程序，文本模式下不能运行。

14. 如何查看磁盘的使用情况？

答：# df -h。

15. /proc目录下，关于系统资源非常有用的文件有哪几个？

/proc/modules、/proc/ioports、/proc/iomen、/proc/devices、/proc/interrupts、/proc/filesystems。

16. 关于内核代码调试时输出打印信息的printk语句有哪些？

如：语句“printk(KERN_DEBUG “Here I am: %s:%i\n”,_FILE_,_LINE_&_);printk(KERN_INFO “ Driver Initional \n”);”等同于“printk(“<6>” “ Driver Initional \n”);printk(“<1> Hello,World! \n”);”。

没有指定优先级的printk语句采用默认日志级别(DEFAULT_MESSAGE_LOGLEVEL)在kernel/printk.c中被指定，根据日志级别，内核可能会把消息输出到当前控制台上。当优先级值小于console_loglevel整数值时，消息才会被显示出来。如果系统同时运行了klogd和syslogd，则无论console_loglevel为何值，都将把内核消息追加到/val/log/messages中。console_loglevel的初始值是DEFAULT_CONSOLE_LOGLEVEL，可以通过文本文件/proc/sys/kernel/printk来读取和修改它及控制台的当前日志级别等。也可以简单地通过输入下面的命令使所有的内核消息得到显示：

```
# echo 8 > /porc/sys/kernel/printk
```

17. 如何查看当前正在运行的进程？

答：# ps。

18. 如何解压缩到指定目录？

答：# tar xvzf linutte.tgz -C /linuette。

19. 当/etc/grub.conf文件中的内容被修改或破坏不能正常启动时，如何在GRUB引导时修改设置使其正常启动？

答：以修改了/etc/grub.conf 文件中的 vga 项使启动时显示器不能显示为例，在 GRUB 启动引导菜单中 Windows XP 和 RedHat Linux(2.4.20-8)两项使用上、下、左、右键选中 Linux 系统，不按 Enter 键，而按 E 键进入菜单项目编辑器，再使用上、下、左、右键选中 Kernel 项，也按 E 键进行编辑，在行的后面输入"vga=791 fb=on"后按 Enter 键，最后按 b 键执行命令，并引导操作系统。

20. 如何使用包管理器 RPM?

以 tmake 为例：

安装：#rpm -ivh tmake-1.7-3mz.noarch.rpm。

升级：#rpm -Uvh tmake-1.7-3mz.noarch.rpm。

查询：#rpm -q tmake。

删除：#rpm -e tmake。

11.1.2　超级终端 Minicom 的使用

1. 启动 Minicom

输入 minicom 启动，或输入 minicom -s 直接进入设置模式。

2. 设　置

选择串口：选择菜单中的 Serial port setup，按回车键，再按 A 键以设置 Serial Device(如果使用串口 1，则输入/dev/ttyS0；如果您使用串口 2，则输入/dev/ttyS1，注意其中的 S 是大写)，按回车键返回。

设置波特率：按 E 键进入设置 bps/par/Bits(波特率)界面，如果按 I 键则设置波特率为 115 200，按回车键返回。

数据流控制：按 F 键设置 Hardware Flow Control 为 NO。

其他为默认设置，然后按回车键到串口设置主菜单，选择 Save setup as dfl，按回车键保存刚才的设置(保存到/etc/minirc.dfl)，再选择 Exit 退出设置模式，回到 minicom 操作模式。

此时就可以像在 Windows 下的超级终端一样使用了。

3. 退出 Minicom

按下 Ctrl+A 键，松开后紧接着再按下 Q 键，在跳出的窗口中，选择 Yes。

查看以前的信息：

➢ 按下 Ctrl+A 键，松开后紧接着再按下 B 键，接着用上下箭头可以翻看前面的信息。

其他有用的功能：

➢ 命令帮助——按 Ctrl+A 键后再按 Z 键；
➢ 清屏——按 Ctrl+A 键后再按 C 键；
➢ 设置——按 Ctrl+A 键后再按 O 键；

➢ 发送文件——按Ctrl+A键后再按S键；
➢ 退出——按Ctrl+A键后再按Q键。

11.1.3　NFS的使用

为什么要使用NFS：网络文件系统(NFS，Network File System)是一种在网络上的计算机间共享文件的方法，通过它可以将计算机上的文件系统导出给另一台计算机。我们在宿主机上编辑、编译好的程序，可以通过它导出到目标板上进行实际的运行。

1. 宿主机配置

从NFS服务器中共享文件又称导出目录，/etc/exports文件控制NFS服务器要导出哪些目录，格式如下：

共享的目录　可以连接的主机(读写权限，其他参数)

如果允许目标板(IP：192.168.0.*)挂载主机的/home目录，则/etc/exports文件的内容如下：

```
/home 192.168.*.*(rw,sync)
```

注：如果出现mount不成功，则可将sync去掉试试。
更改后要使用如下命令重新载入配置文件：

```
#/sbin/services nfs reload 或 #/etc/init.d/nfs reload
```

然后启动NFS服务器，命令如下：

```
/sbin/services nfs start
```

上面两个命令也可以用下面的一条指令完成，如下：

```
/sbin/service nfs restart
```

设置好后也可以通过mount自己来测试NFS服务设置是否成功。如果本机IP地址为192.168.0.1，则可以用"mount 192.168.0.1:/home /mnt"命令进行测试，如果mount成功，则在/mnt目录就可以看到/home目录下面的内容了。

2. 使用mount命令挂载NFS文件系统

下面将宿主机(IP：192.168.0.1)配置的/home目录挂载到(IP：192.168.0.7)目标板上的/mnt目录。

在宿主机启动minicom作为目标板的显示终端，启动目标板的Linux系统，再使用下面命令：

```
mount -o nolock 192.168.0.1:/home /mnt
```

注意：如果没有"-o nolock"选项，而直接使用命令"mount 192.168.0.1:/home/

mnt”时将出现如下错误“portmap：server localhost not responding，timed out”。

目前，笔者都用“mount -t nfs 192.168.0.1:/home /mnt”。

11.2 Ubuntu系统的安装与常用操作

注：对于其他（如RedHat等）的桌面Linux操作系统，下述操作基本上也是可用的。

11.2.1 Ubuntu 14.04的安装

Ubuntu 14.04的安装步骤如下：

1. 下载光盘刻录软件

在Windows桌面系统下，下载光盘刻录软件（如UltraISO Premium Edition 9.6.2.3059），并安装。

2. Ubuntu官网下载ubuntu-14.04.2-desktop-amd64.iso

如果PC是32位的，则可以下载32位的版本。

3. 刻录U盘启动盘

注：如果有些读者在公司的计算机上安装，且公司的计算机上安装有加密软件，此时在加密计算机中刻录，或在加密计算机中使用U盘复制文件，软件会自动给文件加密使之出错，因此不能在加密的系统下制作启动盘。

计算机插入U盘，启动UltraISO软件。在UltraISO软件下选择“文件”→“打开”，打开下载的ubuntu-14.04.2-desktop-amd64.iso文件。

选择“启动”→“写入硬盘映像”，弹出对话框，单击“格式化”按钮将U盘格式化，选择“便捷启动”→“写入新的驱动器引导扇区”→Syslinux（注：如果是公司加密计算机，这步会出错）。

单击“写入”按钮将ubuntu-14.04.2-desktop-amd64.iso文件的内容将刻录到U盘里，这里也可以勾选“写入校验”。

写入完成后返回。

4. 设置BIOS从U盘引导启动并进入Ubuntu安装

以ASUS UEFI BIOS为例，开机时按F2键进入BIOS，选择“启动菜单”→“UEFI：U盘名称”，进入启动选项，可以先选择“检测盘片是否有错误”（如果盘片有错误会安装失败），检测完成没有错误时再选择“安装Ubuntu”。

在“安装”对话框中，不要勾选“安装中下载更新”，这样会使安装过程变慢。

安装类型，如果还没分区，则选择“其他选项”手动分区。

5. 分　区

注：不需要给UEFI分区，否则安装完成后无法启动。

选中已有分区，再按减号“－”可以将已有的分区释放为空闲区，我们将除了已经安装 Win7 系统的 C 盘和另外的 D 盘(即 sda1 和 sda5)保留外，其他都变为空闲区。

然后选中空闲区，再按加号“＋”弹出对话框，分别将空闲区分成以下 4 个区：

逻辑分区	/boot	2048 MByte	ext4	注：这里挂载点选择 /boot
逻辑分区		16483 MByte	swap	注：这里要大于两倍的内存
逻辑分区	/	204800 MByte	EXT4	注：这里挂载点选择“/”
逻辑分区	/home	508888 MByte	EXT4	注：这里挂载点选择 /home

6. 安装语言都选择 English

安装语言都选择 English，接着按各种提示即可完成安装。

11.2.2 Ubuntu 14.04 的基本设置和常用操作

1. Ubuntu 14.04 安装后无法进入 GRUB 引导程序

Ubuntu 14.04 安装完成后重新启动时会直接进入 Ubuntu，而不会进入 GRUB 引导程序让其选择是由 Ubuntu 或是由 Windows 7 启动。解决方法是启动 ubuntu 命令行模式，执行命令：update-grub，扫描计算机中所有操作系统并重新生成/boot/grub/grub.cfg 文件。

2. 终端切换快捷键

命令模式和图形模式的切换：Alt＋Ctrl＋F1～F6 为命令行模式；Alt＋Ctrl＋F7 为图形桌面模式。

3. 设置 root 密码

增加 root 登录命令：sudo passwd root，紧接着分别输入用户密码和 root 密码即可设置好 root 的密码，命令 su root 则切换到 root 模式下。

用户模式下 sudo passwd -l root 禁用 root。

4. 系统启动后默认进入命令行模式的设置

启动后默认为命令行模式：将/etc/default/grub 文件中的

```
GRUB_CMDLINE_LINUX_DEFAULT = "quiet splash"
```

修改为

```
GRUB_CMDLINE_LINUX_DEFAULT = "quiet splash text"
```

注：quiet 代表不显示详细启动过程；splash 代表显示进度条。

有时修改为默认命令行启动时，再按 Ctrl＋Alt＋F7 不能启动图形模式。在命令行下输入命令：service lightdm start 启动图形，之后就可以用 Ctrl＋Alt＋F1 和 F7 切换了。

另外，GRUB_TIMEOUT 为启动等待时间，修改该时间后，还需要屏蔽 GRUP_HIDDEN_TIMEOUT=0；否则会警告，使修改无效。

GRUB_DEFAULT 为默认的启动选项，假设启动界面如图 11.1 所示，则 0 为 Ubuntu，4 为 Windows。

```
*Ubuntu
Advanced options for Ubuntu
Memory test (memtest86+)
Memory test (memtest86+, serial console 115200)
Windows Vista (loader) (on /dev/sda1)
```

图 11.1 启动界面

修改完成后输入命令：update-grub 更新/boot/grub/grub.cfg 后重启，即开机后自动进入 tty1。

5. 网络设置

网络设置需要在图形模式下，右上角选择"上下行"按钮→Edit Connections。

Add→Ethernet→Create，选择 Device MAC address，再选择 IPv4 Settings→Manual，单击 Add 按钮输入：

Address：192.168.1.213

Netmask：255.255.255.0

Gateway：192.168.1.1

DNS servers：202.96.134.133

单击 Save 按钮设置完成。

设置完之后可以选择"上下行"按钮→Connection Information，如果 IP 地址等不正确，可能是无线网络的原因，可以选中 Wired Connection 1，再单击 Delete 按钮将其删除再试，或者在命令行模式下设置 IP 等。

6. 终端启动

在图形模式下，按 Ctrl+Alt+T 键打开终端。

在命令行模式下执行：apt-get install nautilus-open-terminal，安装完成后重启。

7. 终端常用命令

① Tab 键补全。

② 管道机制"|"：这种机制允许把一条命令的输出传送到另一条命令。例如，ls 命令列出当前目录下的所有文件，grep 命令搜索输入其中的指定检索项。

读者可以通过管道机制（"|"字符）把二者结合起来，在当前目录下搜索文件。以下给出的命令（在当前文件夹下）为搜索关键字为"word"的文件：

```
ls | grep word
```

例如：make |more 或 make|less。

③ 通配符"＊"：如"rm really＊name"命令为删除当前文件夹下名为"really long file name"和"really very long file name"的两个文件。

④ 输出重定向"＞"：可以把一条命令的输出重定向到一个文件或另一条命令。

```
ls > file1
```

⑤ 历史记录：Bash能记住以前输入过的命令，上、下方向键可以逐行调出它们。

～ 用来表示当前用户的主目录，如"cd～"等效于"cd /home/name"。

. 当前目录。

.. 当前上一级目录。

- 前一个目录，"cd-"可以返回到前一个工作目录。

⑥ 后台命令

Bash默认情况下会在当前终端下执行键入的每条命令。通常这样是没有问题的，但是如果你想要在启动某个应用后继续使用终端呢？比如通过输入firefox启动火狐浏览器，你的终端将被错误提示等各种信息输出占据，直到关闭火狐浏览器为止。在Bash中可以通过在命令结尾添加"&"操作符来执行后台程序。

```
firefox &
```

⑦ 条件执行

Bash也可以连续执行两条命令。第二条命令仅在第一条命令成功执行后才会开始执行。如此，可以通过键入"&&"，也就是两个"&"字符进行分隔，在同一行输入两条命令。下面给出的命令会在等待5 s后运行gnome-screenshot工具。

```
sleep 5 && gnome - screenshot
```

8. 中文输入法的安装

在桌面系统下，打开Ubuntu Software Center，右上角输入ibus搜索，左下角单击show 64 technical items，找到ibus-table-wubi，以及确定ibus-table-pinying是否已经安装。桌面系统右上角选择En→Text Entry Settings→＋→Chinese(wubi-jidian86)和Chinese(SunPinyin)→Add，可用键＋Space组合键切换，键一般在键盘的左下方或右下方Alt键旁边。

解决命令行模式中文乱码：需要安装zhcon。

在/etc/default/locale文件里设置

```
LANG = "en_US.UTF - 8"
LANGUAGE = "en_US:en"
```

存盘退出，再reboot重启，用env或者locale可查看修改后的结果：

```
apt - get install zhcon
```

安装完后，在命令行直接输入 zhcon 即可开启(每次进入都需要输入 zhcon 开启)。

注：如有黑屏，则需要加载 vgz 驱动和 utf8 支持，输入的命令为：

```
zhcon --utf8 --drv = vga
```

也可以在~/. bashrc 里面加一个别名：vim~/. bashrc 打开文件增加一行

```
alias zhcon = 'zhcon --utf8 --drv = vga'
```

那么每次输入 zhcon 命令都会被认为是 zhcon --utf8 --drv=vga。

注：笔者认为 zhcon 可以正确解析中文，而 LANGUAGE 可以将解析的内容用 en_US：en 英语显示出来。另外，zhcon 也可以切换到输入中文。

9. 屏幕截图软件 shutter：apt-get install Shutter

打开该软件，选择 Selection。

如果不小心将其在左侧的 Launcher 中删除，则需要重新安装，然后在 Launcher 中第一个程序"Search your computer and online source"中输入 Shutter 将其打开，再在 Launcher 中右击 Shutter 程序的图标，可选择 Lock to Launcher 将其锁定在 Launcher 中。

10. Ubuntu Software Center

如果不小心在 Launcher 中将 Ubuntu Software Center 删除了，则在"Search your computer"中查找"Ubuntu Software Center"，看是否可以找到，或者在命令行中输入 software-center，如果没有则重新安装：apt-get install software-center。

安装完后在"Search your computer"中将其打开，然后在 Launcher 中将其锁定。

11. VLC 安装失败

```
add - apt - repository ppa:videolan/master - daily
apt - get update
apt - get install vlc
```

原先已经安装了 VLC1. 16 版本，但采用上述方法安装失败后在软件中心一直找不到原来的 VLC 1. 16 版本，而最新的版本也不能成功安装。这就是乱加 ppa 源的后果，把 https://launchpad. net/～videolan/＋archiv…ster-daily 源给删除。如果是用 apt-add-repository 方式添加的，就用下面的语句删除，执行完成后重新启动软件中心。

```
apt - add - repository - r ppa:videolan/master - daily
```

11.2.3 Ubuntu常用命令

下面这些命令是对上述Ubuntu常用命令的补充。

1. 查看文件夹容量

$df -hT　查看磁盘信息，h表示系统大小用GB、MB、KB为单位输出，T输出文件系统类型。

$du -h --max-depth=1　查看文件及文件夹的容量及使用情况，--max-depth=1表示只查看当前一级。

2. 解　压

$tar -zxvf　*.tgz　*.tar　*.zip　参数z表示解gzip类。

$tar -jxvf　*.bz2　参数j表示解bzip2类。

$cat　*.bz2 | tar jxv-　解多个bz2文件，cat将多个文件合并后通过管道"|"采用stdout格式输出，"-"表示tar采用stdin格式输入。

3. 环境变量

在/home/ccn目录和/root目录都有.bashrc文件，打开后可以通过export添加环境变量。

```
export PATH = $PATH:/home/ccn/bin/gcc-linaro-aarch64-linux-gnu-4.8-2013.10_linux/bin/
```

$source .bashrc　使更改生效。

$echo　$PATH　查看当前用户的环境变量值。

将路径添加到环境变量值后，无论位于哪个目录都可以直接运行该路径下的可执行文件，可以输入aarch后再按Tab键测试，此时可以显示刚添加的路径下的所有以aarch开始的可执行文件。

4. 搜索包含某些字符的文件

$grep -r 'gcc-linaro-aarch64-linux-gnu-4.8-2013.09' *　其中r表示进入子目录。

5. 查找文件

```
$find . -name "set.bin"
```

6. 安　装

安装完Ubuntu系统后第一次安装可以先更新

```
#apt-get update
#apt-get install openjdk-7-jdk
```

7. clone 源码

```
# git clone https://android.googlesource.com/prebuilts/gcc/linux-x86/aarch64/aarch64-linux-android-4.9
```

8. 查看连接的 USB 设备

```
$ lsusb
```

9. 查看 USB 读卡器设备

```
$ ls /dev/sd*
```

10. 将文件写入 TF 卡(用于 TF 启动 U-boot)

```
# dd iflag=dsync oflag=dsync if=set.bin of=/dev/sdb1 seek=1
```

dd 用指定大小的块复制一个文件,并在复制的同时进行指定的转换。

if=文件名,输入文件名,默认为标准输入,即指定源文件<if=input file>。

of=文件名,输出文件名,默认为标准输出,即指定目的文件<of=output file>。

seek=blocks,从输出文件开头跳过 blocks 个块后再开始复制。

11. 十六进制模式及 TF 卡数据查看

```
# hexdump -n 1048576 /dev/sdb1 |more
# hexdump -n 1048576 ./set.bin |more
```

-n 1048576 代表打印出前 1M=1 024 * 1 024=1 048 576 字节的数据。

12. 分区、格式化磁盘

#ls /dev/sd* 查看 U 盘(或 TF 卡)设备名称

```
/dev/sda  /dev/sda1  /dev/sda2  /dev/sda5
```

插入 TF 卡及读卡器后

```
# ls /dev/sd*
/dev/sda  /dev/sda1  /dev/sda2  /dev/sda5  /dev/sdb  /dev/sdb1
```

#fdisk /dev/sdb 磁盘操作,根据提示输入命令:删除分区、增加分区、格式化、写入分区表等。

Command (m for help): m 显示各个命令功能。

Command (m for help): n 增加一个分区。

Partition type:

p primary (0 primary,0 extended,4 free);

e　extended。

Select (default p)：p　选择 primary。

Partition number (1～4,default 1)：1　分区号。

First sector (2 048～7 626 751,default 2 048)：选择该区的起始块号，直接回车选择默认值，Using default value 2 048。

Last sector，+sectors or +size{K,M,G} (2 048～7 626 751,default 7 626 751)：选择该区的结束块号，Using default value 7 626 751。

Command (m for help)：a　标记第一个分区。

Partition number (1～4)：1　分区号为1。

Command (m for help)：w　写入分区表。

```
The partition table has been altered!
Calling ioctl() to re-read partition table.
Syncing disks.
```

#mkfs.vfat /dev/sdb1　将 sdb1 分区格式化为 vfat 文件系统，或 mkfs.ext4 格式化为 ext4。

```
mkfs.fat 3.0.26 (2014-03-07)
```

格式化完成后需 eject 再重新插入后有效，查看：

```
#df -hT
/dev/sdb1    ext3    3.6G    7.4M    3.4G   1%     /media/autel/78c26ecc
```

13. 查看文件格式

```
#file system.img
```

11.2.4　Ubuntu Linux 与 Windows 系统下的文件共享

1. 安装配置 Samba 服务器实现与 Win7 的文件共享目录

Samba 是 Linux 系统上的一种文件共享协议，可以实现 Windows 系统访问 Linux 系统上的共享资源。(注：Linux 和 Linux 间用 NFS，Windows 和 Windows 间用 CIFS。)

① 安装：apt-get install samba samba-common。

② 新建共享目录并设置权限：

```
mkdir /home/share
chmod 777 /home/share
```

③ 修改配置文件：vi /etc/samba/smb.conf，在“max log size = 1000”的下一行增加“security = user”行，说明需要账号和密码才可以访问共享目录。

在文件最后增加：

```
[myshare]
   comment = my share directory
   path = /home/share
   browseable = yes
   writable = yes
```

保存后退出。

④ 新建访问共享资源的用户和密码：

useradd ccn　如果已经存在ccn用户了，则该命令可以省略。

```
smbpasswd -a ccn
```

smbpasswd -a　增加用户（要增加的用户必须已是系统用户）。

smbpasswd -d　冻结用户，这个用户不能再登录了。

smbpasswd -e　恢复用户，解冻用户，让冻结的用户可以再使用。

smbpasswd -n　把用户的密码设置成空，要在global中写入null passwords -true。

smbpasswd -x　删除用户。

⑤ 重启samba服务器：service smbd restart。

⑥ Win7下访问Ubuntu共享目录：

Win7系统下按 +R键打开"运行"窗口，输入：\\192.168.1.213\myshare。

对话框中输入用户名：ccn及密码。

⑦ 利用smbclient使Ubuntu访问Win7下的共享目录。

如果没有安装smbclient，则输入apt-get install smbclient命令。

Win7设置好共享目录后，在Linux下使用smbclient -L //192.168.1.213命令查看Windows下的共享目录，mount -t smbfs -o username=administrator,passwd='dell' //192.168.1.213/linuxsoft /mnt/samba。

2. 安装配置SSH实现Win7利用putty登录Ubuntu

① 安装SSH服务器：apt-get install openssh-server。

② 启动SSH服务器：/etc/init.d/ssh start。

③ 使用putty连接：打开putty，输入Host Name 192.168.1.213，Connection type：ssh。

在login as输入ccn及密码，su root可以切换到root用户。

如果root用户无法登录，在/etc/ssh/sshd_config文件中，将PermitRootLoginl默认值"no或without-password"改为"yes"，修改后需要重启service ssh restart或/etc/init.d/ssh restart。

3. 同一台机器 Linux 挂载 Win7 的磁盘

如果 Win7 加密或其他原因，Ubuntu 桌面不能自动显示 Win7 的磁盘，则可以用 mount 实现。

Linux 命令行模式下，用 fdisk -l 打开要挂载的驱动器/dev/sda2。

```
mkdir -p /mnt/win7
mount -t vfat /dev/sda2 /mnt/win7
```

ls -l /mnt/win7　如果中文显示为乱码，则可以进入图形模式查看/mnt/win7。

11.2.5　Ubuntu Linux 与 Linux 系统下的文件共享

Linux 与 Linux 间的共享使用 NFS，因此需要安装配置 NFS。

1. 服务器安装 NFS：apt-get install nfs-kernel-server

建立 NFS 共享目录 mkdir/home/share(如果已存在则省略)。

在/etc/exports 文件最后一行添加 NFS 服务器共享目录：

```
/home/share *(rw,sync,no_root_squash,no_subtree_check)
```

*：允许所有的网段访问，也可以使用具体的 IP。

rw：挂接此目录的客户端对该共享目录具有读/写权限。

sync：资料同步写入内存和硬盘。

no_root_squash：root 用户具有对根目录的完全管理访问权限。

no_subtree_check：不检查父目录的权限。

启动 NFS 服务器：service nfs-kernel-server start。

本地回环验证：mkdir /mnt/nfs。

挂载：mount -t nfs 127.0.0.1:/home/share /mnt/nfs。

查看：ls /mnt/nfs。

2. 客户端安装(切换到另一台 Linux 计算机)：apt-get install nfs-common

安装后新建一个目录用于挂载：mkdir/mnt/nfs。

查看服务器上已被共享的目录：showmount -e 192.168.1.213。

挂载：mount -t nfs 192.168.1.213:/home/share/mnt/nfs/。

查看：ls /mnt/nfs。

11.2.6　超级终端 Minicom 的使用

1. Ubuntu 下 USB 转串口芯片驱动程序安装 cp210x

由于笔者使用的 USB 串口转换器的芯片是 cp210x(是目前最常见的 USB 转串口驱动芯片之一)所以需要安装该芯片的驱动。

当插入USB转串口板后，会出现/dev/ttyUSB0，如图11.2所示。

```
ccn@ccn01: ~
ccn@ccn01:~$ ls /dev/ttyUSB*
ls: cannot access /dev/ttyUSB*: No such file or directory
ccn@ccn01:~$ ls /dev/ttyUSB*
/dev/ttyUSB0
ccn@ccn01:~$
```

图11.2　ttyUSB0

2. 安装设置Minicom

安装apt-get install minicom。

选择minicom -s命令进入设置模式，选择Serial port setup后按回车键，然后按A键在Serial Device下选择/dev/ttyUSB0后按回车键，再按F键使硬件流控为No后，按回车键退出，再选择Save setup as dfl，将刚才的设置保存到/etc/minicom/minirc.dfl下，再选择Exit退出。

如果提示"Cannot open /dev/ttyUSB0: Permission denied"，则需要用root登录。

3. Minicom常用功能键

退出：按Ctrl+A键松开后按Q键。

查看信息：按Ctrl+A键松开后按B键，接着用上下箭头可以翻看前面的信息。

帮助：按Ctrl+A键松开后按Z键。

清屏：按Ctrl+A键松开后按C键。

设置：按Ctrl+A键松开后按O键。

发送文件：按Ctrl+A键松开后按S键。

11.3　Linux下的应用编程

《GNU/Linux编程指南》和《Unix高级环境编程》都是学习Linux应用编程的经典书籍。它们详细介绍了Linux编程方面的很多基础知识和源码实例，即使读者将来的工作不是应用而只想学习驱动开发，也要先学习它们。阿南当初就是没有学习它们，而直接去看《Linux设备驱动程序》，结果很多基础知识都不理解；学习它们之后，看驱动的书就非常轻松了。

下面是阿南在嵌入式Linux应用编程实践过程中对进程的一点理解(注：也可以在PC机上测试)。

11.3.1 进程间隔定时器

1. 概 念

所谓间隔定时器(Interval Timer,简称 itimer)就是指定时器采用"间隔"值(interval)来作为计时方式,当定时器启动后,间隔值 interval 将不断减小。当 interval 值减到 0 时,我们就说该间隔定时器到期。它主要应用在用户进程上,每个 Linux 进程都有三个相互关联的间隔定时器:

① 真实间隔定时器 ITIMER_REAL:这种间隔定时器在启动后,不管进程是否运行,每个时钟滴答都将其间隔计数器减 1。当减到 0 值时,内核向进程发送 SIGALRM 信号。

② 虚拟间隔定时器 ITIMER_VIRT:也称为进程的用户态间隔定时器。当虚拟间隔定时器启动后,只有当进程在用户态下运行时,一次时钟滴答才能使间隔计数器当前值 it_virt_value 减 1。当减到 0 值时,内核向进程发送 SIGVTALRM 信号(虚拟时钟信号),并将 it_virt_value 重置为初值 it_virt_incr。

③ PROF 间隔定时器 ITIMER_PROF:当一个进程的 PROF 间隔定时器启动后,则只要该进程处于运行中,而不管是在用户态还是核心态下执行,每个时钟滴答都使间隔计数器 it_prof_value 值减 1。当减到 0 值时,内核向进程发送 SIGPROF 信号,并将 it_prof_value 重置为初值 it_prof_incr。

定时器在初始化时,被赋予一个初始值,随时间递减至 0 后发出信号,同时恢复初始值。在任务中我们可以使用其中一种或全部三种定时器,但同一时刻同一类型的定时器只能使用一个。

2. 数据结构

Linux 在 include/linux/time.h 头文件中为上述三种进程间隔定时器定义了索引标识,如下所示:

```
#define ITIMER_REAL 0
#define ITIMER_VIRTUAL 1
#define ITIMER_PROF 2
```

虽然,在内核中间隔定时器的间隔计数器是以时钟滴答次数为单位的,但是让用户以时钟滴答为单位来指定间隔定时器的间隔计数器的初值显然是不太方便的,因为用户习惯的时间单位是秒、毫秒或微秒等。所以 Linux 定义了数据结构 itimerval 来让用户以秒或微秒为单位指定间隔定时器的时间间隔值。其定义如下(include/linux/time.h):

```
struct itimerval {
struct timeval it_interval;//timer interval
struct timeval it_value;//current value
};
```

其中,it_interval 成员表示间隔计数器的初始值,而 it_value 成员表示间隔计数器的当前值,它们都是 timeval 结构类型的变量。

```
struct timeval {
time_t tv_sec;//seconds
suseconds_t tv_usec;//microseconds
};
```

3. 操作函数

```
int getitimer(int which,struct itimerval * value);
```

getitimer()函数得到间隔计时器的时间值并保存在 value 中。它有两个参数:① which,指定查询调用进程的哪一个间隔定时器,其取值可以是 ITIMER_REAL、ITIMER_VIRT 和 ITIMER_PROF 三者之一;② value 指针,指向用户空间中的一个 itimerval 结构,用于接收查询结果。


```
setitimer(int which,struct itimerval * value,struct itimerval * ovalue);
```

setitimer()不仅设置调用进程的指定间隔定时器,而且还返回该间隔定时器的原有信息。它有三个参数:① which,含义与 sys_getitimer()中的参数相同;② 输入参数 value,指向用户空间中的一个 itimerval 结构,含有待设置的新值;③ 输出参数 ovalue,指向用户空间中的一个 itimerval 结构,用于接收间隔定时器的原有信息。

```
int sigaction(int signum,const struct sigaction * act,struct sigaction * oldact);
```

该函数为 signum 所指定的信号设置信号处理器。sigaction 结构(act 和 oldact)描述了信号的部署。它在<signal. h>中的完整定义为:

```
struct sigaction{
void ( * sa_handler)(int);//规定了当 signum 中的信号产生后,要调用的处理函数
sigset_t sa_mask;//定义了在执行处理函数期间应该阻塞的其他信号集合的信号掩码
int sa_flags;
void ( * sa_restorer)(void);
};
```

4. 测试程序

```
#include <stdio.h>
#include <signal.h>
#include <sys/time.h>
#include <asm/param.h>//#define HZ
struct timeval tpstart,tpend;
float timeuse;
```

```
/* signal process */
static timer_count = 0;
void prompt_info(int signo){
    if ((++timer_count)%100 == 0){
    time_t t = time(NULL);//获取秒数表示的系统当前时间
    printf("[%d]prompt_info called",timer_count);
    printf("      current time %s",ctime(&t));//将秒数转换成字符串输出
    gettimeofday(&tpend,NULL);
    timeuse = 1000000 * (tpend.tv_sec - tpstart.tv_sec) + tpend.tv_usec - tpstart.
                     tv_usec;
//    timeuse /= 1000000;
    printf("      Used Time: %f\n",timeuse);
    }
}
void init_sigaction(void){
    struct sigaction act;
    act.sa_handler = prompt_info;
    act.sa_flags   = 0;
sigemptyset(&act.sa_mask);//初始化一个空的信号集合 act.sa_mask
sigaction(SIGALRM,&act,NULL);//为 SIGALRM 信号设置信号处理函数
    gettimeofday(&tpstart,NULL);
}
void init_time(void){
    struct itimerval val;
    val.it_value.tv_sec = 0;//设置间隔定时器的当前值
    val.it_value.tv_usec = 10000;//取值要大,等于时钟滴答的周期,否则仍为时钟
                                  //滴答的时间
    val.it_interval = val.it_value; //间隔计数器的初始值
    setitimer(ITIMER_REAL,&val,NULL);//设置真实间隔定时器,定时结束后将发出
                                      //SIGALRM 信号
}
int main(void){
    printf("clock tick frequency is %d\n",HZ);//输出时钟滴答的频率
    init_sigaction();
    init_time();
//    printf("clock tick frequency is %d\n",HZ);
    while(1);
    exit(0);
}
```

注：单独运行该测试程序时，几乎看不到什么误差，但在运行比较复杂程序的同时，再运行该测试程序，误差就会显示出来，而且比想象中的要恶劣得多，所以在实际

应用时一定要注意！

11.3.2　关于进程的体会

进程能少创建一个就少创建一个，因为它会浪费很多资源，如内核调度需要时间；子进程会对父进程的内存进行复制而花费大量的内存；特别是进程之间不共享全局变量，之间需要通过进程间的系列通信机制才能交互等。

1. 进程间不共享变量

在程序中定义了全局变量，在main函数中创建了一个子进程，但在父进程中对该变量所做的修改不会影响到子进程，同时对子进程的修改也不会对父进程的该变量产生任何影响。在创建子进程时，子进程对父进程的内存区做了完全的复制，且相互独立地占用了两个不同的内存区，所以子进程的该变量值只是刚创建时父进程的当前值，以后就相互独立互不影响了。可以用下述代码进行测试：

```
#include <stdio.h>
#include <unistd.h>
#include <sys/types.h>
unsigned char val,val1;
int main(){
    pid_t pid;
    val1 = 0x44;
    if ( (pid = fork()) == 0){
        while (1){
            val = 0x77;
            printf("child:val = %02x,val1 = %02x\n",val,val1);
            sleep(1);
            val = 0x55;
            printf("child:val = %02x,val1 = %02x\n",val,val1);
            sleep(1);
        }
    }
    else{
        while (1){
            sleep(1);
            val1 = 0x99;
            printf("father:val = %02x,val1 = %02x\n",val,val1);
            sleep(1);
            val1 = 0xaa;
            printf("father:val = %02x,val1 = %02x\n",val,val1);
        }
    }
```

```
}
```

运行结果如下：

```
child:val = 77,val1 = 44
father:val = 00,val1 = 99
child:val = 55,val1 = 44
father:val = 00,val1 = aa
child:val = 77,val1 = 44
father:val = 00,val1 = 99
```

2. 进程间通信——信号的使用

信号定义系统调用：

```
#include <signal.h>
void ( * signal(int signo,void ( * func)(int)))(int);
```


返回：成功则指向以前的信号处理函数，若出错则为 SIG_ERR。

signo 参数为信号名，func 为函数指针(即产生该信号时将调用的信号处理函数)。

上述函数原型太过复杂，可以用 typedef 定义使其简单化。

```
typedef void Sigfunc(int);
```

然后 signal 函数原型可写成：

```
Sigfunc * signal(int signo,Sigfunc * func);
```

使用例子：

```
#define SIG_HARDRESET 34//定义信号名
typedef void Sigfunc(int);
Sigfunc * signal(int signo,Sigfunc * func){
    struct sigaction act,oact;
    act.sa_handler = func;
    sigemptyset(&act.sa_mask);
    act.sa_flags = 0;
    if (signo == SIGALRM){
#ifdef SA_INTERRUPT
        act.sa_flags |= SA_INTERRUPT;
#endif
    }
    else{
#ifdef    SA_RESTART
        act.sa_flags |= SA_RESTART;
#endif
    }
```

```
    if (sigaction(signo,&act,&oact) < 0){
        return(SIG_ERR);
    }
    return(oact.sa_handler);
}
Sigfunc * Signal(int signo,Sigfunc * func){
    Sigfunc * sigfunc;
    if ( (sigfunc = signal(signo,func)) == SIG_ERR){
        perror("signal error");
    }
    return(sigfunc);
}
void sig_softreset(int signo){ //信号处理函数
    static pid_t pid_UserSystem = 0;
    if (signo == SIG_SOFTRESET){
        if (pid_UserSystem){
            kill(pid_UserSystem,SIGKILL);
            pid_UserSystem = 0;
            printf("kill UserSystem\n");
        }
        if ( (pid_UserSystem = fork()) == 0){
          ......
            kill(getppid(),SIG_SOFTRESET);//在子进程中发信号调用处理函数
        }
    }
    return;
}
int main(int argc,char * argv[]){
    int ret;
    pid_t pid_InputElement,pid_TimeTick,pid_IrQueue,pid_UserSystem;
    Signal(SIG_SOFTRESET,sig_softreset);
    kill(getpid(),SIG_SOFTRESET);//在父进程中发信号调用处理函数
    ......
}
```

3. 防止僵死进程

当因某种原因使父进程终止时，子进程将会变成僵死进程继续工作，而浪费各种资源，这不是我们所期望的。

在上述例子中加入一个信号处理函数，以及在main函数中增加信号定义，如下：

```
void sig_chld(int signo){//信号处理，当一个子进程结束时被调用
    pid_t pid;
```

```
    int stat;
    while ( (pid = waitpid( - 1,&stat,WNOHANG)) > 0){
        printf("child % d terminated\n",pid);
    }
    return;
}
Signal(SIGCHLD,sig_chld);//加入到上述的 main()函数中,且在创建子进程前调用
```

11.4 Linux下的驱动程序设计

《Linux设备驱动程序》一书,也是Linux驱动程序设计的权威书籍,建议在编写驱动之前先看看。下面两个实验是笔者当初学习时的实验,虽然简单,但当时还是出现了不少问题,希望对读者有所帮助。

11.4.1 模块编程实验

参考《Linux设备驱动程序》P25页,hello.c源程序如下:

```
#define MODULE
#include <linux/module.h>
t init_module(void){
    printk ("<1>Hello,world\n");
    return 0;
}
void cleanup_module(void){
    printk("<1>Goodbye cruel world\n");
}
```

用以下命令进行编译:

```
#gcc -c hello.c
#insmod ./hello.o
hello,world
#rmmod hello                //注:不是#rmmod hello.o
Goodbye cruel world
```

出现的问题:在执行"#insmod ./hello.o"时并没出打印出"hello,world"而是出现了下述错误:

```
./hello:kernel - module version mismatch
hello.o was compiled for kernel version 2.4.20
while this kernel is version 2.4.20 - 8
```

原因:模块和内核版本不匹配,即编译内核的编译器与现在编译模块的编译器

版本不一致。

解决方法：

① 将/usr/include/linux/version. h 文件中的 # define…“2. 4. 20”修改成 # define…“2. 4. 20 - 8”，再重新编译。

② 用 insmod 的-f(force，强制)选项强行装入模块 insmod -f . /hello. o。

③ 因为用 vi /usr/include/linux/version. h 查看到定义的内核版本是 2. 4. 20，而在内核源代码树下/usr/src/linux-2. 4/include/linux/version. h 中定义为 2. 4. 20-8 版本，所以用如下命令进行编译：gcc -c -I/usr/src/linux-2. 4/include hello. c，再装载就 OK 了！

注：可查看/proc/modules 文件。

11.4.2　简单的字符设备驱动实验

驱动程序源代码如下：

```
/ * CharDriver.c * /
#define  _NO VERSION
#include <linux/module.h>
#include <linux/version.h>
char kernel_version[] = UTS_RELEASE;
#define  KERNEL
#include <linux/types.h>
#include <linux/fs.h>
#include <linux/mm.h>
#include <linux/errno.h>
#include <asm/segment.h>
#define  SUCCESS 0

static int device_read(struct file * file,char * buf,size_t count,loff_t * f_pos);
static int device_open(struct inode * inode,struct file * file);
static void device_release(struct inode * inode,struct file * file);

struct file_operations tdd_fops = {
read: device_read,
open: device_open,
release: device_release,
};
#define DEVICE_NAME "char_dev"
static int Device_Open = 0;
unsigned int test_major = 0;
//static char Messaege[1024];
```

```
static int device_open(struct inode * inode,struct file * file){
ifdef DEBUG
printk ("device_open( % p)\n",file);
# endif
if (Device_Open)
return - EBUSY;
Device_Open ++ ;
MOD_INC_USE_COUNT;
return SUCCESS;
}
static void device_release(struct inode * inode,struct file * file){
# ifdef DEBUG
printk ("device_release( % p, % p)\n",inode,file);
# endif
Device_Open -- ;
MOD_DEC_USE_COUNT;
}
static int device_read(struct file * file,char * buf,size_t count,loff_t * f_pos){
int left;
if (verify_area(VERIFY_WRITE,buf,count) == - EFAULT)
return - EFAULT;
for (left = count; left > 0; left -- ){
put_user(1,buf);
buf ++ ;
}
return count;
}
int init_module(void){
int result;
result = register_chrdev(0,"char_dev",&tdd_fops);
if (result < 0){
printk("char_dev:can't get major number\n");
return result;
}
if (test_major == 0)
test_major = result;
printk ("Hello,I'm in kenel mode\n");
return 0;
}
void cleanup_module(void){
printk("Hello,I'm goint to out\n");
unregister_chrdev(test_major,"char_dev");
```

```
}
```

用下述命令进行编译：

```
#gcc -O2 -DMODULE -D__KERNEL__ -c CharDriver.c
```

出现下述错误：

```
CharDriver.c 18: Tvariable 'fops' has initializer but incomplete type
CharDriver.c 19: unknown field 'read' specified in initializer
CharDriver.c 20: unknown field 'open' specified in initializer
CharDriver.c 20: unknown field 'release' specified in initializer
CharDriver.c 53: storage size of 'fops' isn't known
```

原因：在用gcc进行编译时，默认搜索的include文件路径为/usr/include，但/usr/include/linux/fs.h中没有定义file_operations结构体（可以打开该文件看一下），file_operations是在源代码目录树/usr/src/linux-2.4/include下linux/fs.h中定义的，而且用了#ifdef __KERNEL__条件编译。因此在编译时必须指定该目录和定义__KERNEL__。编译命令如下：

```
#gcc -O2 -DMODULE -D__KERNEL__ -I/usr/src/linux-2.4/include -c CharDriver.c
```

经过编译得到的文件CharDriver.o就是设备驱动程序，执行以下命令把它安装到系统中：

```
#insmod CharDriver.o
```

如果安装成功，用#vi /proc/devices打开文件将看到设备char_dev和主设备号。如果需要卸载，则可运行#rmmod char_dev。

然后就需要创建设备文件了，执行以下命令：mknod /dev/char_dev c 254 0，其中"/dev/char_dev"是要生成的设备名及目录，选项c是指字符设备，254是生成的主设备号。可用上面方法查看到，也可能是253或其他，从设备号设成0就可以。另外可以用rm -f /dev/char_dev删除mknod命令创建的设备文件。可用下述测试程序源代码进行测试。

驱动的测试程序如下：

```
/*DriverTest.c*/
#include <stdio.h>
#include <sys/types.h>
#include <sys/stat.h>
#include <fcntl.h>
int main(int argc,char *argv[]){
int testdev;
int i;
char buf[10];
```

```
testdev = open("/dev/char_dev",O_RDWR);
if (testdev == -1){
printf("Cann't open file \n");
exit(0);
}
read(testdev,buf,10);
for (i = 0; i < 10; i++){
printf("%d\n",buf[i]);
}
close(testdev);
}
```

需要注意的是，将驱动程序数据传给用户程序时是使用函数 put_user(1,buf)，而不是 put_user(1,buf,1)；否则会出现错误。还有 copy_to_user(buf,Buffer++,1)函数的 Buffer 是指针变量，而不是具体的数据，如果写成 copy_to_user(buf,1,1)，也会出错。

第12章

嵌入式 Linux 开发环境

12.1 概　述

12.1.1 Linux 开发环境概述

当有了在 PC 上 Linux 的各方面基础后，就可以开始基于嵌入式处理器 2410A、AM335x 等的 Linux 学习与开发了，而嵌入式的开发环境、工具也是最为重要及首先应该学习的。由于嵌入式系统本身资源的缺乏，Linux 通常采用宿主机＋目标机的交叉编译调试方式。宿主机就是安装有嵌入式 Linux 开发环境的 PC，利用它强大的功能资源可高效地完成目标系统的开发。嵌入式开发通常需要有交叉编译、调试等系列工具，目标板的编程下载工具，还需有目标处理器及平台的整个内核源码树等。有些读者往往想着要自己去组建、移植这所有的一切，但是考虑到我们只是应用工程师，也不可能是第一个应用该处理器平台的工程师，因此从零开始是没有必要的。通常原厂或第三方都会发布这些资源。由于我们选用的是 TI AM335x Starter Kit 硬件平台，相应的 TI 也提供了其 Linux 软件 SDK 开发包，该平台除了工具及内核源码等资源外，还包含了英文版的文档，本章的很多内容也都是从其中摘录和总结出来的，读者也可以自行去参考阅读。

12.1.2 TI 官方 AM335x Linux SDK 资源及参考文档

AM335x Sitara 处理器的 SDK 主页：http://www.ti.com.cn/tool/cn/processor-sdk-am335x。

AM335x Sitara 处理器 PROCESSOR-SDK-LINUX-AM335X 及相关资源软件下载主页：http://software-dl.ti.com/processor-sdk-linux/esd/AM335X/latest/index_FDS.html。

PROCESSOR-SDK-LINUX-AM335X 一些有帮助的维基：

- http://processors.wiki.ti.com/index.php/Processor_Linux_SDK_How_To_Guides。
- http://processors.wiki.ti.com/index.php/Sitara_Linux_SDK_Training。

➢ http://processors.wiki.ti.com/index.php/Processor_SDK_Linux_Release_Notes。

PROCESSOR-SDK-LINUX-AM335X 软件开发指南：http://processors.wiki.ti.com/index.php/Processor_SDK_Linux_Software_Developer's_Guide。

PROCESSOR-SDK-LINUX-AM335X 入门指南：http://processors.wiki.ti.com/index.php/Processor_SDK_Linux_Getting_Started_Guide。

PROCESSOR-SDK-LINUX-AM335X 安装指南：http://processors.wiki.ti.com/index.php/Processor_SDK_Linux_Installer。

12.1.3 PROCESSOR-SDK-LINUX-AM335x 概述

PROCESSOR-SDK-LINUX-AM335x 是 TI 基于 AM335x 嵌入式处理器的统一的 Linux 软件平台(即 Sitara Linux SDK，包括 Bootloader 引导程序、Linux 内核、文件系统和交叉编译工具等)，它设置简单，提供开箱即用的基准测试和演示，且完全免费。下面摘自官方主页上介绍的主要特点。

1. Linux 亮点

➢ 长期稳定（LTS）主线 Linux 内核支持；

➢ U-boot 引导加载程序支持；

➢ Linaro GNU 编译器集（GCC）工具链；

➢ 兼容 Yocto Project OE Core 的文件系统。

2. Linaro 工具链支持

Linaro 工具链包括强大的商用级工具，这些工具专为 Cortex-A 处理器进行过优化。此工具链得到了 TI 和整个 Linaro 社区的全力支持，包括来自 Linaro 内部工程师、成员公司开发者以及开源社区的其他人员的支持。Linaro 工具、软件和测试程序在最新版的 TI 处理器 SDK 中提供。

3. Yocto Project 支持

Yocto 项目是由 Linux 基金会设立的开源协作项目，旨在简化构建嵌入式 Linux 软件发行版的框架。TI 通过 Arago 发行方式提供。Arago 项目提供经验证和测试并受支持的软件包子集，并且是用免费和开放的工具链构建而成的。有关 Yocto 项目和 TI Arago 发行版的其他资源，请访问 arago-project.org。

12.2 PC 宿主机环境的创建

12.2.1 安装基本的软件开发工具

在 PC 上安装 Ubuntu 14.04 或其他版本的桌面 Linux 操作系统，当完成后基本的开发工具也就安装完成，主要有：binutils、gcc、gcc-c++、glibc-kernheaders、glibccommon、glibc、glibc-devel、patch、make、minicom 等。

如果在开发过程中，特别是编译过程中提示缺少某些工具或库等错误时，可以利用 apt-get install 安装。在 TI 的 Processor SDK Building The SDK 中要求安装的软件："$ sudo apt-get install git build-essential python diffstat texinfo gawk chrpath dos2unix wget unzip socat doxygen libc6:i386 libncurses5:i386 libstdc++6:i386 libz1:i386"，且需要配置"dash:"，即"$ sudo dpkg-reconfigure dash"。

实际上，在后续的 SDK 安装中有说明，SDK 软件开发所需的很多工具软件包会通过 setup.sh 自动安装完成，所以这里安装完 Ubuntu 14.04 LTS 后无需太担心将来需要安装哪些软件包。

12.2.2　下载安装 Sitara Linux SDK for AM335x

我们可以从 TI 的官网上下载最新的基于 AM335x Starter Kit 的免费的 Sitara Linux SDK 开发包。进入 TI 的 AM335x 资源页面：http://www.ti.com.cn/tool/cn/processor-sdk-am335x，或者 http://software-dl.ti.com/processor-sdk-linux/esd/AM335X/latest/index_FDS.html。

下载最新的 Linux 开发包：ti-processor-sdk-linux-am335x-evm-01.00.00.03-Linux-x86-Install.bin。

注：可能版本会有所不同，2016 年最新版本为 ti-processor-sdk-linux-am335x-evm-03.00.00.04-Linux-x86-Install.bin。

如图 12.1 所示为 2016 年最新版本的 PROCESSOR-SDK-AM335X。

图 12.1　PROCESSOR-SDK-AM335X

单击“获得软件”按钮可以进入SDK的下载页面，并找到SDK的安装文件，如图12.2所示。

图12.2　PROCESSOR-SDK-AM335X安装文件下载

需要注意：该版本为适应64位的PC主机和Linux操作系统，如果用户是基于32位的系统需下载32位的版本。

同时，在开发包下载页面里，还列出了AM335x Linux SDK Documentation，如图12.3所示。

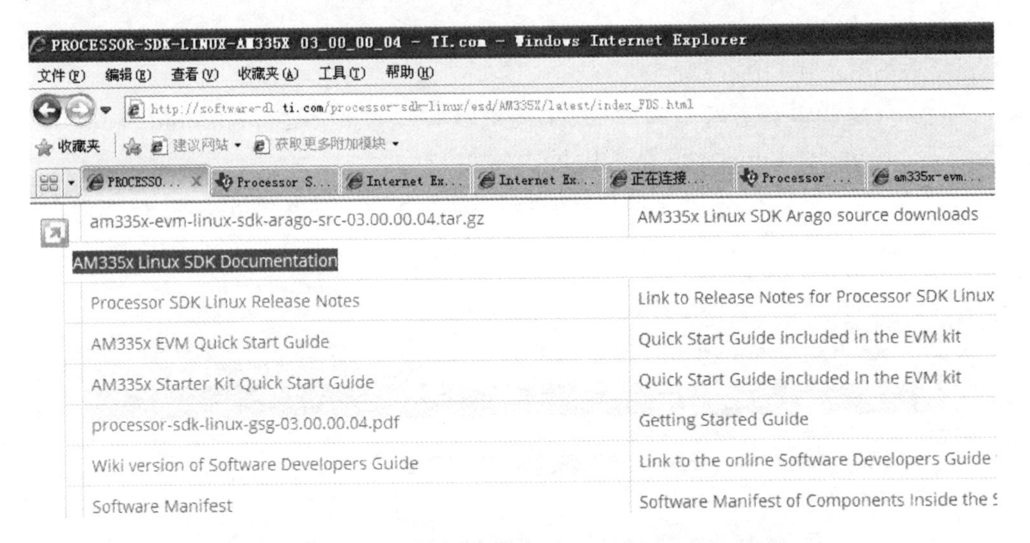

图12.3　AM335x Linux SDK Documentation

打开Processor SDK Linux Release Notes链接：http://processors.wiki.ti.com/index.php/Processor_SDK_Linux_Release_Notes，会得到很多有用的信息，如SDK包含哪些组件及对应的版本号等。如图12.4所示为Release_Notes中列出的Processor SDK Linux包含的主要内容。

如图12.5所示为01.00.00.03版本发布笔记中列出的各组件版本。

如图12.6所示为03.00.00.04版本发布笔记中列出的各组件版本。

另外，可以在Installation and Usage项找到Software Developer's Guide链接：

- Bootloaders & Filesystems
- SDK Installer
- Setup Scripts
- Makefiles
- Matrix Application Launcher
- Example Applications
- WLAN support (Wilink 8)
- Code Composer Studio v6

图 12.4 Processor SDK Linux 包含的主要内容

- Using 3.14.43 LTS based Kernel
- Using 2014.07 U-boot
- Validated SDK scripts on Ubuntu 14.04 LTS
- Support for AM437x-SK
- Upgraded GStreamer infrastructure from 0.10 to 1.2
- build-essentials (gcc, make, git, etc) added to target filesystem

图 12.5 01.00.00.03 SDK 包含的组件及版本

SDK Components & Versions

Component	Version
Linux Kernel	4.4.12+ (2016 LTS)
U-Boot	2016.05
Yocto Project	2.1 (Krogoth)
Linaro Toolchain (gcc)	5.3 2016.02 hard-float
Qt	5.6
OpenCL	1.1.9
OpenCV	3.1
Wayland	1.9

图 12.6 03.00.00.04 SDK 包含的组件及版本

http://processors.wiki.ti.com/index.php/Processor_SDK_Linux_Software_Developer's_Guide,进入其主页后可看到其列出的各项链接,如图 12.7 所示。

我们可以先阅读 Getting Started Guide,了解如何开始工作。

Processor SDK Linux	
Getting Started Guide <-- **Start Here**	How To Guides
Supported Platforms and Versions	Linux Software Stack
Directory Structure Overview	Building the SDK
Release Notes	Migration Guide
GPLv3 Disclaimer	MCSDK to Processor SDK Migration Guide
Technical Support	

Foundational Components (more information on each piece of the distribution)			
U-Boot	Boot Monitor	Kernel	Filesystem
Tools	OpenCL	OpenCV	Graphics & Display
Multimedia	Examples, Demos	PRU-ICSS	

图 12.7 Software Developer's Guide

接着，可参考安装指南：http://processors.wiki.ti.com/index.php/Processor_SDK_Linux_Installer。

在 PC 的 Linux 宿主机，启动 Ubuntu 操作系统，进入 SDK 所在目录，执行修改文件的可执行权限命令：

```
chmod +x ./ti-processor-sdk-linux-am335x-evm-01.00.00.03-Linux-x86-Install.bin
```

再执行安装命令：

```
./ti-processor-sdk-linux-am335x-evm-01.00.00.03-Linux-x86-Install.bin
```

注：如果是在 Ubuntu 的图形界面下，可以直接进入 SDK 目录，双击文件：ti-processor-sdk-linux-am335x-evm-01.00.00.03-Linux-x86-Install.bin。

另外，下述命令可以列出附加的帮助信息：

```
./ti-processor-sdk-linux-am335x-evm-01.00.00.03-Linux-x86-Install.bin --help
```

SDK 安装命令终端信息，如图 12.8 所示。

```
root@ccn:/home/ccn/Downloads# ls -lh
total 4.3G
-rw------- 1 ccn ccn 1.8G  7月 29 05:28 ti-processor-sdk-linux-am335x-evm-01.00.00.03-Linux-x86-Install.bin
-rw------- 1 ccn ccn 2.6G 10月  5 06:35 ti-processor-sdk-linux-am335x-evm-03.00.00.04-Linux-x86-Install.bin
root@ccn:/home/ccn/Downloads# chmod +x ./ti-processor-sdk-linux-am335x-evm-01.00.00.03-Linux-x86-Install.bin
root@ccn:/home/ccn/Downloads# ls -lh
total 4.3G
-rwx--x--x 1 ccn ccn 1.8G  7月 29 05:28 ti-processor-sdk-linux-am335x-evm-01.00.00.03-Linux-x86-Install.bin
-rw------- 1 ccn ccn 2.6G 10月  5 06:35 ti-processor-sdk-linux-am335x-evm-03.00.00.04-Linux-x86-Install.bin
root@ccn:/home/ccn/Downloads# ./ti-processor-sdk-linux-am335x-evm-01.00.00.03-Linux-x86-Install.bin
```

图 12.8 SDK 安装命令

执行安装命令后，会弹出一系列安装对话框，可以选择默认方式。如图 12.9 所示为选择默认的安装路径/home/sitara/ti-processor-sdk-linux-am335x-evm-01.00.00.03。

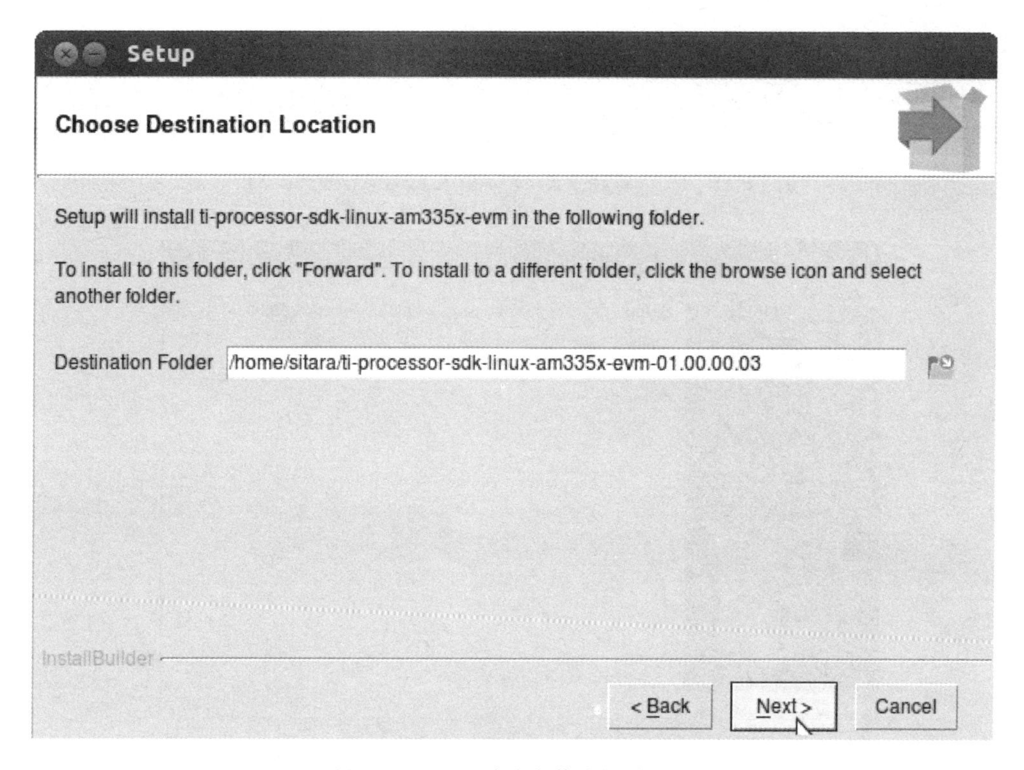

图 12.9　SDK 默认安装路径对话框

需要特别注意：ti-processor-sdk-linux-am335x-evm-01.00.00.03-Linux-x86-Install.bin 是基于 PC 主机的 32 位的 Linux 系统，如果在 64 位 Linux 系统下执行安装则会没有任何反应(如图 12.10 所示，Ubuntu 下可以使用 uname -a 查看到系统版本，i386 代表 32 位系统，x86_64 代表 64 位系统)，需要下载基于 64 位的安装文件，如：ti-processor-sdk-linux-am335x-evm-03.00.00.04-Linux-x86-Install.bin。

```
-rw------- 1 ccn ccn 1.8G  7月 29 05:28 ti-processor-sdk-linux-am335x-evm-01.00.00.03-Linux-x86-Install.bin
-rw------- 1 ccn ccn 2.6G 10月  5 06:35 ti-processor-sdk-linux-am335x-evm-03.00.00.04-Linux-x86-Install.bin
root@ccn:/home/ccn/Downloads# chmod +x ./ti-processor-sdk-linux-am335x-evm-01.00.00.03-Linux-x86-Install.bin
root@ccn:/home/ccn/Downloads# ls -lh
total 4.3G
-rwx--x--x 1 ccn ccn 1.8G  7月 29 05:28 ti-processor-sdk-linux-am335x-evm-01.00.00.03-Linux-x86-Install.bin
-rw------- 1 ccn ccn 2.6G 10月  5 06:35 ti-processor-sdk-linux-am335x-evm-03.00.00.04-Linux-x86-Install.bin
root@ccn:/home/ccn/Downloads# ./ti-processor-sdk-linux-am335x-evm-01.00.00.03-Linux-x86-Install.bin
root@ccn:/home/ccn/Downloads# 
```

图 12.10　64 位系统安装 32 位 SDK 的情况

64 位 SDK 下载完成后，执行安装命令，如图 12.11 所示。

执行安装命令后会弹出 Setup 对话框，如图 12.12 所示。

单击 Next 按钮，如图 12.13 所示。

```
root@ccn:/home/ccn/Downloads# chmod +x ti-processor-sdk-linux-am335x-evm-03.00.00.04-Linux-x86-Install.bin
root@ccn:/home/ccn/Downloads# ls -lh
total 4.3G
-rwx--x--x 1 ccn ccn 1.8G  7月 29 05:28 ti-processor-sdk-linux-am335x-evm-01.00.00.03-Linux-x86-Install.bin
-rwx--x--x 1 ccn ccn 2.6G 10月  5 06:35 ti-processor-sdk-linux-am335x-evm-03.00.00.04-Linux-x86-Install.bin
root@ccn:/home/ccn/Downloads# ./ti-processor-sdk-linux-am335x-evm-03.00.00.04-Linux-x86-Install.bin
```

图12.11 64位系统安装64位SDK版本03.00.00.04

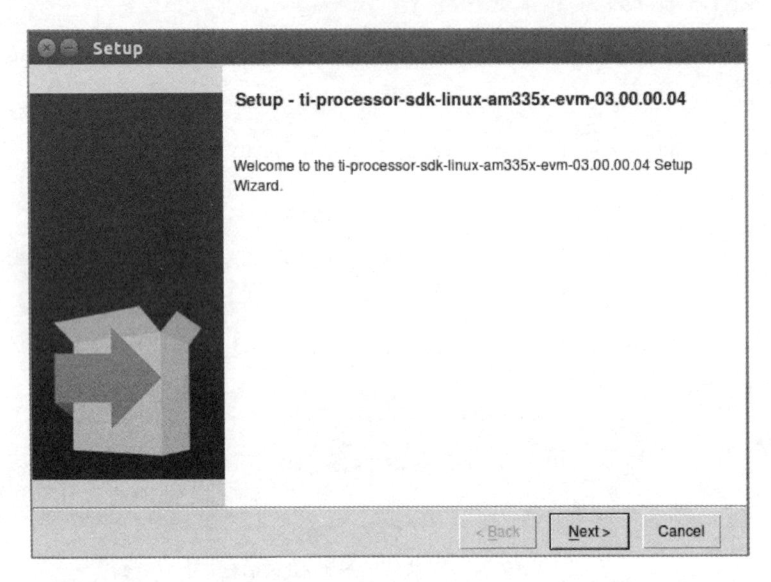

图12.12 64位SDK版本03.00.00.04 Setup对话框(一)

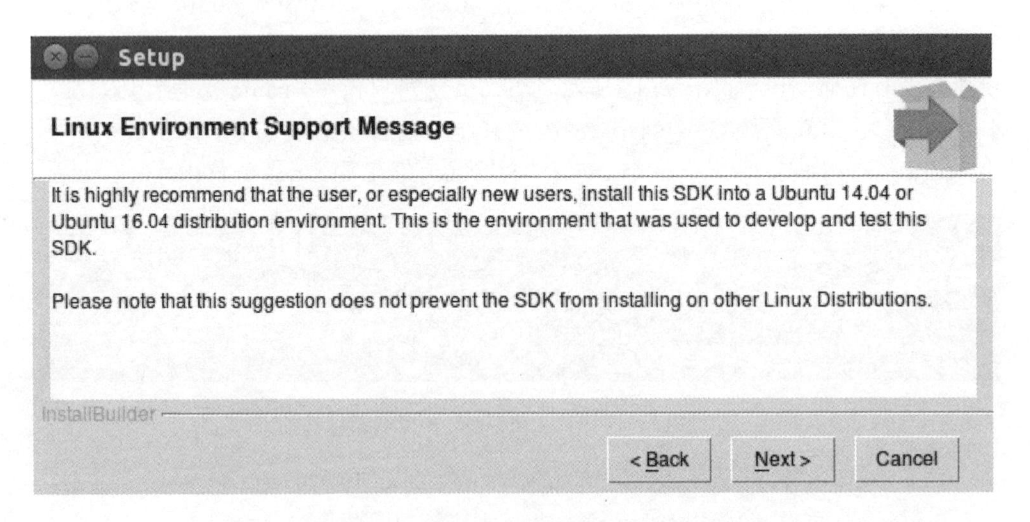

图12.13 64位SDK版本03.00.00.04 Setup对话框(二)

单击Next按钮,如图12.14所示。

单击Next按钮,选择默认的安装路径,如图12.15所示。

图 12.14　64 位 SDK 版本 03.00.00.04 Setup 对话框(三)

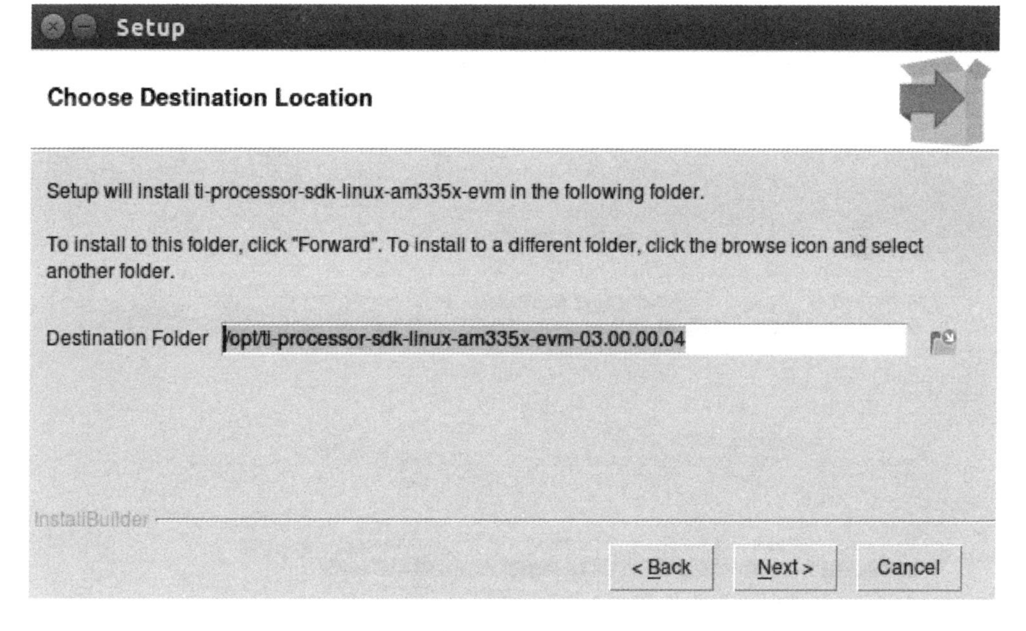

图 12.15　64 位 SDK 版本 03.00.00.04 Setup 对话框(四)

单击 Next 按钮,显示安装路径,如图 12.16 所示。

单击 Next 按钮,设置完毕将准备安装,如图 12.17 所示。

单击 Next 按钮,进入安装状态,如图 12.18 所示,需要等待相当长的时间。

安装完成后,如图 12.19 所示,单击 Finish 按钮完成。

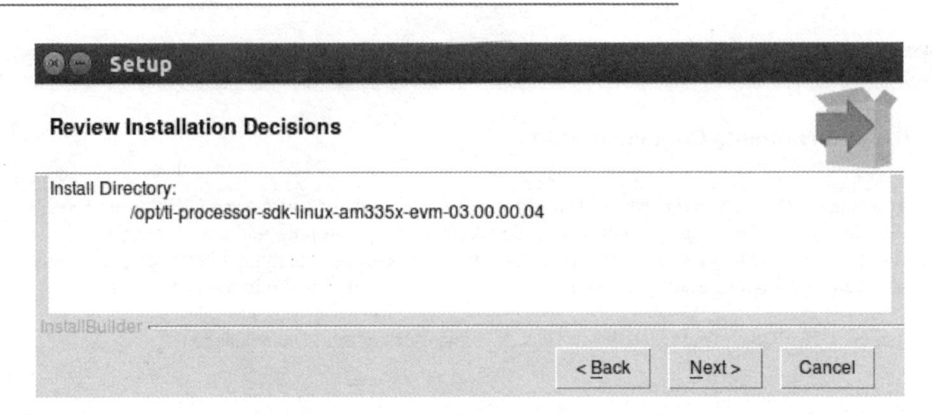

图 12.16　64 位 SDK 版本 03.00.00.04 Setup 对话框(五)

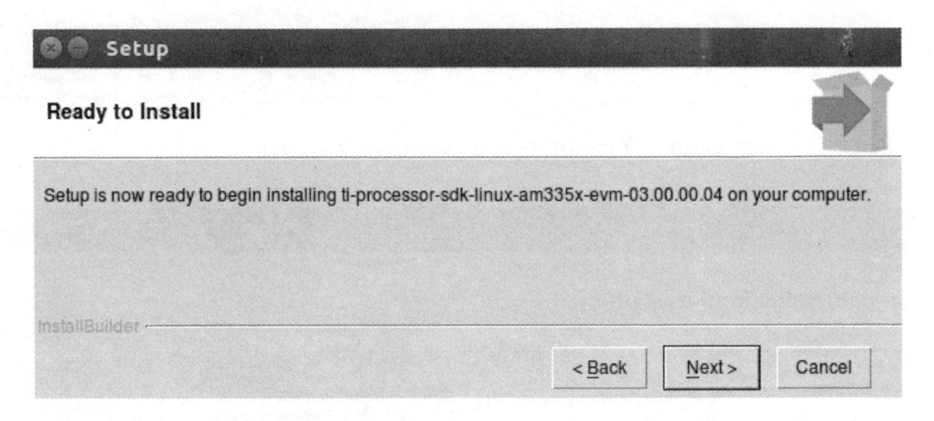

图 12.17　64 位 SDK 版本 03.00.00.04 Setup 对话框(六)

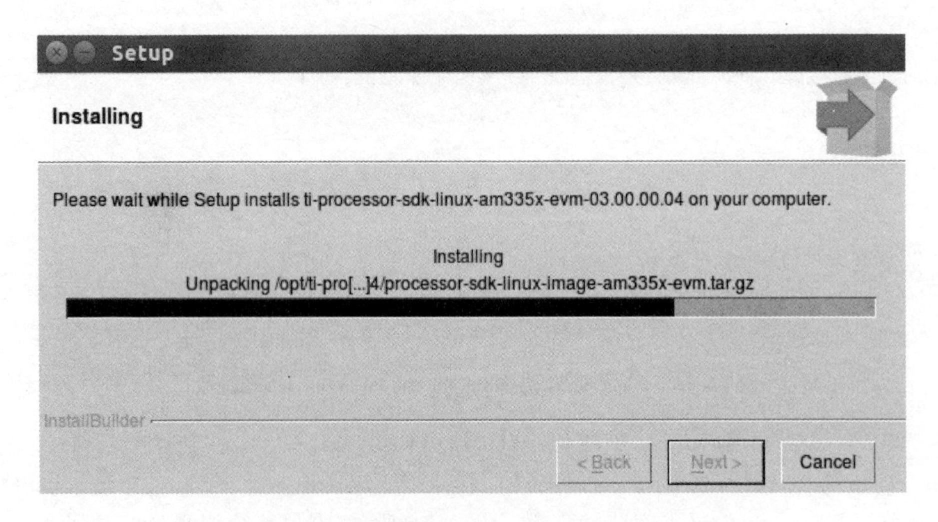

图 12.18　64 位 SDK 版本 03.00.00.04 Setup 对话框(七)

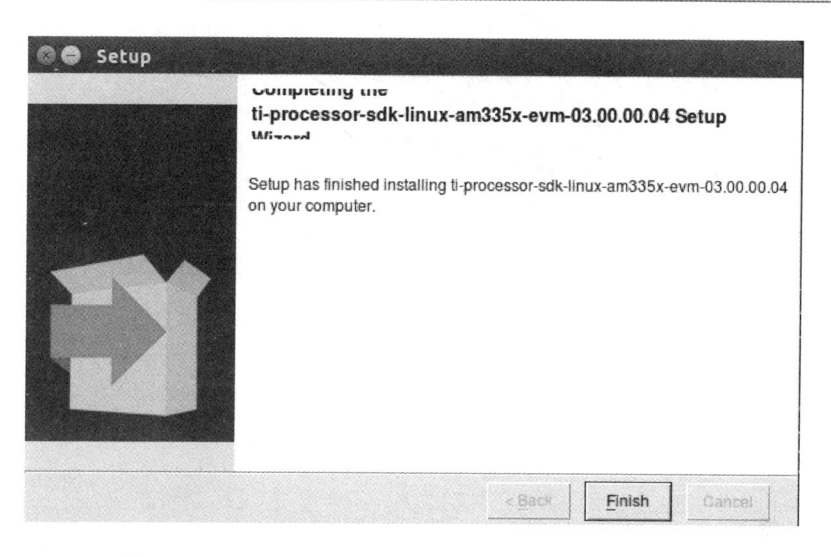

图 12.19　64 位 SDK 版本 03.00.00.04 Setup 对话框(八)

12.2.3　Sitara Linux SDK for AM335x 目录结构和软件架构

当 SDK 安装成功后，可以进入到默认的安装目录：/opt/ti-processor-sdk-linux-am335x-evm-03.00.00.04，如图 12.20 所示为目录下包含的文件及文件夹。

```
root@ccn:/home/ccn/Downloads# cd /opt/ti-processor-sdk-linux-am335x-evm-03.00.00.04/;ls -lh
total 52K
drwxr-xr-x  2 1001 1001 4.0K 10月  5 07:35 bin
drwxr-xr-x  6 1001 1001 4.0K  7月  9 09:13 board-support
drwxr-xr-x  3 1001 1001 4.0K  7月  9 09:14 docs
drwxrwxr-x 11 1001 1001 4.0K  7月  9 09:12 example-applications
drwxr-xr-x  2 1001 1001 4.0K  7月  9 09:16 filesystem
drwxr-xr-x  3 root root 4.0K 10月  5 07:35 linux-devkit
-rwxr-xr-x  1 1001 1001  16K  7月  9 07:28 Makefile
-rwxr-xr-x  1 1001 1001 1.3K 10月  5 07:35 Rules.make
-rwxr-xr-x  1 1001 1001 4.1K  7月  9 07:28 setup.sh
root@ccn:/opt/ti-processor-sdk-linux-am335x-evm-03.00.00.04#
```

图 12.20　64 位 SDK 版本 03.00.00.04 Setup 安装后的目录结构

关于 SDK 目录结构的说明，也有 TI 官方维基可以参考：http://processors.wiki.ti.com/index.php/Processor_SDK_Linux_Directory_Structure。

如图 12.21 所示为目录顶层结构。

bin：包含一些有用的 sh 脚本，用于自动配置主机和目标器件，其大部分都是通过由 SDK 顶层目录下的 setup.sh 调用执行的。

board-support：包含需要移植到定制平台时所需要修改的 SDK 的组件，包括 boot loaders 引导程序、Linux kernel 内核和 tree drivers 驱动设备树等的源代码目录。

docs：包含各种 SDK 文档，如软件清单和附加的软件工具使用指南。

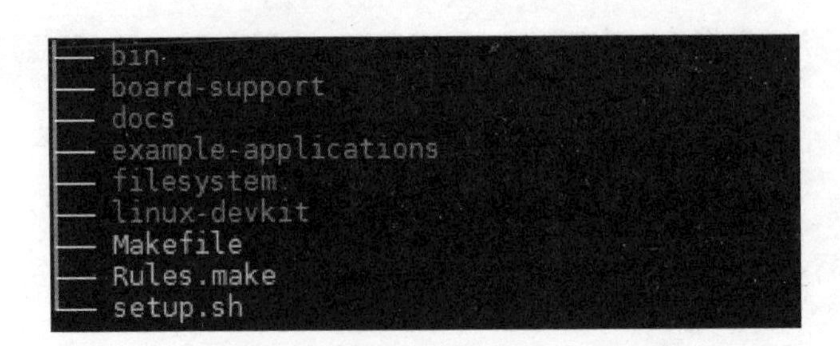

图 12.21　The Processor SDK for Linux 顶层目录和文件

example-applications：包含 TI 提供的可以作为评估的一些应用例子源码。

filesystem：包含可参考的文件系统。它们当中，有最小的只有基本功能的文件系统，也有 SDK 全功能特性的文件系统。

linux-devkit：包含交叉编译工具链和库，用于加速目标器件的开发。

Makefile：SDK 顶层的 Makefile 文件，用于配置编译 SDK 的各个组件。

Rules.make：为顶层 Makefile，以及各子目录的 Makefiles 提供各种设置的默认参数值。

setup.sh：配置安装脚本。

SDK 安装根目录下各文件夹中包含的文件，如图 12.22 所示。

```
root@ccn:/home/ccn/Downloads# cd /opt/ti-processor-sdk-linux-am335x-evm-03.00.00.04/;ls -lh
total 52K
drwxr-xr-x  2 1001 1001 4.0K 10月  5 07:35 bin
drwxr-xr-x  6 1001 1001 4.0K  7月  9 09:13 board-support
drwxr-xr-x  3 1001 1001 4.0K  7月  9 09:14 docs
drwxrwxr-x 11 1001 1001 4.0K  7月  9 09:12 example-applications
drwxr-xr-x  2 1001 1001 4.0K  7月  9 09:16 filesystem
drwxr-xr-x  3 root root 4.0K 10月  5 07:35 linux-devkit
-rwxr-xr-x  1 1001 1001  16K  7月  9 07:28 Makefile
-rwxr-xr-x  1 1001 1001 1.3K 10月  5 07:35 Rules.make
-rwxr-xr-x  1 1001 1001 4.1K  7月  9 07:28 setup.sh
root@ccn:/opt/ti-processor-sdk-linux-am335x-evm-03.00.00.04# ls bin/
add-to-group.sh  create-sdcard.sh  setup-host-check.sh  setup-package-install.sh  setup-tftp.sh       un
common.sh        create-ubifs.sh   setup-minicom.sh     setup-targetfs-nfs.sh     setup-uboot-env.sh
root@ccn:/opt/ti-processor-sdk-linux-am335x-evm-03.00.00.04# ls board-support/
extra-drivers  linux-4.4.12+gitAUTOINC+3639bea54a-g3639bea54a  prebuilt-images  u-boot-2016.05+gitAUTOIN
root@ccn:/opt/ti-processor-sdk-linux-am335x-evm-03.00.00.04# ls docs/
licenses                     Processor_SDK_Linux_Getting_Started_Guide.txt  software_manifest.htm
Multimedia_Data_Files_Manifest.doc  processor-sdk-linux-gsg-03.00.00.04.pdf        software_manifest.txt
root@ccn:/opt/ti-processor-sdk-linux-am335x-evm-03.00.00.04# ls example-applications/
am-sysinfo-git      matrix-gui-2.0          omapconf-1.72      pru-icss-4.0.2       ti-crypto-examples-git
arm-benchmarks-1.3  matrix-gui-browser-2.0  oprofile-example  refresh-screen-2.0
root@ccn:/opt/ti-processor-sdk-linux-am335x-evm-03.00.00.04# ls filesystem/
arago-base-tisdk-image-am335x-evm.tar.gz  tisdk-rootfs-image-am335x-evm.tar.gz
arago-base-tisdk-image-am335x-evm.ubi     tisdk-rootfs-image-am335x-evm.ubi
root@ccn:/opt/ti-processor-sdk-linux-am335x-evm-03.00.00.04# ls linux-devkit/
environment-setup  site-config-cortexa8hf-neon-linux-gnueabi  sysroots  version-cortexa8hf-neon-linux-gn
root@ccn:/opt/ti-processor-sdk-linux-am335x-evm-03.00.00.04#
```

图 12.22　SDK 根目录下各文件夹的内容

SDK 整体的软件多层架构，可以参考 TI 的官方维基：http://processors.wiki.ti.com/index.php/Processor_SDK_Linux_Software_Stack。

如图 12.23 所示为 SDK 软件多层架构，从图中我们可以看到 SDK 在各个层中都提供了哪些应用。

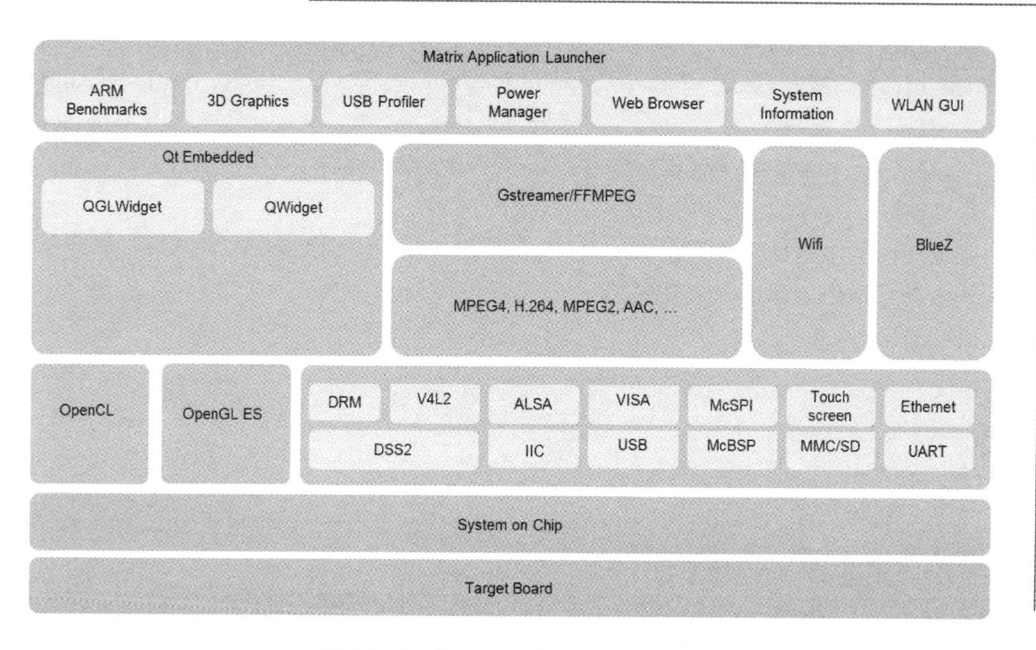

图 12.23 Processor SDK Linux stack

位于最顶层的 Matrix Application Launcher 为基于 GUI 图形界面的应用级别的软件，提供了如 3D Graphics（3D 动画图像加速处理）、Power Manager（电源管理）、Web Browser（WEB 网页服务器）等评估软件。

第二层有嵌入式 QT 图形库，这说明 Matrix Application Launcher 就是基于 QT 而开发的软件。另外还有非常主流的，用于构建流媒体应用的开源多媒体框架（framework）：Gstreamer 和 FFMPEG，它们提供了非常强大的音/视频编解码，可以大大减少音/视频应用程序的开发，顶层的音视频多媒体播放软件就是基于 Gstreamer 或 FFMPEG 库而开发的。而 Gstreamer 和 FFMPEG 又是基于 MPEG4、H.264、MPEG2、AAC 等视频/音频编解码库的。第二层还提供了 Wi-Fi 和 BlueZ 等应用库，所以在开发实现 Wi-Fi 和蓝牙通信互联方面的应用也就非常简单了。

第三层就是如 OpenGL、IIC、USB 等常用的处理器外设接口，笔者认为应该就是指 Linux 下的外设驱动了。

第四层简单地显示为 System on Chip（片上系统），指某个处理器，如 AM335x 系列，或者是 TI 的其他处理器。

第五层 Target Board 指特定的目标板，有 TI 提供支持的官方 EVM、Starter Kit 或 BeagleBone，可由用户自己基于上述平台修改定义的特定的用户产品。

12.2.4 Sitara Linux SDK for AM335x 环境配置

在 SDK 的安装根目录下，我们会发现有一个 setup.sh 脚本，它会调用 SDK 安装根目录下 bin 文件夹中的各个 sh 脚本执行，它就是用于配置 SDK 软件开发所需要的 Linux 主机下的各种环境，详细介绍可参考 TI 官方维基：

- ➢ http://processors.wiki.ti.com/index.php/Processor_SDK_Linux_Setup_Script。
- ➢ http://processors.wiki.ti.com/index.php/Processor_SDK_Linux_Getting_Started_Guide。

上述两个维基中都强调了在安装完 SDK 后，应先运行该 setup.sh 脚本。

由于脚本中的很多配置需要管理员权限，所以我们使用 root 用户登录 Ubuntu 主机(或使用 sudo)去执行 setup.sh 脚本，执行命令如下：

```
#./setup.sh
```

或者

```
$ sudo ./setup.sh
```


如图 12.24 所示为 SDK 主机环境设置。

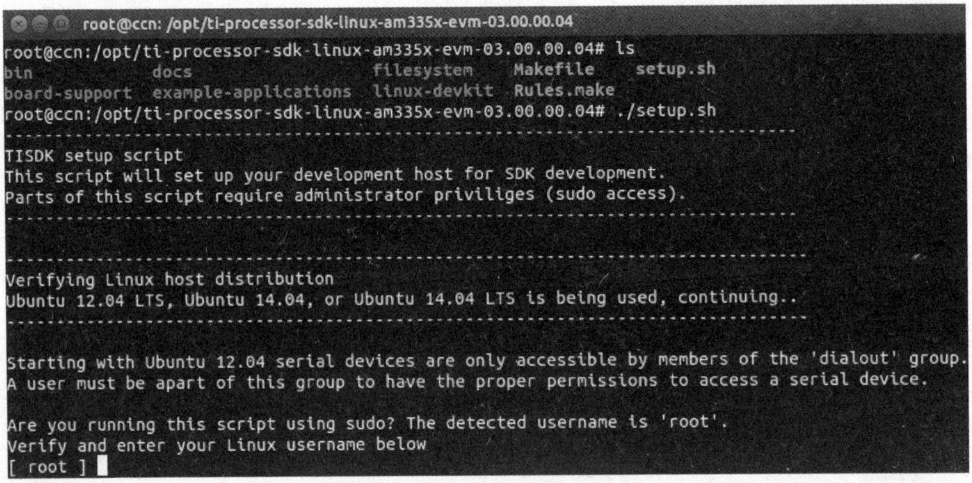

```
root@ccn: /opt/ti-processor-sdk-linux-am335x-evm-03.00.00.04
root@ccn:/opt/ti-processor-sdk-linux-am335x-evm-03.00.00.04# ls
bin            docs                  filesystem    Makefile    setup.sh
board-support  example-applications  linux-devkit  Rules.make
root@ccn:/opt/ti-processor-sdk-linux-am335x-evm-03.00.00.04# ./setup.sh
--------------------------------------------------------------------------------
TISDK setup script
This script will set up your development host for SDK development.
Parts of this script require administrator priviliges (sudo access).
--------------------------------------------------------------------------------

--------------------------------------------------------------------------------
Verifying Linux host distribution
Ubuntu 12.04 LTS, Ubuntu 14.04, or Ubuntu 14.04 LTS is being used, continuing..
--------------------------------------------------------------------------------

Starting with Ubuntu 12.04 serial devices are only accessible by members of the 'dialout' group.
A user must be apart of this group to have the proper permissions to access a serial device.

Are you running this script using sudo? The detected username is 'root'.
Verify and enter your Linux username below
[ root ]
```

图 12.24 SDK 主机环境设置

除了参考 TI 官方维基外，当然也可以直接用文本或 vi 编辑器打开 setup.sh 或其他各 sh 文件进行参考，详细了解它们做了哪些操作。从 setup.sh 中可以看到，它先后调用了 bin 文件夹下的下述脚本：

- ➢ common.sh；
- ➢ setup-host-check.sh；
- ➢ add-to-group.sh；
- ➢ setup-package-install.sh；
- ➢ setup-targetfs-nfs.sh；
- ➢ setup-tftp.sh；

- ➢ setup-minicom. sh；
- ➢ setup-uboot-env. sh。

因此，bin 文件夹下还没有被调用的只剩下：

- ➢ create-sdcard. sh；
- ➢ create-ubifs. sh；
- ➢ unshallow-repositories. sh。

参考脚本调用及维基的解释，我们知道执行 setup. sh 后做了如下操作：

① 校验所运行的 Linux 主机是否是本 SDK 推荐的 Ubuntu LTS 版本(Ubuntu 12.04 及以上)，从图 12.24 也能看出来。

② 下载安装 SDK 软件开发要求主机所具备的工具、软件包如下：

- ➢ telnet 远程登录工具；
- ➢ menuconfig 内核配置的工具；
- ➢ NFS 网络文件系统挂载工具；
- ➢ TFTP 网络上传工具；
- ➢ minicom 串口调试终端；
- ➢ U-boot 编译工具及重新编译 U-boot。

如执行下述命令：

```
# apt - get install xinetd tftpd nfs - kernel - server minicom build - essential libncurses5 - dev uboot - mkimage autoconf automake
```

注：实际可能会因主机原来是否已经安装某些工具软件包而有所不同。

如图 12.25 所示为工具包安装。

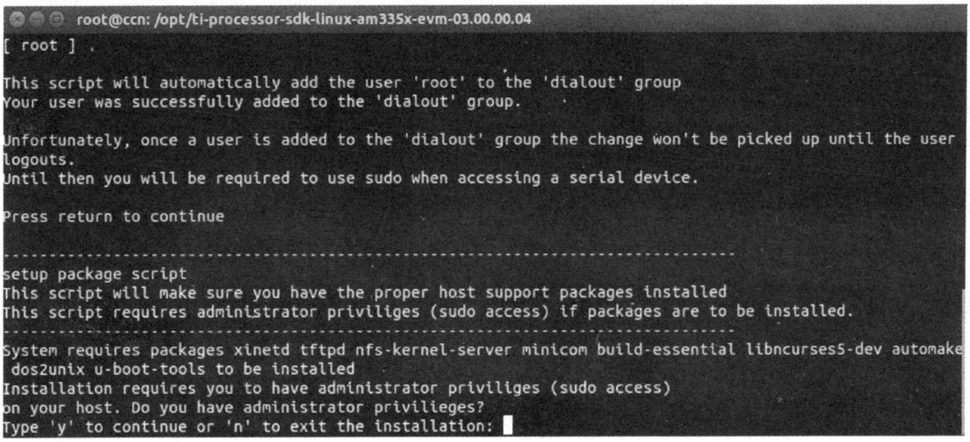

```
root@ccn: /opt/ti-processor-sdk-linux-am335x-evm-03.00.00.04
[ root ] .

This script will automatically add the user 'root' to the 'dialout' group
Your user was successfully added to the 'dialout' group.

Unfortunately, once a user is added to the 'dialout' group the change won't be picked up until the user logouts.
Until then you will be required to use sudo when accessing a serial device.

Press return to continue

--------------------------------------------------------------------------------
setup package script
This script will make sure you have the proper host support packages installed
This script requires administrator priviliges (sudo access) if packages are to be installed.
--------------------------------------------------------------------------------
System requires packages xinetd tftpd nfs-kernel-server minicom build-essential libncurses5-dev automake
 dos2unix u-boot-tools to be installed
Installation requires you to have administrator priviliges (sudo access)
on your host. Do you have administrator privilieges?
Type 'y' to continue or 'n' to exit the installation:
```

图 12.25　工具包安装

③ Target FileSystem installation。

在默认的路径下安装目标文件系统：/opt/ti-processor-sdk-linux-am335x-evm-

03.00.00.04/targetNFS，如图 12.26 所示。

将 SDK 根目录下的 filesystem/tisdk-rootfs-image-am335x-evm.tar.gz 文件系统解压到网络文件系统 targetNFS 目录下。

```
root@ccn: /opt/ti-processor-sdk-linux-am335x-evm-03.00.00.04
This step will extract the target filesystem to /opt/ti-processor-sdk-linux-am335x-evm-03.00.00.04/targe
tNFS

Note! This command requires you to have administrator priviliges (sudo access)
on your host.
Press return to continue

Successfully extracted tisdk-rootfs-image-am335x-evm.tar.gz to /opt/ti-processor-sdk-linux-am335x-evm-03
.00.00.04/targetNFS
--------------------------------------------------------------------------------

--------------------------------------------------------------------------------
This step will set up the SDK to install binaries in to:
    /opt/ti-processor-sdk-linux-am335x-evm-03.00.00.04/targetNFS/home/root/am335x-evm

The files will be available from /home/root/am335x-evm on the target.

This setting can be changed later by editing Rules.make and changing the
EXEC_DIR or DESTDIR variable (depending on your SDK).

Press return to continue
```


图 12.26　目标文件系统安装

④ NFS 配置：/etc/exports。

⑤ TFTP 配置。

默认的 TFTP 根目录为/tftpboot，创建及配置：/etc/xinetd.d/tftp，如图 12.27 所示。

图 12.27　TFTP 配置

⑥ Minicom 配置。

默认选择 USB 转串口驱动设备：/dev/ttyUSB0，保存到配置文件：home/user/.minirc.dfl，如图 12.28 所示。

⑦ U-boot 设置。

```
root@ccn: /opt/ti-processor-sdk-linux-am335x-evm-03.00.00.04
--------------------------------------------------------------------------------

--------------------------------------------------------------------------------"
This step will set up minicom (serial communication application) for
SDK development


For boards that contain a USB-to-Serial converter on the board such as:
        * BeagleBone
        * Beaglebone Black
        * AM335x EVM-SK
        * AM57xx EVM
        * K2H, K2L, and K2E EVMs

the port used for minicom will be automatically detected. By default Ubuntu
will not recognize this device. Setup will add a udev rule to
/etc/udev/ so that from now on it will be recognized as soon as the board is
plugged in.

For other boards, the serial will defualt to /dev/ttyS0. Please update based
on your setup.

--------------------------------------------------------------------------------


NOTE: If your using any of the above boards simply hit enter
and the correct port will be determined automatically at a
later step.  For all other boards select the serial port
that the board is connected to.
Which serial port do you want to use with minicom?
[ /dev/ttyS0 ]
```

图 12.28　Minicom 配置

创建引导所用目标板必需的 U-boot 命令，如检测、显示主机所用的 IP 地址，从哪个媒介（如 TFTP 或 SD 卡）挂载内核和文件系统等，如图 12.29 所示。

```
root@ccn: /opt/ti-processor-sdk-linux-am335x-evm-03.00.00.04
--------------------------------------------------------------------------------

--------------------------------------------------------------------------------
This step will set up the U-Boot variables for booting the EVM.

Autodetected the following ip address of your host, correct it if necessary
[ 192.168.1.101 ]

Select Linux kernel location:
 1: TFTP
 2: SD card

[ 1 ] 2

Select root file system location:
 1: NFS
 2: SD card

[ 1 ] 2
--------------------------------------------------------------------------------
Would you like to create a minicom script with the above parameters (y/n)?
[ y ]
```

图 12.29　U-boot 配置

⑧ U-boot 脚本装载。

创建一个 Minicom 脚本或 uEnv. txt 脚本，用于放置在 SD 卡的/boot 分区。

最后，我们可以从 TI 官方维基 Processor SDK Linux Setup Script 中看到如图 12.30 所示的整个开发环境的结构示意图。

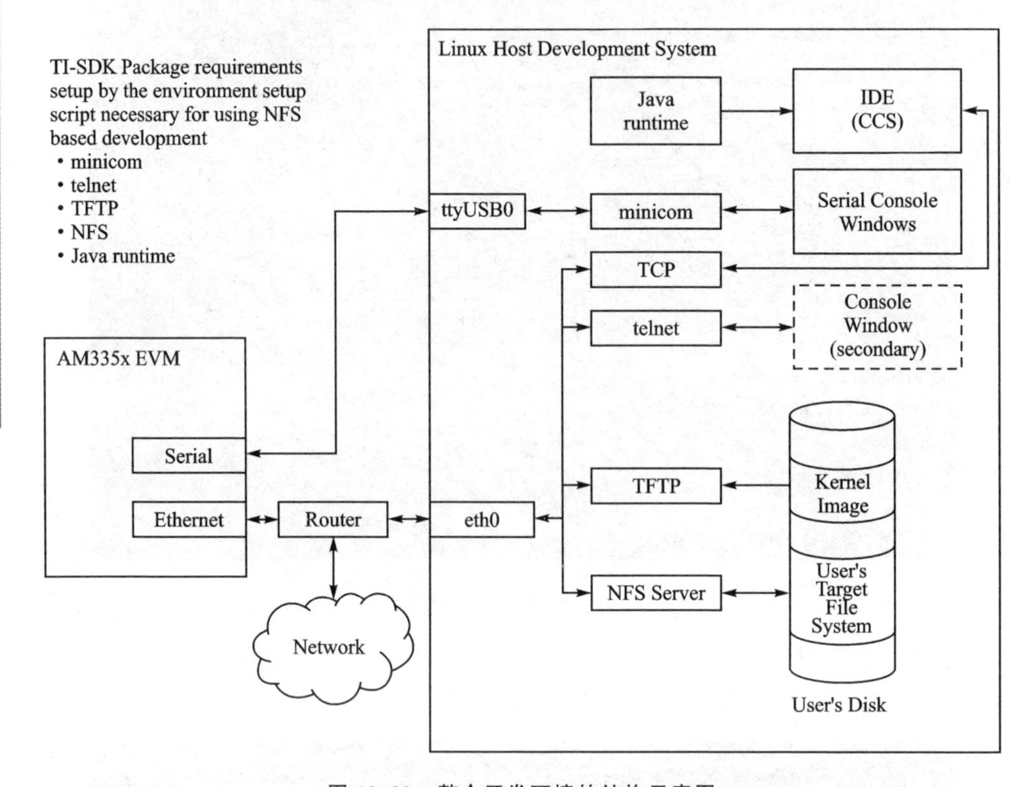

图 12.30　整个开发环境的结构示意图

图 12.30 是从软件、工具的角度概括，如果从具体的 PC 宿主机和目标开发板等实物概括，整个开发环境如图 12.31 所示，其中目标开发板以 BeagleBone Black 展示，用户可以采用 AM335x Starter Kit 或其他开发板。

12.2.5　交叉编译工具链的安装与配置

按照 *Processor SDK Linux Software Developer's Guide* 中 Building the SDK 的链接 Processor SDK Building The SDK，http://processors.wiki.ti.com/index.php/Processor_SDK_Building_The_SDK 中的说明似乎需要下载、安装 AM335x 的交叉编译工具链 Linaro，如图 12.32 所示。

```
$ wget https://releases.linaro.org/components/toolchain/binaries/5.3-2016.02/arm-linux-gnueabihf/gcc-linaro-5.3-2016.02-x86_64_arm-linux-gnueabihf.tar.xz
$ tar -Jxvf gcc-linaro-5.3-2016.02-x86_64_arm-linux-gnueabihf.tar.xz -C $HOME
```

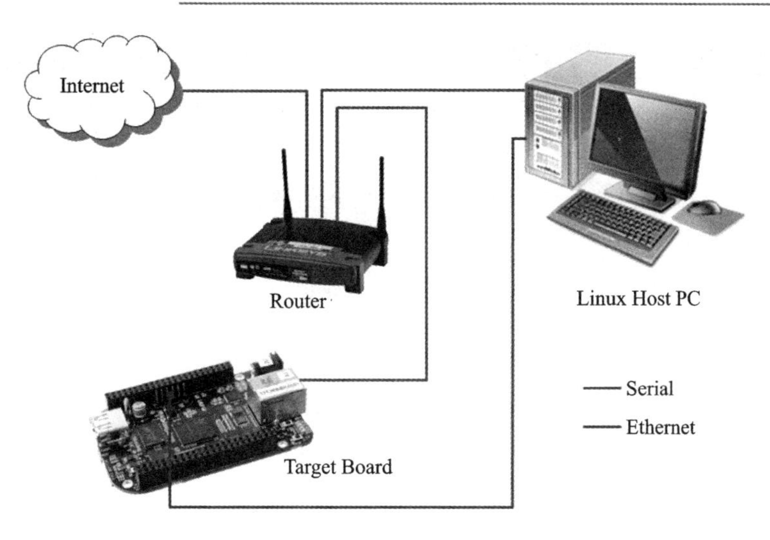

图 12.31　PC 宿主机和目标开发板

Cross-compile Toolchain

Run the following commands to install the Linaro Toolchain.

```
$ wget https://releases.linaro.org/components/toolchain/binaries/5.3-2016.02/arm-linux-gnueabihf/gc
$ tar -Jxvf gcc-linaro-5.3-2016.02-x86_64_arm-linux-gnueabihf.tar.xz -C $HOME
```

图 12.32　Building The SDK

但实际上，正如上一小节中提到的，在 SDK 的安装目录的顶层——linux-devkit 目录下，已经包含了交叉编译工具链：ti-processor-sdk-linux-am335x-evm-03.00.00.04\linux-devkit\sysroots\x86_64-arago-linux，且在 SDK 安装根目录下的 Rules.make 中查看，也明确指明了交叉编译工具链的路径，如图 12.33 所示。

因此，我们安装完 ti-processor-sdk-linux-am335x-evm-03.00.00.04-Linux-x86-Install.bin SDK 后，无需再手动安装交叉编译工具链。但为编译方便，还是需要在环境变量中添加其交叉编译工具链的路径：

```
vi /etc/profile
```

在打开的文件末尾添加：

```
export PATH = "/opt/ti - processor - sdk - linux - am335x - evm - 03.00.00.04/linux - devkit/sysroots/x86_64 - arago - linux/usr/bin: $ PATH"
export ARCH = arm
export CROSS_COMPILE = arm - linux - gnueabi -
```

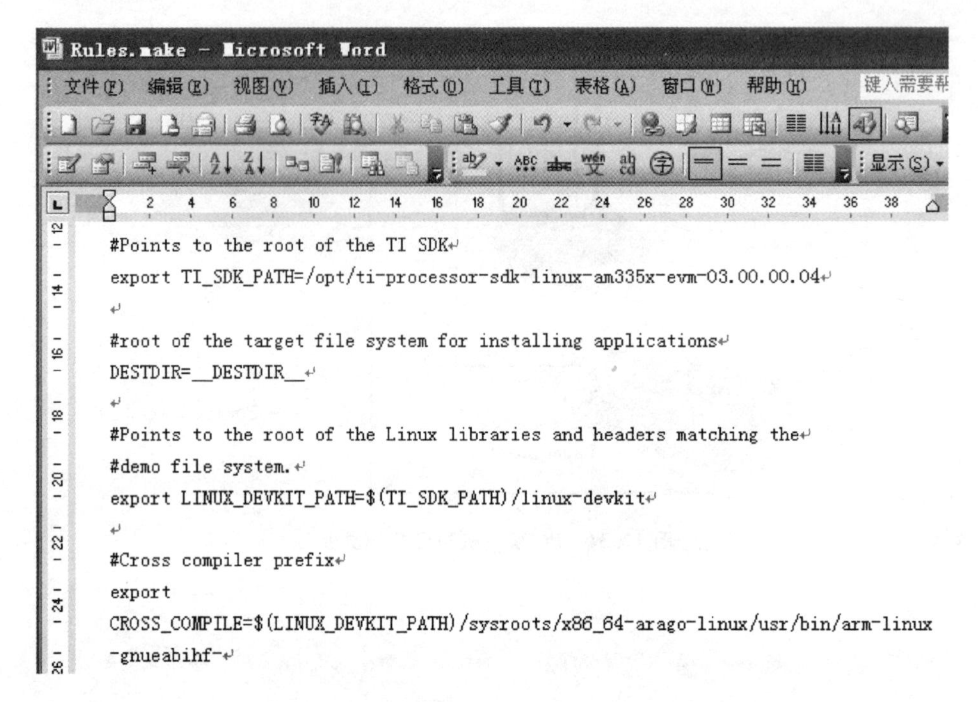

图12.33　Rules.make——CROSS_COMPILE

退出profile文件后执行下述命令，不用重启计算机使环境变量生效。

```
source /etc/profile
```

实际上，设置环境变量的方法很多，除了在/etc/profile文件内设置外，还可以在/etc/environment中设置，或者在～.bashrc、～.profile（如果是ccn用户，即为/home/ccn/.profile；如果是root用户，即为/root/.profile）等文件中设置。

经测试，在/etc/profile文件中添加环境变量后，启动后默认用户的环境变量生效，命令行中输入arm，再连续按两次Tab键后会出现安装好的arm-linux-gnueabihf-gcc等众多的编译工具。但在root用户下环境变量设置没有生效，每次进入root用户后都需要执行一下source /etc/profile才能生效，因此尝试在/etc/environment中添加，如图12.34所示。

图12.34　environment文件内容

12.3　嵌入式 Linux 系统的配置和编译

嵌入式 Linux 软件系统由 Bootloader、kernel、root filesysytem 和可选的 user filesystem 构成，如下：

Bootloader(一次固化)+内核(多次更新)+根文件系统(多次更新)[+用户文件系统]

Sitara Linux SDK for AM335x 中 Bootloader 使用的是 U-boot。

12.3.1　SDK 根目录下编译 U-boot 和 Linux 内核

参考 *Processor SDK Linux Getting Started Guide*(http://processors.wiki.ti.com/index.php/Processor_SDK_Linux_Getting_Started_Guide)的最后，如图 12.35 所示。

7. **Rebuild sources** using the top-level makefile in the SDK root directory. For example:
   - `make all` rebuilds all components in the SDK
   - `make linux` configures and builds the kernel
   - `make u-boot-spl` builds u-boot and u-boot-spl

   The file system can be rebuilt following these instructions.

图 12.35　Rebuild sources

也可以直接参考 SDK 根目录的顶层 makefile 文件，make all 重新编译 SDK 的所有组件，make linux 配置和编译 Linux 内核，make u-boot-spl 编译 U-boot 的第一阶段 u-boot-spl 和 U-boot 的后半断，通常就是指 U-boot，如图 12.36 和图 12.37 所示。

12.3.2　Bootloader 的配置和编译

PROCESSOR-SDK-AM335x 使用的 Bootloader 是目前使用最为广泛的跨平台的 U-boot，源代码包位于 SDK 安装的根目录下的 board-support/。

正如前面所述，可以直接在 SDK 根目录下使用 make u-boot-spl 编译，也可以进入到 U-boot 源代码目录下进行配置和编译。

进入 U-boot 源代码目录后，通常可以像 S3C2410 使用的 vivi 那样执行配置命令

```
#make menuconfig
```

以进入图形化配置界面，如图 12.38 所示。

但 AM335x 的 U-boot 通常不采用这种方法，详细的配置、编译，可参考 TI 官方维基关于 U-boot 的使用手册，如下：

```
root@ccn: /opt/ti-processor-sdk-linux-am335x-evm-03.00.00.04
-include Rules.make

MAKE_JOBS ?= 1

all: linux matrix-gui arm-benchmarks am-sysinfo oprofile-example matrix-gui-browser
refresh-screen u-boot-spl ti-crypto-examples linux-dtbs cryptodev ti-sgx-ddk-km pru-
icss
clean: linux_clean matrix-gui_clean arm-benchmarks_clean am-sysinfo_clean oprofile-e
xample_clean matrix-gui-browser_clean refresh-screen_clean u-boot-spl_clean ti-crypt
o-examples_clean linux-dtbs_clean cryptodev_clean ti-sgx-ddk-km_clean pru-icss_clean

install: linux_install matrix-gui_install arm-benchmarks_install am-sysinfo_install
oprofile-example_install matrix-gui-browser_install refresh-screen_install u-boot-sp
l_install ti-crypto-examples_install linux-dtbs_install cryptodev_install ti-sgx-ddk
-km_install pru-icss_install
# Kernel build targets
linux: linux-dtbs
        @echo =================================
        @echo     Building the Linux Kernel
        @echo =================================
        $(MAKE) -C $(LINUXKERNEL_INSTALL_DIR) ARCH=arm CROSS_COMPILE=$(CROSS_COMPILE
) $(DEFCONFIG)
        $(MAKE) -j $(MAKE_JOBS) -C $(LINUXKERNEL_INSTALL_DIR) ARCH=arm CROSS_COMPILE
=$(CROSS_COMPILE) zImage
        $(MAKE) -j $(MAKE_JOBS) -C $(LINUXKERNEL_INSTALL_DIR) ARCH=arm CROSS_COMPILE
=$(CROSS_COMPILE) modules

linux_install: linux-dtbs_install
        @echo ===================================
        @echo     Installing the Linux Kernel
        @echo ===================================
        @if [ ! -d $(DESTDIR) ] ; then \
                echo "The extracted target filesystem directory doesn't exist."; \
                echo "Please run setup.sh in the SDK's root directory and then try a
gain."; \
                exit 1; \
        fi
        install -d $(DESTDIR)/boot
"Makefile" 367L, 15400C                                       1,1           Top
```

图 12.36　Make Linux

- ➢ http://processors.wiki.ti.com/index.php/Processor_SDK_Linux_U-Boot。
- ➢ http://processors.wiki.ti.com/index.php/Processor_SDK_Linux_U-Boot_Release_Notes。
- ➢ http://processors.wiki.ti.com/index.php/Linux_Core_U-Boot_User%27s_Guide。

由 Processor SDK Linux U-Boot Release Notes 可知，该 SDK 下的 U-boot 支持 AM335x EVM SK。其中，关于 AM335x 处理器的配置如图 12.39 所示。

AM335x EVM SK 和 AM335x EVM 使用相同的配置文件，为 am335x_evm_defconfig 或 am335x_evm_config。在 SDK 根目录的顶层，可以打开 Rules.make 文件查看，发现 U-boot 使用的是 am335x_evm_config。

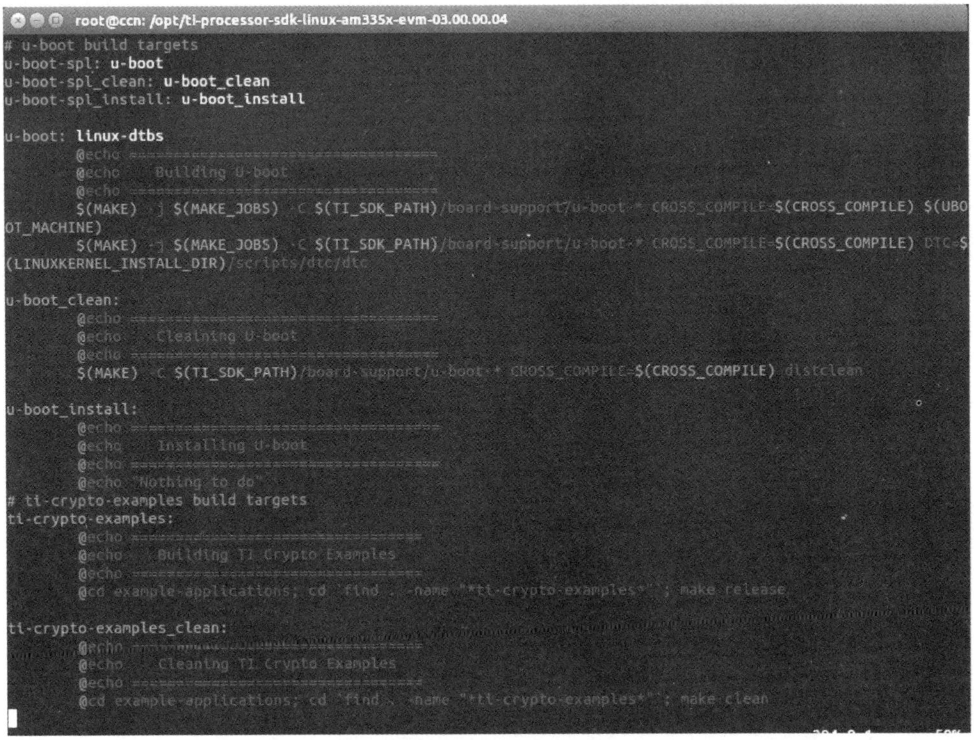

```
root@ccn: /opt/ti-processor-sdk-linux-am335x-evm-03.00.00.04
# u-boot build targets
u-boot-spl: u-boot
u-boot-spl_clean: u-boot_clean
u-boot-spl_install: u-boot_install

u-boot: linux-dtbs
        @echo =====================================
        @echo    Building U-boot
        @echo =====================================
        $(MAKE) -j $(MAKE_JOBS) -C $(TI_SDK_PATH)/board-support/u-boot-* CROSS_COMPILE=$(CROSS_COMPILE) $(UBO
OT_MACHINE)
        $(MAKE) -j $(MAKE_JOBS) -C $(TI_SDK_PATH)/board-support/u-boot-* CROSS_COMPILE=$(CROSS_COMPILE) DTC=$
(LINUXKERNEL_INSTALL_DIR)/scripts/dtc/dtc

u-boot_clean:
        @echo =====================================
        @echo    Cleaining U-boot
        @echo =====================================
        $(MAKE) -C $(TI_SDK_PATH)/board-support/u-boot-* CROSS_COMPILE=$(CROSS_COMPILE) distclean

u-boot_install:
        @echo =====================================
        @echo    Installing U-boot
        @echo =====================================
        @echo "Nothing to do"
# ti-crypto-examples build targets
ti-crypto-examples:
        @echo =================================
        @echo    Building TI Crypto Examples
        @echo =================================
        @cd example-applications; cd `find . -name "*ti-crypto-examples*"`; make release

ti-crypto-examples_clean:
        @echo =================================
        @echo    Cleaning TI Crypto Examples
        @echo =================================
        @cd example-applications; cd `find . -name "*ti-crypto-examples*"`; make clean
```

图 12.37 Make U-boot

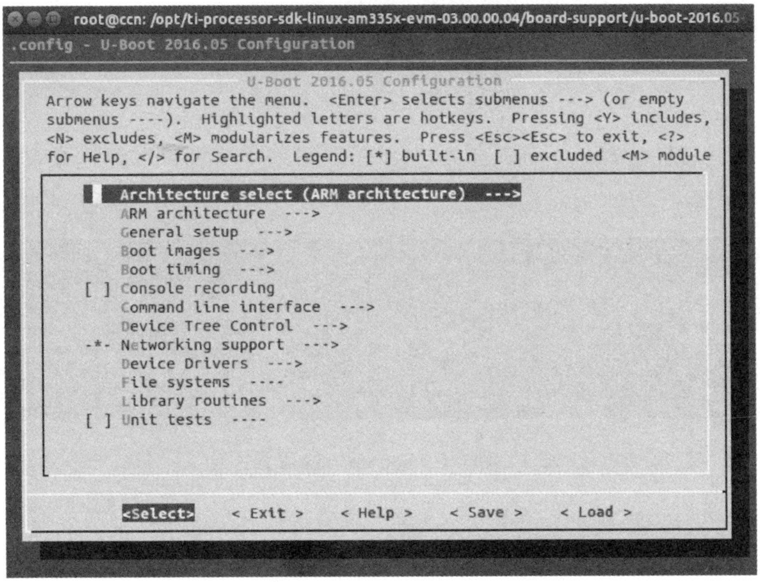

```
root@ccn: /opt/ti-processor-sdk-linux-am335x-evm-03.00.00.04/board-support/u-boot-2016.05
.config - U-Boot 2016.05 Configuration
 ─────────────────── U-Boot 2016.05 Configuration ───────────────────
  Arrow keys navigate the menu.  <Enter> selects submenus ---> (or empty
  submenus ----).  Highlighted letters are hotkeys.  Pressing <Y> includes,
  <N> excludes, <M> modularizes features.  Press <Esc><Esc> to exit, <?>
  for Help, </> for Search.  Legend: [*] built-in  [ ] excluded  <M> module

          Architecture select (ARM architecture)  --->
          ARM architecture  --->
          General setup  --->
          Boot images  --->
          Boot timing  --->
      [ ] Console recording
          Command line interface  --->
          Device Tree Control  --->
      -*- Networking support  --->
          Device Drivers  --->
          File systems  ----
          Library routines  --->
      [ ] Unit tests  ----

        <Select>    < Exit >    < Help >    < Save >    < Load >
```

图 12.38 U-boot make menuconfig

Board configs built:

- am43xx_evm_qspiboot_defconfig
- am43xx_hs_evm_qspiboot_defconfig
- am335x_igep0033_defconfig
- am335x_evm_norboot_defconfig
- am43xx_evm_ethboot_defconfig
- am335x_evm_spiboot_defconfig
- am43xx_hs_evm_ethboot_defconfig
- am335x_boneblack_defconfig
- am335x_evm_nor_defconfig
- am335x_evm_usbspl_defconfig
- am335x_baltos_defconfig
- k2l_evm_defconfig
- k2g_evm_defconfig
- am57xx_hs_evm_defconfig
- am43xx_evm_rtconly_defconfig
- am335x_boneblack_vboot_defconfig
- am57xx_evm_defconfig
- k2hk_evm_defconfig
- k2e_evm_defconfig
- am43xx_evm_usbhost_boot_defconfig
- am335x_evm_defconfig
- am335x_evm_nandboot_defconfig

图 12.39　U-boot 配置文件

在 U-boot 源代码包 configs 目录下查找到的关于 am335x 的所有配置文件，如图 12.40 所示。

```
root@ccn:/opt/ti-processor-sdk-linux-am335x-evm-03.00.00.04/board-support/u-boot-2016.05
+gitAUTOINC+b4e185a8c3-gb4e185a8c3# ls configs/am335x*
configs/am335x_baltos_defconfig             configs/am335x_evm_norboot_defconfig
configs/am335x_boneblack_defconfig          configs/am335x_evm_nor_defconfig
configs/am335x_boneblack_vboot_defconfig    configs/am335x_evm_spiboot_defconfig
configs/am335x_evm_defconfig                configs/am335x_evm_usbspl_defconfig
configs/am335x_evm_nandboot_defconfig       configs/am335x_igep0033_defconfig
configs/am335x_evm_nodt_defconfig           configs/am335x_sl50_defconfig
```

图 12.40　U-boot am335x 配置文件

根据 *Linux Core U-Boot User's Guide* 介绍，清除上一次的配置及编译结果的命令为

```
$ make CROSS_COMPILE = arm - linux - gnueabihf -  distclean
```

如果上述命令执行过程中出现arm-linux-gnueabihf-gcc的交叉工具链编译器没有找到的错误(见图12.41),则说明$PATH环境变量没有包含编译器的安装路径。我们交叉编译器的安装路径为

```
/opt/ti-processor-sdk-linux-am335x-evm-03.00.00.04/linux-devkit/sysroots/x86_64-arago-linux/usr/bin/arm-linux-gnueabihf-
```

```
root@ccn:/opt/ti-processor-sdk-linux-am335x-evm-03.00.00.04/board-support/u-boot-2016.05
+gitAUTOINC+b4e185a8c3-gb4e185a8c3# make CROSS_COMPILE=arm-linux-gnueabihf- distclean
make: arm-linux-gnueabihf-gcc: Command not found
/bin/sh: 1: arm-linux-gnueabihf-gcc: not found
dirname: missing operand
Try 'dirname --help' for more information.
root@ccn:/opt/ti-processor-sdk-linux-am335x-evm-03.00.00.04/board-support/u-boot-2016.05
+gitAUTOINC+b4e185a8c3-gb4e185a8c3#
```

图12.41 arm-linux-gnueabihf-gcc命令没有找到的错误

因此需要执行echo $PATH命令,查看$PATH环境变量值,以及重新设置$PATH指定工具链编译器的路径:

```
export PATH=/opt/ti-processor-sdk-linux-am335x-evm-03.00.00.04/linux-devkit/sysroots/x86_64-arago-linux/usr/bin/:$PATH
```

指定之后再重新用echo $PATH查看,重新执行编译命令,如图12.42所示。

```
+gitAUTOINC+b4e185a8c3-gb4e185a8c3# export PATH=/opt/ti-processor-sdk-linux-am335x-evm-0
3.00.00.04/linux-devkit/sysroots/x86_64-arago-linux/usr/bin:$PATH
root@ccn:/opt/ti-processor-sdk-linux-am335x-evm-03.00.00.04/board-support/u-boot-2016.05
+gitAUTOINC+b4e185a8c3-gb4e185a8c3# make CROSS_COMPILE=arm-linux-gnueabihf- distclean
root@ccn:/opt/ti-processor-sdk-linux-am335x-evm-03.00.00.04/board-support/u-boot-2016.05
+gitAUTOINC+b4e185a8c3-gb4e185a8c3# echo $PATH
/opt/ti-processor-sdk-linux-am335x-evm-03.00.00.04/linux-devkit/sysroots/x86_64-arago-li
nux/usr/bin:/usr/local/sbin:/usr/local/bin:/usr/sbin:/usr/bin:/sbin:/bin:/usr/games:/usr
/local/games
root@ccn:/opt/ti-processor-sdk-linux-am335x-evm-03.00.00.04/board-support/u-boot-2016.05
+gitAUTOINC+b4e185a8c3-gb4e185a8c3# make CROSS_COMPILE=arm-linux-gnueabihf- distclean
root@ccn:/opt/ti-processor-sdk-linux-am335x-evm-03.00.00.04/board-support/u-boot-2016.05
+gitAUTOINC+b4e185a8c3-gb4e185a8c3#
```

图12.42 指定PATH环境变量

彻底删除编译结果的目录(假设编译时,使用'O=am335x_evm'编译输出目标目录,且目录为am335x_evm):

```
$ rm -rf ./am335x_evm
```

编译生成MLO和u-boot:

```
$ make CROSS_COMPILE=arm-linux-gnueabihf- O=am335x_evm am335x_evm_config all
```

上述选择的配置目标为am335x_evm_config,可以通过include/configs/am335x_evm.h实现配置。

上述编译还可能会出现一些莫名其妙的错误,并提示执行make mrproper清除

之后再编译，那么就执行 make mrproper 后，再重新执行编译命令。

make mrproper 命令会删除所有的编译生成文件、内核配置文件（.config 文件）和各种备份文件，所以一般只在第一次执行内核编译前才用这条命令。

可能还会提示错误，如图 12.43 所示。

```
  LDS     u-boot.lds
  LD      u-boot
  OBJCOPY u-boot-nodtb.bin
../scripts/dtc-version.sh: line 17: dtc: command not found
../scripts/dtc-version.sh: line 18: dtc: command not found
*** Your dtc is too old, please upgrade to dtc 1.4 or newer
make[2]: *** [checkdtc] Error 1
make[1]: *** [__build_one_by_one] Error 2
make[1]: Leaving directory `/opt/ti-processor-sdk-linux-am335x-evm-03.00.00.04/
335x_evm'
make: *** [sub-make] Error 2
```

图 12.43　dtc 命令没有找到

dtc 是 device-tree-compiler 的缩写，即设备树编译器，说明系统中没有安装这个编译器，或者是该编译器版本太低，所以需要重新安装：

```
#apt-get install device-tree-compiler
```

编译完成后会在 U-boot 目录下生成 am335x_evm 文件夹，同时会生成 U-boot 和 SPL 等目标镜像。

12.3.3　Linux 内核的配置和编译

ti-processor-sdk-linux-am335x-evm-03.00.00.04 使用的是 Linux 内核，可参考 TI 官方维基关于 Linux 内核的使用手册，如下：

- http://processors.wiki.ti.com/index.php/Processor_SDK_Linux_Kernel。
- http://processors.wiki.ti.com/index.php/Processor_SDK_Linux_Kernel_Release_Notes。
- http://processors.wiki.ti.com/index.php/Linux_Kernel_Users_Guide。
- http://processors.wiki.ti.com/index.php/Processor_SDK_Linux_Kernel_Performance_Guide。

清除上一次的配置和编译结果：

```
make ARCH=arm CROSS_COMPILE=arm-linux-gnueabihf- distclean
```

配置内核：

```
make ARCH=arm CROSS_COMPILE=arm-linux-gnueabihf- mrproper
make ARCH=arm CROSS_COMPILE=arm-linux-gnueabihf- tisdk_am335x-evm_defconfig
make ARCH=arm CROSS_COMPILE=arm-linux-gnueabihf- menuconfig
```

执行了上述命令之后，会出现配置窗口，如图12.44所示。

```
.config - Linux/arm 3.1.0 Kernel Configuration
                    Linux/arm 3.1.0 Kernel Configuration
  Arrow keys navigate the menu.  <Enter> selects submenus --->.
  Highlighted letters are hotkeys.  Pressing <Y> includes, <N> excludes,
  <M> modularizes features.  Press <Esc><Esc> to exit, <?> for Help, </>
  for Search.  Legend: [*] built-in  [ ] excluded  <M> module  < >

      [ ] Patch physical to virtual translations at runtime
          General setup  --->
      [*] Enable loadable module support  --->
      -*- Enable the block layer  --->
          System Type  --->
          Bus support  --->
          Kernel Features  --->
          Boot options  --->
          CPU Power Management  --->
          Floating point emulation  --->
      v(+)

              <Select>    < Exit >    < Help >
```

图12.44　内核配置界面

可以按目标板实际需要选择内核组件进行配置，如修改驱动配置则往下选择Device Drivers，如图12.45所示，选择的配置结果将保存在.config文件中。

```
.config - Linux/arm 3.1.0 Kernel Configuration
                    Linux/arm 3.1.0 Kernel Configuration
  Arrow keys navigate the menu.  <Enter> selects submenus --->.
  Highlighted letters are hotkeys.  Pressing <Y> includes, <N> excludes,
  <M> modularizes features.  Press <Esc><Esc> to exit, <?> for Help, </>
  for Search.  Legend: [*] built-in  [ ] excluded  <M> module  < >
  ^(-)
          Power management options  --->
      [*] Networking support  --->
          Device Drivers  --->
          File systems  --->
          Kernel hacking  --->
          Security options  --->
      -*- Cryptographic API  --->
          Library routines  --->
      ---
          Load an Alternate Configuration File
  v(+)

              <Select>    < Exit >    < Help >
```

图12.45　Device Drivers

编译内核镜像：

```
make ARCH = arm CROSS_COMPILE = arm - linux - gnueabihf -  zImage
```

编译结果将位于内核目录arch/arm/boot/，生成zImage文件。

Linux从Kernel 3.8开始，每个TI EVM板的Linux内核都有对应的一个设备树文件，因此需要为所用的目标板编译、安装对应的设备树dtb文件。所有的设备树都位于arch/arm/boot/dts/目录下，AM335x Starter Kit开发板对应的设备树文件为am335x-evmsk.dts。

单独编译一个AM335x Starter Kit设备树文件：

```
make ARCH = arm CROSS_COMPILE = arm - linux - gnueabihf -  am335x - evmsk.dtb
```

编译的结果将位于arch/arm/boot/dts目录。

默认情况下，Linux设备驱动是编译集成到Linux内核镜像zImage里，但也可以编译成模块形式，使其可以动态加载：

```
make ARCH = arm CROSS_COMPILE = arm - linux - gnueabihf -  modules
```

内核镜像zImage和设备树dtb的安装，是安装到SD卡的rootfs分区：

```
cd /opt/ti - processor - sdk - linux - am335x - evm - 03.00.00.04/board - support/linux *
cp arch/arm/boot/zImage /media/ccn/rootfs/boot
cp arch/arm/boot/dts/am335x - evmsk.dtb /media/ccn/rootfs/boot
```


内核模块的安装，将模块安装到SD卡的rootfs分区：

```
make ARCH = arm INSTALL_MOD_PATH = /media/ccn/rootfs modules_install
```

注：对SD卡进行格式化时需要有rootfs分区，因为文件系统需要存放在该分区。这里的/media/ccn/rootfs可能并不是读者插入SD卡的真实路径，需要读者根据自己的实际情况修改。详情请参考“12.4.2　Ubuntu系统下AM335x Linux SDK SD卡的创建”。

12.3.4　文件系统

Linux少不了文件系统，TI也不例外，它提供了基于Arago的开源嵌入式文件系统。在TI提供的SDK开发包里(即SDK安装目录：/opt/ti-processor-sdk-linux-am335x-evm-03.00.00.04/filesystem)，主要有两种文件系统的支持：一种是完整版文件系统(tisdk-rootfs-image-am335x-evm.tar.gz)的支持，包括QT、3D、Gstreamers以及各种DEMO程序等；另一种是空的文件系统(arago-base-tisdk-image-am335x-evm.tar.gz)，用户可以自己进行添加、更改、使用。TI提供了上述两种编译好的文件系统，而且不建议重新进行编译，因为编译一次需要花费很多时间，况且开发人员完全可以基于TI提供的这两种文件系统去进行开发、移植等，完全可以满足产品需求，大大缩短开发时间。

12.4　目标板 Linux 系统的创建

12.4.1　Windows 系统下 AM335x Linux SDK SD 卡的创建

参考：http://software-dl. ti. com/processor-sdk-linux/esd/AM335X/latest/index_FDS. html 主页下的 PROCESSOR - SDK - LINUX - AM335X Product downloads→AM335x Linux SDK SD Card Creation→Windows SD Card Creation Wiki。

参考 TI 官方维基：http://processors. wiki. ti. com/index. php/Processor_SDK_Linux_Creating_a_SD_Card_with_Windows。

虽然，最终我们会使用自己编译的 Linux 镜像用于创建 SD 卡的 Linux 系统。但是，为了验证 SD 卡 Linux 系统创建方法的正确性，还是建议先去官网下载现成的 Linux 镜像，验证成功后再使用自己编译的镜像，这样可以在出现问题的情况下排除编译本身出现的问题。

Linux 镜像下载链接：http://software-dl. ti. com/sitara_linux/esd/processor-sdk/PROCESSOR-SDK-LINUX-AM335X/latest/index_FDS. html。

找到 AM335x Linux SDK SD Card Creation 项，如 am335x-evm-linux-03. 00. 00. 04. img. zip，或更高版本 am335x-evm-linux-03. 01. 00. 06. img. zip 镜像文件，再单击下载，并解压成目标镜像 am335x-evm-linux-03. 01. 00. 06. img。

注：官方提示，该镜像只用于在 Windows 系统下创建 SD 卡。

从 http://sourceforge. net/projects/win32diskimager 地址链接下载开源的，可以将 img 镜像烧入到 SD 卡的工具 Win32 Disk Imager，下载完成后为 Win32Disk-Imager-0. 9. 5-install. exe，双击该文件安装到默认目录下，并在桌面创建快捷方式。

双击启动桌面的 Win32 Disk Imager，在 Image File 选项中，单击如图 12.46 所示的光标处。

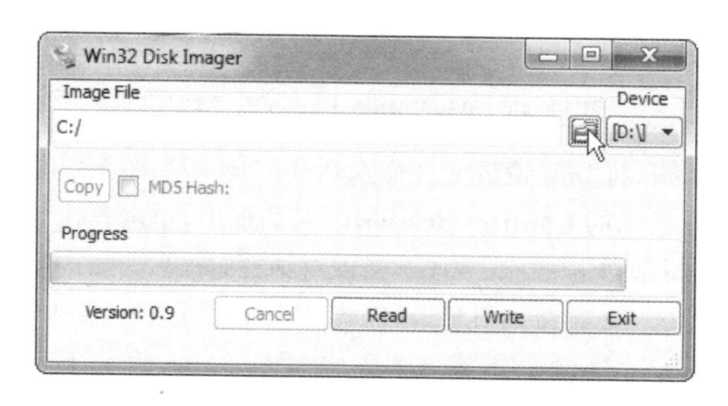

图 12.46　Win32 Disk Imager 对话框(一)

打开 am335x-evm-linux-03.01.00.06.img 镜像文件，如图 14.47 所示。

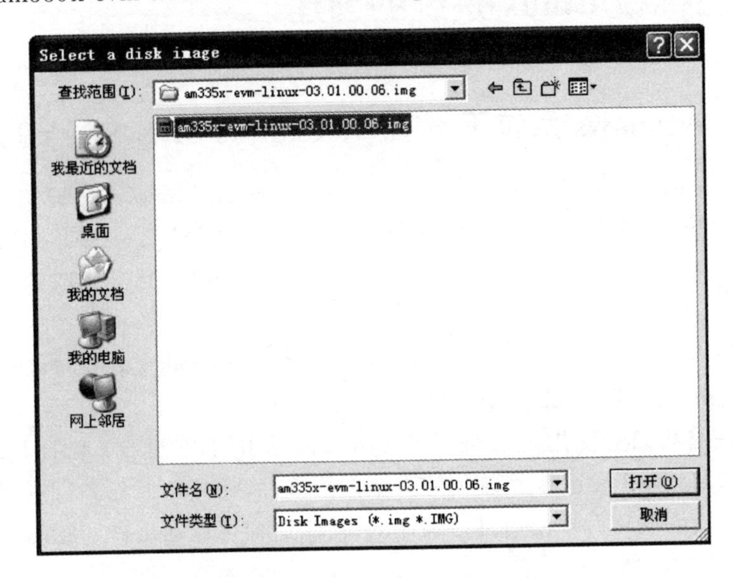

图 12.47　打开 am335x-evm-linux-03.01.00.06.img

将 TF 卡通过 USB 读卡器插入 PC，并格式化 TF 卡，格式化时在卷标项填 boot。格式化完成后，在 Win32 Disk Imager 工具的 Device 选项中，选择可移动磁盘所在的盘符 H:\，如图 12.48 所示。

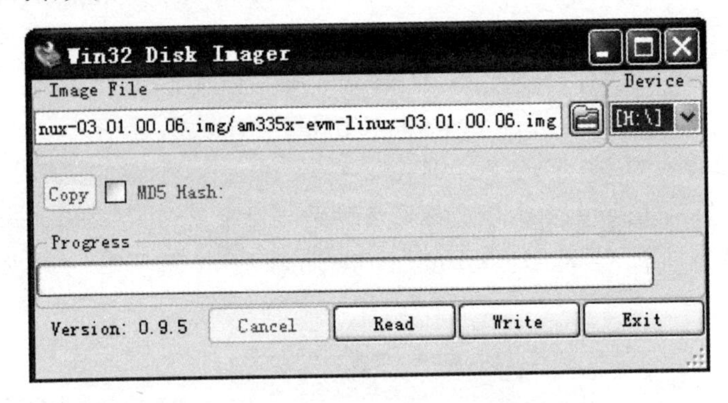

图 12.48　Win32 Disk Imager 对话框(二)

单击 Write 按钮，将 img 镜像文件烧入到 TF 卡中，如图 12.49 所示。

在如图 12.50 所示的 Confirm overwrite 对话框中，单击 Yes 按钮确认。但没有顺利地进行烧写，而是弹出 Write Error 错误对话框，如图 12.51 所示。

原因是使用了 4 GB 容量的 TF 卡，因为 TF 卡会存在一些坏块，及其他如文件表等的占用，所以实际容量会不满 4 GB。如图 12.52 所示 H 盘的实际容量为 3.68 GB，而要烧写的 am335x-evm-linux-03.01.00.06.img 镜像文件有 3.90 GB(见图 12.53)，因此会出现磁盘没有足够空间的错误。

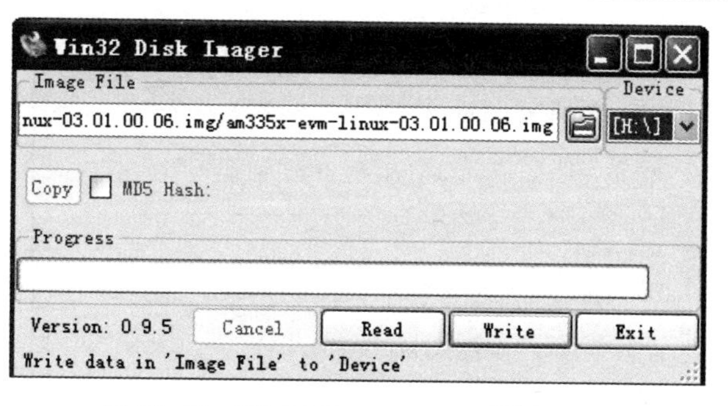

图 12.49 Win32 Disk Imager 对话框(三)

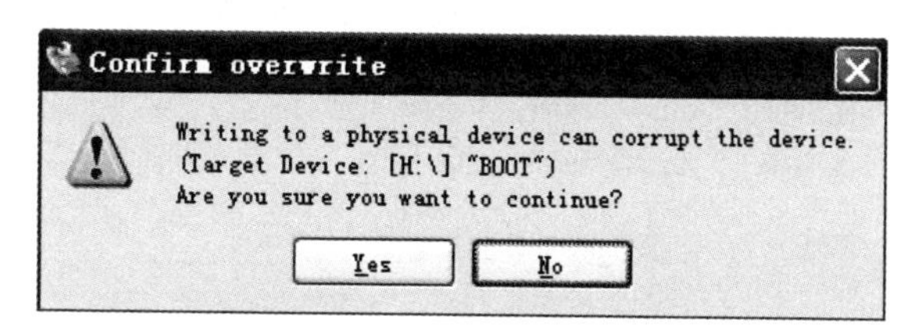

图 12.50 Confirm overwrite 对话框

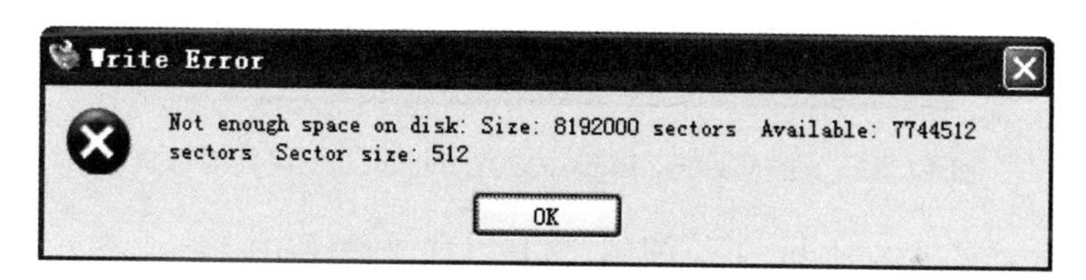

图 12.51 Write Error 对话框

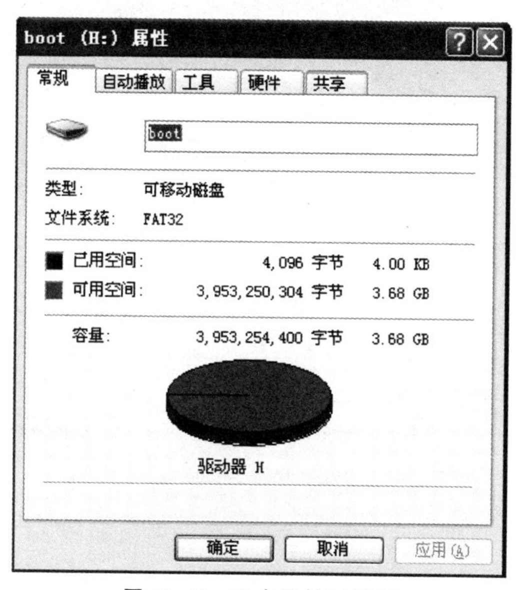

图 12.52 H 盘属性对话框

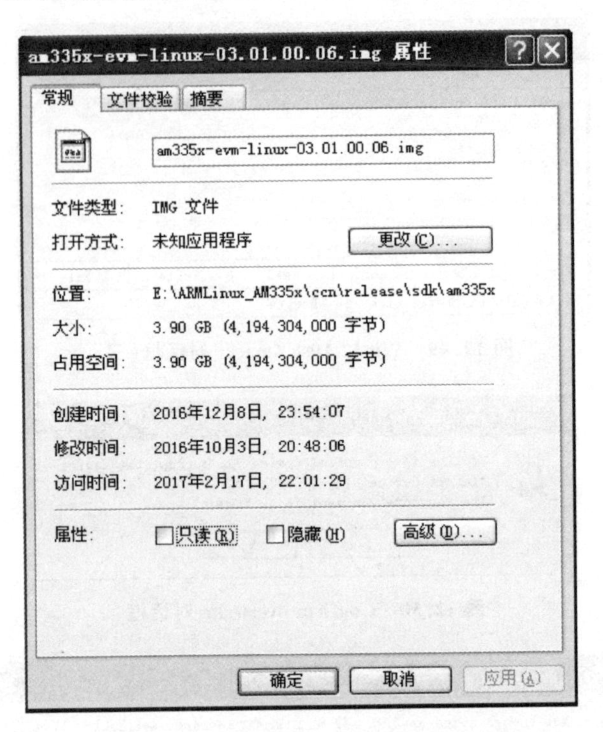

图 12.53 am335x-evm-linux-03.01.00.06.img 属性对话框

更换一张容量更大的 TF 卡后测试，没有出现问题，如图 12.54 所示。

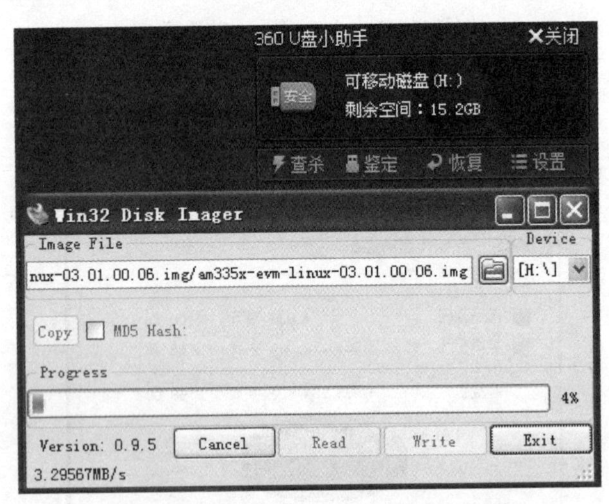

图 12.54 Win32 Disk Imager 对话框(四)

烧写有点慢，需要一些时间，当烧写完成后会弹出如图 12.55 所示的完成对话框，单击 OK 按钮结束，同时单击 Win32 Disk Imager 对话框的 Exit 按钮退出烧写工具。

图 12.55　Complete 对话框

但奇怪的是，将镜像文件烧写到 TF 卡完成后，查看 TF 卡属性发现，居然是如图 12.56 所示的容量属性。

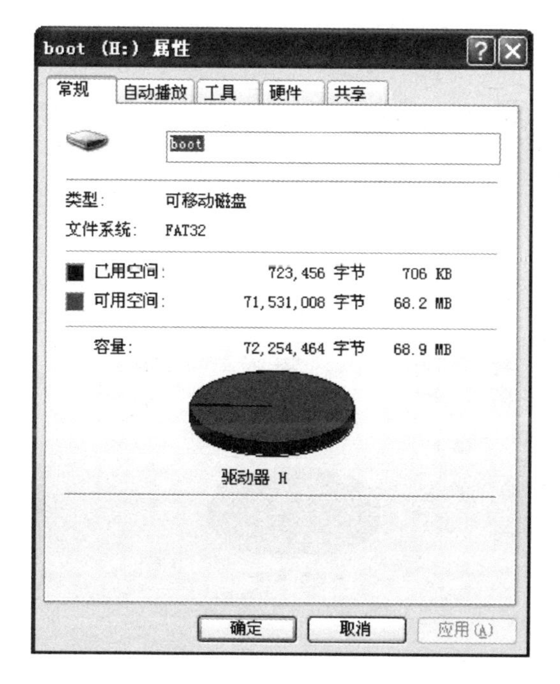

图 12.56　16 GB TF 卡烧写完镜像后的容量属性

查看 TF 卡，发现磁盘里只有如图 12.57 所示的 MLO 和 u-boot.img 两个文件。

最重要的是，将该 TF 卡插入 Starter Kit 开发板的 TF 卡槽后，开发板上电后再按 PWRON 按键，LCD 居然没有出现正常的启动界面。那么到底是什么问题呢？这里有个疑点，TF 卡里为何没有 Linux 内核和文件系统。我们带着疑问去尝试其他的方法是否可以实验成功。

另外，可以下载老版本的镜像 sitara_linux_sdk_image_am335x.img.zip（可能需要 Google 或百度查找一下）实验，解压后为 sitara_linux_sdk_am335x.img，只有 1 024 000 KB，即 1 GB 的容量，完全可以烧写到一张 4 GB 的 TF 卡上，烧写后如图 12.58 所示。

查看 TF 卡，磁盘里也只有如图 12.59 所示的 MLO 和 u-boot.img 两个文件。

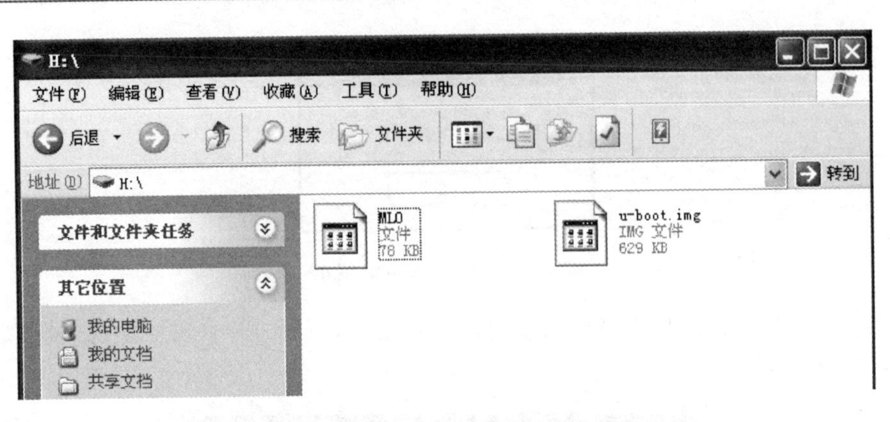

图 12.57 16 GB TF 卡烧写完镜像后包含的文件

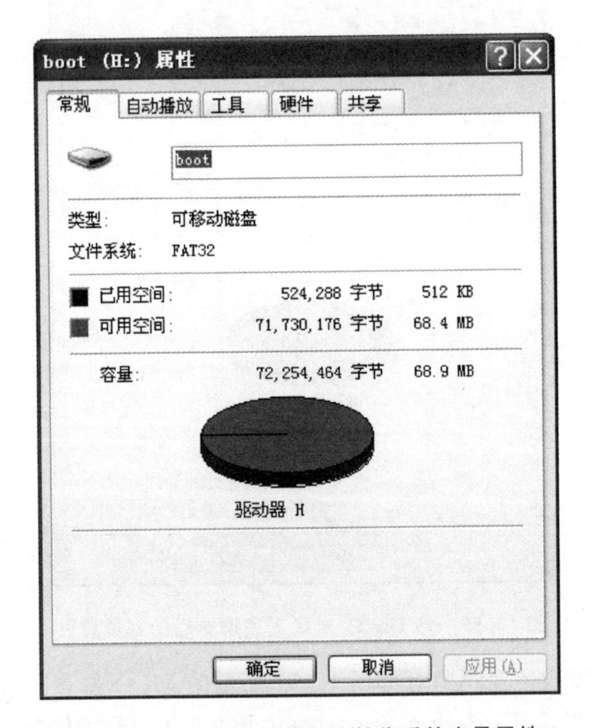

图 12.58 4 GB TF 卡烧写完镜像后的容量属性

MLO 为之前章节提到过的启动引导程序，而 u-boot.img 为对应的 APP 应用程序。在整个嵌入式 Linux 的软件架构中，MLO 算是 U-boot 的前一级引导程序，而 U-boot 就是通用的 Linux 系统的 Bootloader 引导程序，那么 Linux 内核 uImage 和文件系统 ramdisk 在哪里呢？非常庆幸的是，该 4 GB 容量的 TF 卡放入 Starter Kit 开发板后，上电居然可以正常启动，不仅有 LED 指示灯闪烁，LCD 屏上也出现了 TI 的开机启动界面并有进度条显示，接着进入触摸屏校准程序，如图 12.60 和图 12.61 所示。

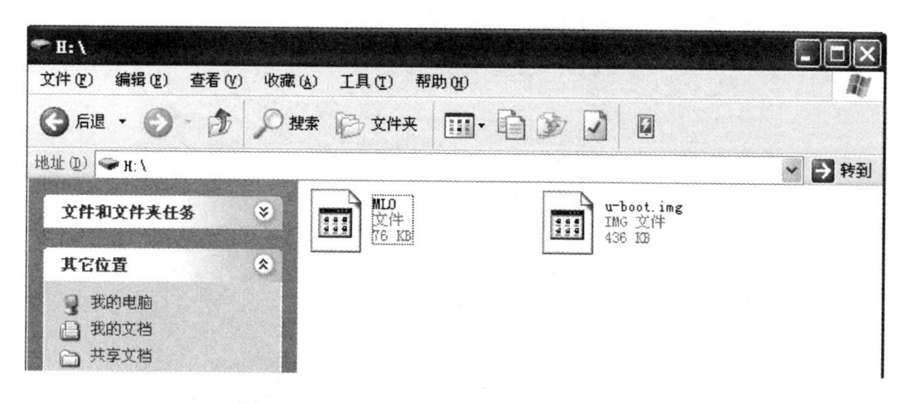

图 12.59　4 GB TF 卡烧写完镜像后的文件

图 12.60　4 GB TF 卡烧写完镜像后的开机启动画面

图 12.61　4 GB TF 卡烧写完镜像后的触摸屏校准程序

依次单击五个方位的十字坐标，校准成功后进入菜单画面，如图 12.62 所示。

如果通过 USB 将开发板接入 PC 且启动超级终端，将进入开发板的 Linux 系统，如图 12.63 所示。

图 12.62 4 GB TF 卡烧写完镜像后的菜单画面

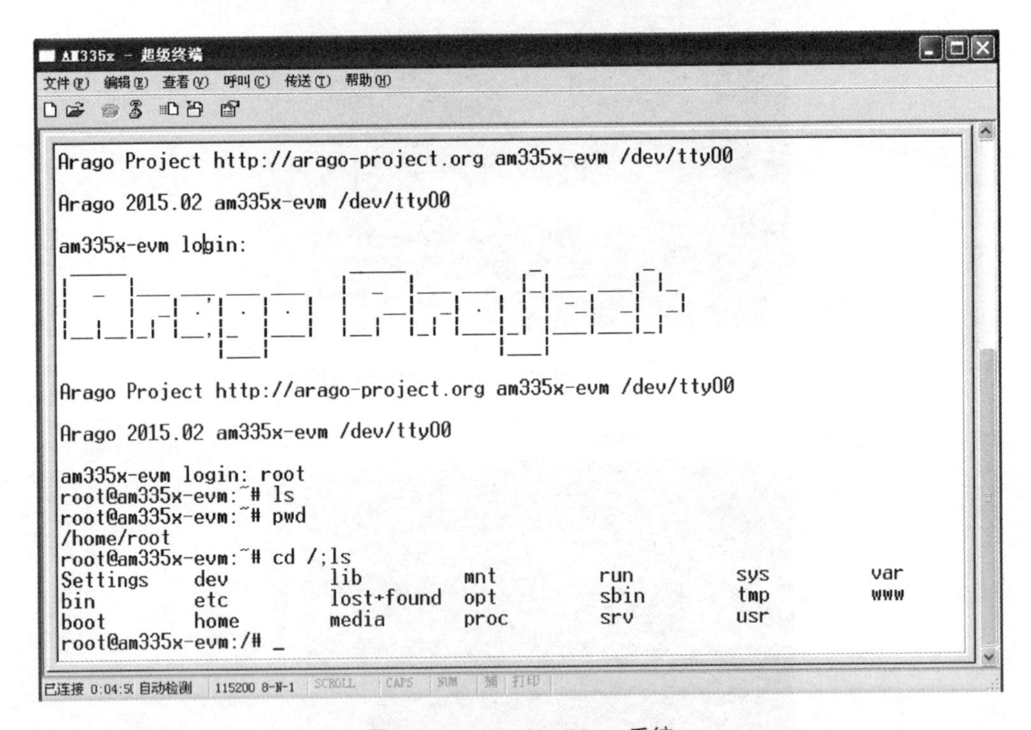

图 12.63 AM335x Linux 系统

也就是说 4 GB 的 TF 卡镜像及烧写过程都是正确的，TF 卡中显示的文件属性和只有 MLO 和 u-boot.img 两个文件也是正常的。为了进一步验证真实性，我们可以打开 Starter Kit 配套的两张 TF 卡进行比较，结果也和我们烧写的 TF 卡属性类似，如图 12.64 所示。

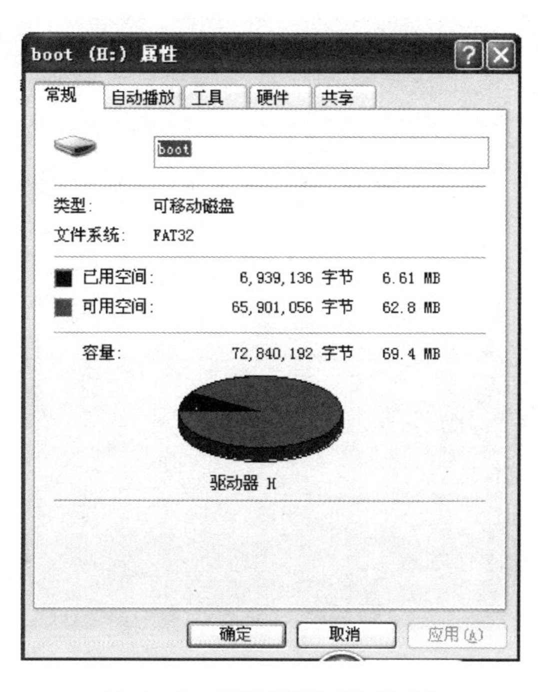

图 12.64 原厂配带的TF卡属性

但是,在内容上除了包含MLO和u-boot.img外,比我们烧写的多出了uImage内核镜像,如图12.65所示为Android系统的TF卡的内容。

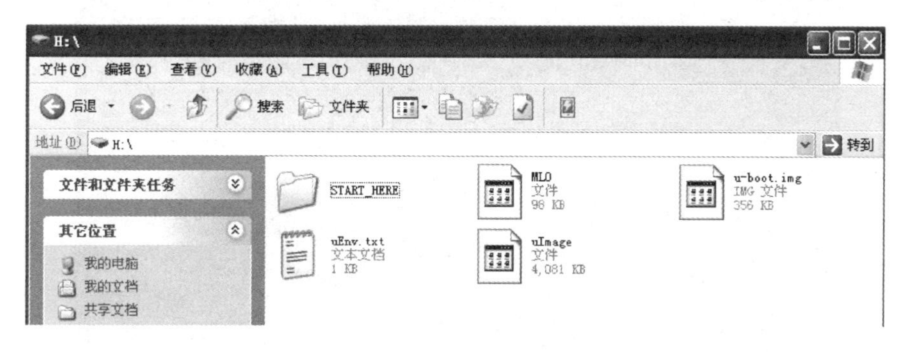

图 12.65 原厂配带Android系统TF卡内容

在正常情况下,Android系统也是会有ramdisk根文件系统和system用户文件系统等其他文件的。

那么是什么原因让我们看不到其他文件呢?其实,切换到Ubuntu下查看TF卡就全明白了。

如图12.66所示为在Ubuntu下查看到的我们烧写的4 GB TF卡。

如图12.67所示为在Ubuntu下查看到的原厂附带的Android系统的TF卡。

```
root@ccn:/home/ccn# df -hT
Filesystem     Type      Size  Used Avail Use% Mounted on
/dev/sda10     ext4       92G   18G   70G  20% /
none           tmpfs     4.0K     0  4.0K   0% /sys/fs/cgroup
udev           devtmpfs  1.9G  4.0K  1.9G   1% /dev
tmpfs          tmpfs     376M  1.4M  374M   1% /run
none           tmpfs     5.0M     0  5.0M   0% /run/lock
none           tmpfs     1.9G   76K  1.9G   1% /run/shm
none           tmpfs     100M   36K  100M   1% /run/user
/dev/sdb1      vfat       69M  512K   69M   1% /media/ccn/boot
/dev/sdb2      ext3      883M  609M  229M  73% /media/ccn/rootfs
root@ccn:/home/ccn# ls /media/ccn -h
boot  rootfs
root@ccn:/home/ccn# ls /media/ccn/boot/ -lh
total 512K
-rw-r--r-- 1 ccn ccn  76K  2月 10  2015 MLO
-rw-r--r-- 1 ccn ccn 436K  2月 10  2015 u-boot.img
root@ccn:/home/ccn# ls /media/ccn/rootfs/ -lh
total 84K
drwxr-sr-x  2 root root 4.0K  2月 11  2015 bin
drwxr-sr-x  2 root root 4.0K  2月 11  2015 boot
drwxr-sr-x  2 root root 4.0K  2月 11  2015 dev
drwxr-sr-x 35 root root 4.0K  2月 11  2015 etc
drwxr-sr-x  3 root root 4.0K  2月 11  2015 home
drwxr-sr-x  8 root root 4.0K  2月 11  2015 lib
drwx------  2 root root  16K  2月 11  2015 lost+found
drwxr-sr-x  2 root root 4.0K  2月 11  2015 media
drwxr-sr-x  3 root root 4.0K  2月 11  2015 mnt
drwxr-sr-x  3 root root 4.0K  2月 11  2015 opt
drwxr-sr-x  2 root root 4.0K  2月 11  2015 proc
drwxr-sr-x  2 root root 4.0K  2月 11  2015 run
drwxr-sr-x  2 root root 4.0K  2月 11  2015 sbin
drwxr-sr-x  3 root root 4.0K  2月 11  2015 srv
drwxr-sr-x  2 root root 4.0K  2月 11  2015 sys
lrwxrwxrwx  1 root root    8  1月  1  2000 tmp -> /var/tmp
drwxr-sr-x 11 root root 4.0K  2月 11  2015 usr
drwxr-sr-x  8 root root 4.0K  2月 11  2015 var
drwxr-sr-x  3 root root 4.0K  2月 11  2015 www
root@ccn:/home/ccn# ls /media/ccn/rootfs/
bin  boot  dev  etc  home  lib  lost+found  media  mnt  opt  proc  run  sbin  srv  sys  tmp  usr  var  www
root@ccn:/home/ccn#
```

图 12.66　Ubuntu 下 4 GB TF 卡

```
root@ccn:/home/ccn# df -hT
Filesystem     Type      Size  Used Avail Use% Mounted on
/dev/sda10     ext4       92G   18G   70G  20% /
none           tmpfs     4.0K     0  4.0K   0% /sys/fs/cgroup
udev           devtmpfs  1.9G  4.0K  1.9G   1% /dev
tmpfs          tmpfs     376M  1.4M  374M   1% /run
none           tmpfs     5.0M     0  5.0M   0% /run/lock
none           tmpfs     1.9G   76K  1.9G   1% /run/shm
none           tmpfs     100M   44K  100M   1% /run/user
/dev/sdb1      vfat       70M  6.7M   63M  10% /media/ccn/boot
/dev/sdb3      ext4      6.3G  5.6G  297M  96% /media/ccn/devkit
/dev/sdb2      ext4      448M  188M  233M  45% /media/ccn/rootfs
root@ccn:/home/ccn# ls /media/ccn/boot/
MLO  START_HERE  u-boot.img  uEnv.txt  uImage
root@ccn:/home/ccn# ls /media/ccn/rootfs/
acct          fstab.am335xevm         lost+found  ueventd.am335xevm.rc
cache         init                    mnt         ueventd.goldfish.rc
config        init.am335xevm.rc       proc        ueventd.rc
d             init.am335xevm.usb.rc   sbin        usbdrive
data          init.goldfish.rc        sdcard      vendor
default.prop  init.rc                 storage
dev           init.trace.rc           sys
etc           init.usb.rc             system
root@ccn:/home/ccn# ls /media/ccn/devkit/
app            dontpanic   resource-cache
app-asec       drm         ssh
app-lib        local       system
app-private    lost+found  TI_Android_JB_4.2.2_DevKit_4.1.1.tar.gz
backup         media       user
dalvik-cache   misc
data           property
root@ccn:/home/ccn#
```

图 12.67　Ubuntu 下 Andriod TF 卡

如图 12.68 所示为在 Ubuntu 下查看到的原厂附带的 Linux 系统的 TF 卡。

```
root@ccn:/home/ccn# df -hT
Filesystem     Type      Size  Used Avail Use% Mounted on
/dev/sda10     ext4       92G   18G   70G  20% /
none           tmpfs     4.0K     0  4.0K   0% /sys/fs/cgroup
udev           devtmpfs  1.9G  4.0K  1.9G   1% /dev
tmpfs          tmpfs     376M  1.4M  374M   1% /run
none           tmpfs     5.0M     0  5.0M   0% /run/lock
none           tmpfs     1.9G   76K  1.9G   1% /run/shm
none           tmpfs     100M   52K  100M   1% /run/user
/dev/sdb1      vfat       70M  5.4M   65M   8% /media/ccn/AM335X SDK
/dev/sdb3      ext3      2.9G  2.7G   52M  99% /media/ccn/START_HERE
/dev/sdb2      ext3      679M  471M  173M  74% /media/ccn/rootfs
root@ccn:/home/ccn# ls /media/ccn/AM335X\ SDK/
AM335x-EVM-quickstartguide.pdf     pointercal
AM335x-EVM-SK-quickstartguide.pdf  u-boot.img
autorun.inf                        uImage
Beaglebone-quickstartguide.pdf     windows_users
MLO
root@ccn:/home/ccn# ls /media/ccn/START_HERE/
AM335x-EVM-SK-quickstartguide.pdf
beaglebone-quickstartguide.pdf
CCS
ccs_install.sh
quickstartguide.pdf
setup.htm
START_HERE.sh
ti-sdk-am335x-evm-06.00.00.00-Linux-x86-Install
root@ccn:/home/ccn# ls /media/ccn/rootfs/
bin   dev  home  media  opt   sbin      srv  [illegible]  usr  www
boot  etc  lib   mnt    proc  Settings  sys  tmp          var
root@ccn:/home/ccn# 
```

图 12.68　Ubuntu 下 Linux TF 卡

从图 12.66～图 12.68 对应的 3 张 TF 卡可以看出 TF 卡(/dev/sdb *,被挂载到目录/media/ccn)内有 2 种分区：vfat 和 ext3。而其中 ext3 在 Windows 系统下不可见,只能看到 vfat 分区,所以我们只看到了 vfat 下的 MLO、u-boot. img,或者 MLO、u-boot. img 和 uImage。

12.4.2　Ubuntu 系统下 AM335x Linux SDK SD 卡的创建

参考：http://software-dl. ti. com/processor-sdk-linux/esd/AM335X/latest/index_FDS. html 主页下的 PROCESSOR-SDK-LINUX-AM335X Product downloads→AM335x Linux SDK SD Card Creation→Linux SD Card Creation Wiki。

参考：http://processors. wiki. ti. com/index. php/Processor_SDK_Linux_create_SD_card_script。在 TI Linux SDK 安装目录下有现成的 SD 卡制作脚本：

```
/opt/ti-processor-sdk-linux-am335x-evm-03.00.00.04/bin/create-sdcard.sh
```

当时,在安装完 SDK 后的环境配置 setup. sh 中并没有调用 create-sdcard. sh 脚本,它需要插入 TF 卡后以 root 权限手动执行该脚本去完成 AM335x Linux SDK SD 卡的创建。维基上介绍可以为 TF 卡创建如下 3 种镜像：Linux SDK 目录下默认的 images 镜像、定制的镜像、tarballs 分区的镜像。

以 root 身份登录并进入到脚本目录下：

```
#cd /opt/ti-processor-sdk-linux-am335x-evm-03.00.00.04/bin/
#./create-sdcard.sh
```

注：如果提示 root permissions 错误,需要切换到 root 权限。

脚本的第一步要求我们选择想要格式化的 TF 卡代表的驱动器,当提示我们输入对应的 TF 卡的驱动器号时,我们输入 1,选择驱动器 sdb,如图 12.69 所示。

```
Availible Drives to write images to:

#  major   minor    size   name
1:   8       16    7761920 sdb

Enter Device Number:
```

图 12.69 选择插入 TF 卡对应的驱动器

脚本的第二步是要求选择是否重新分区 TF 卡,如图 12.70 所示。

```
Would you like to re-partition the drive anyways [y/n] :
```

图 12.70 选择是否重新分区 TF 卡

y：选择重新分区。

n：如果 TF 卡已经分区完成,则选择 n 跳过此项。

第三步是选择需要分区的数,2 或者 3,如图 12.71 所示。

```
Number of partitions needed [2/3] :
```

图 12.71 选择分区数

2：多数情况下选择 2,可以给 root 根文件系统提供更多的空间。

3：只用于为 EVM 制作 SD 卡时选择,当制作 tarballs 分区的镜像时需要选择。

第四步是安装 SD 卡内容。SD 卡被分区完成后,将提示是否要继续安装文件系统或退出脚本。

y：选择 y 将开始安装 SD 卡内容。此操作将擦除 SD 卡现有的数据。

可选择默认、定制和 tarballs 分区中的一种镜像方式给 SD 卡创建。

n：选择 n 将不安装镜像,而只完成之前的 SD 卡分区,并离开。

第五步,使用 AM335x Linux SDK 默认镜像创建 SD 启动卡(Create the SD

Card Using Default Image)。

先决条件：① AM335x Linux SDK 已安装在 Linux 主机上；② 要被制作的 SD 卡已经插入到 Linux 主机上，且有足够大的存储空间用于保存 Bootloaders、kernel 内核和 root 根文件系统；③ 继续执行上述脚本。

选择安装预编译镜像，如图 12.72 所示。

```
################################################################################

        Choose file path to install from

        1 ) Install pre-built images from SDK
        2 ) Enter in custom boot and rootfs file paths

################################################################################

Choose now [1/2] :
```

图 12.72 选择预安装镜像文件的路径

选择 1)使用 SDK 中预编译镜像去创建 SD 卡。

如果在 SDK 目录下执行该脚本，则脚本会自动确认 SDK 路径，并将所有镜像复制到 SD 卡，一旦所有镜像复制完成后将退出安装脚本。

如果在安装的 SDK 目录外执行该脚本(如复制该脚本到其他目录下，再执行)，那么应注意：选项 1)将只识别默认安装的 SDK 目录名，如果我们手动修改了安装目录，则需要选择选项 2)进入定制 boot 和 rootfs 文件路径。

为 K2G 选择 tarball rootfs。

如果是 K2G 处理器(即 66AK2G01 Multicore DSP＋ARM KeyStone Ⅱ System-on-Chip (SoC))，而不是我们使用的 AM335x，则进入 rootfs Tarballs，如图 12.73 所示。

```
################################################################################

   Multiple rootfs Tarballs found

################################################################################

        1:tisdk-server-extra-rootfs-image-k2g-evm.tar.gz
        2:tisdk-server-rootfs-image-k2g-evm.tar.gz

Enter Number of rootfs Tarball:
```

图 12.73 选择 rootfs Tarballs

选项 1 使用 K2G SDK 里完整的文件系统镜像去创建 SD 启动卡。

选项 2 是在 SD 卡没有足够空间的情况下，使用只包含基本文件系统功能的小尺寸的镜像去创建 SD 启动卡。

在执行脚本安装时，如果找不到 SDK 的安装路径，则会提示，如图 12.74 所示。

```
no SDK PATH found
Enter path to SDK :
```

图 12.74　no SDK PATH found

此时，需进入正确的 SDK 安装路径，例如 sitara 安装在 home 目录下，将进入 /home/sitara/ti-processor-sdk-linux-<machine>-<version>格式去配置安装路径，而后将用默认镜像去创建 SD 启动卡，完成后退出安装脚本。

第六步，使用定制镜像创建 SD 启动卡。

在开发过程中，时常会分别使用 TFTP 和 NFS 去传输 kernel 内核镜像和引导 root 根文件系统，在使用这种开发方式时，还可能会想着将这些镜像放到 SD 卡中，以便于它们在不需要网络连接到服务器时能够单独使用。

先决条件：① AM335x Linux SDK 已安装在 Linux 主机上；② 要被制作的 SD 卡已经插入到 Linux 主机上，且有足够大的存储空间用于保存 Bootloaders、kernel 内核和 root 根文件系统；③ 继续执行上述脚本。

选择进入定制引导和根文件系统镜像，如图 12.75 所示。

```
################################################################################

        Choose file path to install from

        1 ) Install pre-built images from SDK
        2 ) Enter in custom boot and rootfs file paths

################################################################################

Choose now [1/2] :
```

图 12.75　选择定制引导和根文件系统文件的路径

选择 2)进入定制 boot 引导和 rootfs 根文件系统文件路径去创建 SD 卡。

选择 boot 引导分区：此时将提示提供引导分区文件位于的路径，提示将解释该文件需要放置在哪个路径，基本的选项有：

① Point to a tarball containing all of the files you want placed on the boot partition. This would include the boot loaders and the kernel image as well as any optional files like uEnv. txt

② Point to a directory containing the files for the boot partition like those in the first option.

该脚本能够自动识别出我们提供的是 tarball 还是文件夹路径，且对应地复制这些文件。我们提供复制的文件列表，如果不正确将会给出选项去改变路径。

选择 root 根文件系统分区：此时将提示提供 root 根文件系统分区文件位于的路径，这些提示将解释该文件需要放置在哪个路径，基本的选项如图 12.76 所示。

```
1.  Point to a tarball of the root file system you want to use
2.  Point to a directory containing the root file sysetm such as an NFS share directory.
```

图 12.76　选择引导分区文件的路径

该脚本能够自动识别出我们提供的是 tarball 还是文件夹路径，且对应地复制这些文件。我们提供复制的文件列表，如果不正确将会给出选项去改变路径。

第七步，使用 Partition Tarballs 创建 SD 启动卡。

该选项意味着是板厂商去创建为 EVM 包装配带的 SD 启动卡，它需要访问 3 个 tarballs 对应的 EVM 配带的 SD 卡分区。

先决条件：① AM335x Linux SDK 已安装在 Linux 主机上；② 要被制作的 SD 卡已经插入到 Linux 主机上，且有足够大的存储空间用于保存 Bootloaders、kernel 内核和 root 根文件系统；③ 继续执行上述脚本。

提供 tarball 位置：SD 卡被分区后将提示，如图 12.77 所示。

```
Enter path where SD card tarballs were downloaded :
```

图 12.77　进入 SD 卡 tarballs 的下载路径

如图 12.78 所示为指向包含 tarball 文件的目录。

- **boot_partition.tar.gz**
- **rootfs_partition.tar.gz**
- **start_here_partition.tar.gz**

图 12.78　tarballs 文件目录

该脚本将显示目录内容，并要求校验当前目录下的 tarballs 内容。而后，包含 tarballs 内容的 SD 卡可作为 EVM 配套附件被发行。

总结脚本执行的整个过程(4 GB TF 卡及 USB 读卡器已经插入计算机)，如图 12.79 所示。

TF 卡及读卡器为/dev/sdb＊设备，所以执行后提示写入镜像到 sdb，选择 1，如图 12.80 所示。

选择 2 开始分区，如图 12.81 所示。

分区完成后，开始安装文件系统，如图 12.82 所示。

分区完成，选择 y；继续安装文件系统，选择 1，将 SDK 安装目录下镜像复制到 TF 卡，如图 12.83 所示。

```
root@ccn: /opt/ti-processor-sdk-linux-am335x-evm-03.00.00.04/bin
root@ccn:/opt/ti-processor-sdk-linux-am335x-evm-03.00.00.04/bin# pwd
/opt/ti-processor-sdk-linux-am335x-evm-03.00.00.04/bin
root@ccn:/opt/ti-processor-sdk-linux-am335x-evm-03.00.00.04/bin# ls /dev/sd*
/dev/sda   /dev/sda10  /dev/sda5  /dev/sda7  /dev/sda9  /dev/sdb1
/dev/sda1  /dev/sda2   /dev/sda6  /dev/sda8  /dev/sdb   /dev/sdb2
root@ccn:/opt/ti-processor-sdk-linux-am335x-evm-03.00.00.04/bin# ./create-sdcard.sh


################################################################################

This script will create a bootable SD card from custom or pre-built binaries.

The script must be run with root permissions and from the bin directory of
the SDK

Example:
 $ sudo ./create-sdcard.sh

Formatting can be skipped if the SD card is already formatted and
partitioned properly.

################################################################################


Available Drives to write images to:

#  major   minor    size   name
1:   8       16    3872256 sdb

Enter Device Number or n to exit: 
```

图 12.79　SD 启动卡制作脚本执行过程(一)

```
root@ccn: /opt/ti-processor-sdk-linux-am335x-evm-03.00.00.04/bin
################################################################################


Available Drives to write images to:

#  major   minor    size   name
1:   8       16    3872256 sdb

Enter Device Number or n to exit: 1

sdb was selected

/dev/sdb is an sdx device
Unmounting the sdb drives
 unmounted /dev/sdb1
 unmounted /dev/sdb2
Current size of sdb1 71680 bytes
Current size of sdb2 934912 bytes

################################################################################

        Select 2 partitions if only need boot and rootfs (most users).
        Select 3 partitions if need SDK & other content on SD card.  This is
        usually used by device manufacturers with access to partition tarballs.

        ****WARNING**** continuing will erase all data on sdb

################################################################################

Number of partitions needed [2/3] : 2
```

图 12.80　SD 启动卡制作脚本执行过程(二)

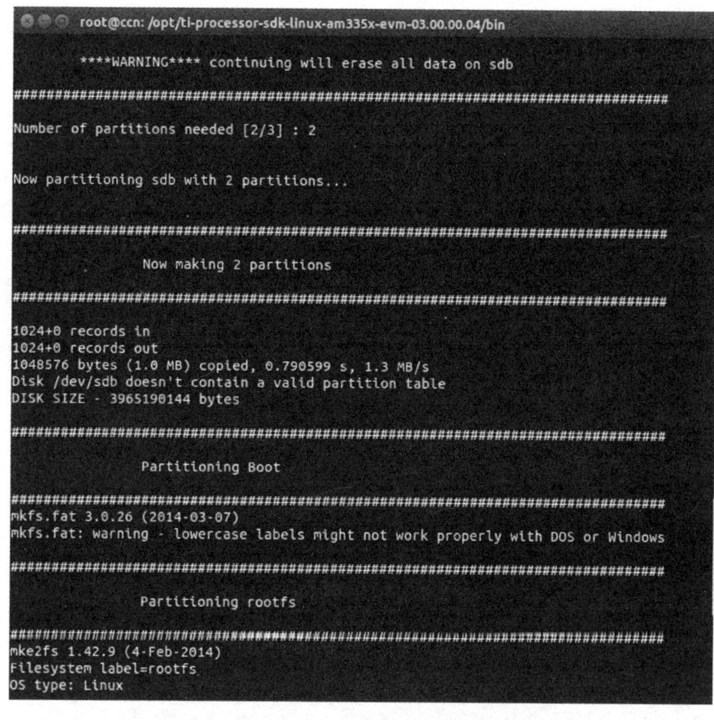

图 12.81　SD 启动卡制作脚本执行过程(三)

```
root@ccn: /opt/ti-processor-sdk-linux-am335x-evm-03.00.00.04/bin
################################################################################

                Partitioning rootfs

################################################################################
mke2fs 1.42.9 (4-Feb-2014)
Filesystem label=rootfs
OS type: Linux
Block size=4096 (log=2)
Fragment size=4096 (log=2)
Stride=0 blocks, Stripe width=0 blocks
236640 inodes, 945920 blocks
47296 blocks (5.00%) reserved for the super user
First data block=0
Maximum filesystem blocks=968884224
29 block groups
32768 blocks per group, 32768 fragments per group
8160 inodes per group
Superblock backups stored on blocks:
        32768, 98304, 163840, 229376, 294912, 819200, 884736

Allocating group tables: done
Writing inode tables: done
Creating journal (16384 blocks): done
Writing superblocks and filesystem accounting information: done


################################################################################

   Partitioning is now done
   Continue to install filesystem or select 'n' to safe exit

   **Warning** Continuing will erase files any files in the partitions

################################################################################

Would you like to continue? [y/n] :
```

图 12.82　SD 启动卡制作脚本执行过程(四)

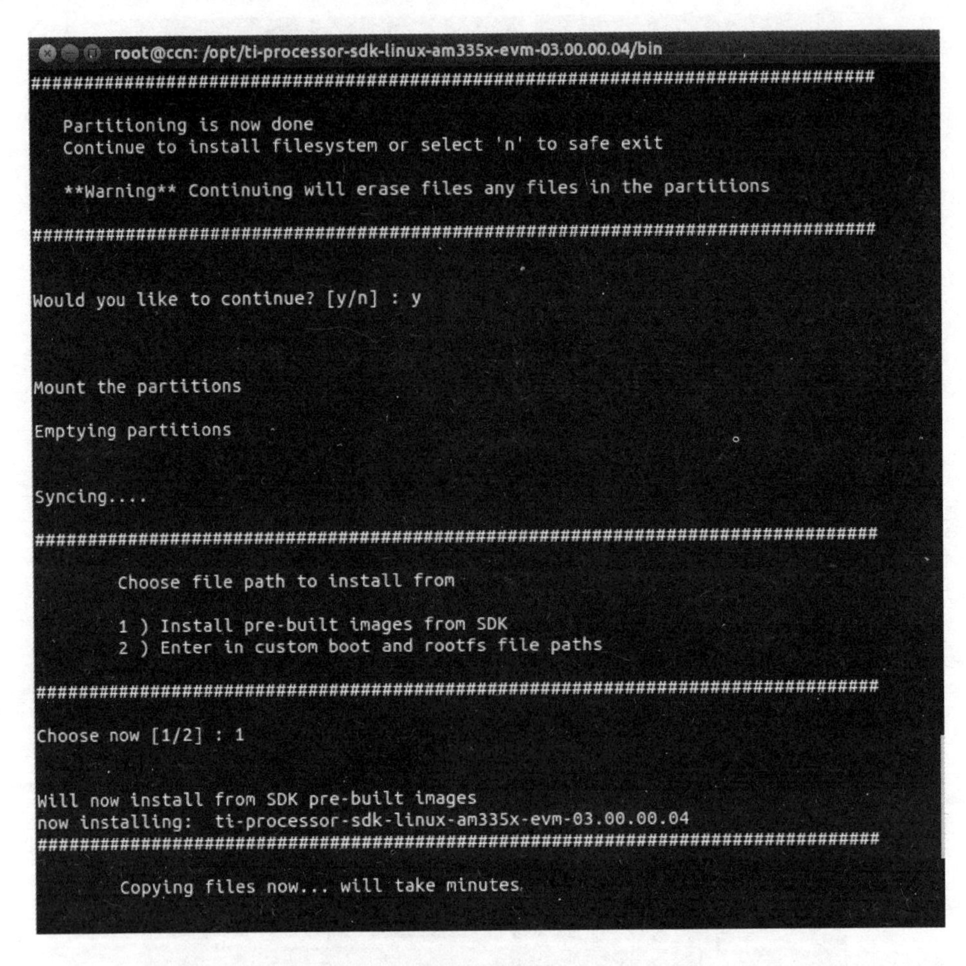

图 12.83　SD启动卡制作脚本执行过程(五)

执行到这步后需要花费些时间,复制完成后将退出脚本的执行,如图12.84所示。

TF启动卡制作完成后,将看到两个分区/dev/sdb1和/dev/sdb2,分别挂载对应的/media/ccn/boot和/media/ccn/rootfs目录,可以查看目录下的文件内容,如图12.85所示。

4 GB容量的TF启动卡制作完成后,将其放入Starter Kit开发板,上电启动,LED指示灯将闪烁,一定时间后进入触摸屏校准程序,如图12.86所示。

触摸屏校验完成后,将进入菜单界面,如图12.87所示。

```
root@ccn: /opt/ti-processor-sdk-linux-am335x-evm-03.00.00.04/bin
        Copying files now... will take minutes

################################################################################

Copying boot partition

MLO copied


u-boot.img copied


Copying rootfs System partition
Written 100%

Syncing...

Un-mount the partitions

Remove created temp directories

Operation Finished

root@ccn:/opt/ti-processor-sdk-linux-am335x-evm-03.00.00.04/bin# ls /dev/sd*
/dev/sda   /dev/sda10  /dev/sda5  /dev/sda7  /dev/sda9  /dev/sdb1
/dev/sda1  /dev/sda2   /dev/sda6  /dev/sda8  /dev/sdb   /dev/sdb2
root@ccn:/opt/ti-processor-sdk-linux-am335x-evm-03.00.00.04/bin# df -hT
Filesystem     Type      Size  Used Avail Use% Mounted on
/dev/sda10     ext4       92G   18G   70G  20% /
none           tmpfs     4.0K     0  4.0K   0% /sys/fs/cgroup
udev           devtmpfs  1.9G  4.0K  1.9G   1% /dev
tmpfs          tmpfs     376M  1.5M  374M   1% /run
none           tmpfs     5.0M     0  5.0M   0% /run/lock
none           tmpfs     1.9G   76K  1.9G   1% /run/shm
none           tmpfs     100M   60K  100M   1% /run/user
/dev/sda6      fuseblk   197G  183G   14G  93% /media/ccn/0003A054000BC55C
/dev/sdb2      ext3      3.5G  1.2G  2.2G  35% /media/ccn/rootfs
/dev/sdb1      vfat       69M  703K   69M   1% /media/ccn/boot
root@ccn:/opt/ti-processor-sdk-linux-am335x-evm-03.00.00.04/bin#
```

图 12.84　SD 启动卡制作脚本执行过程(六)

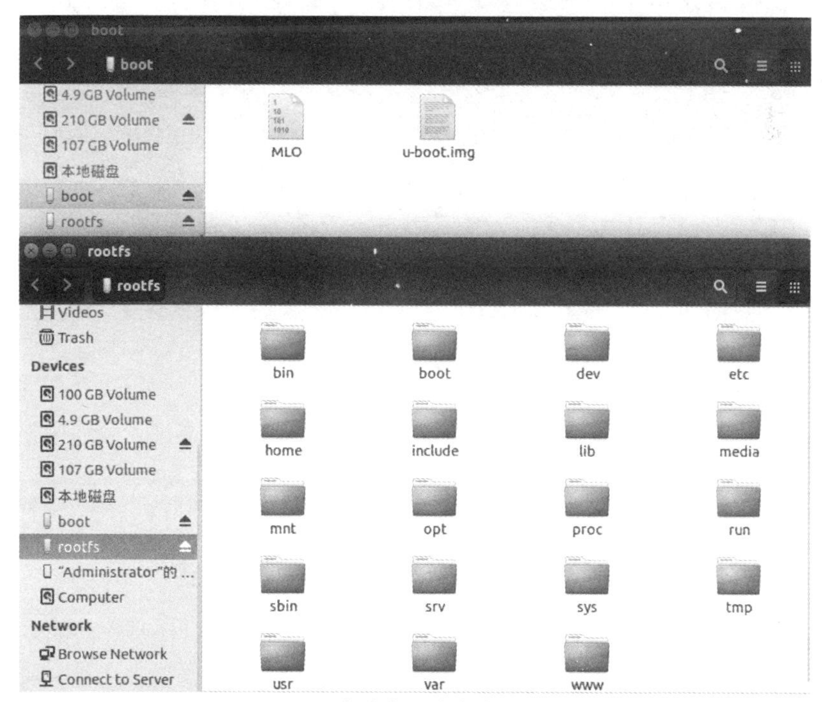

图 12.85　SD 启动卡制作脚本执行过程(七)

图 12.86 触摸屏校验程序

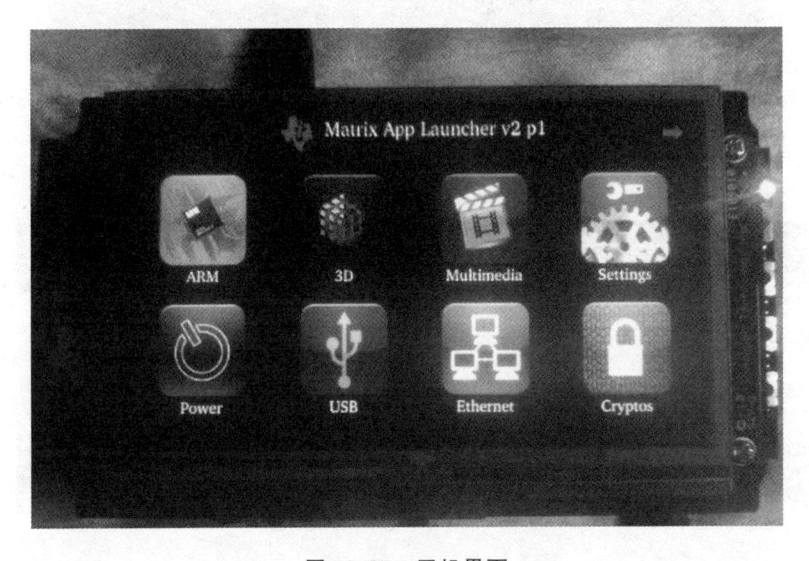

图 12.87 开机界面

12.5 嵌入式 Linux 平台测试

12.5.1 串口调试终端 Minicom 和以太网测试

前面已经介绍过通常情况下的 Minicom 设置及使用,这里针对 TI SDK 包及 TI 开发板再次介绍。

AM335x Starter Kit 板上带有 USB 转串口芯片,因此当使用 USB 线将其连接到计算机后，会出现/dev/ttyUSB0 和/dev/ttyUSB1，后者为 USB 转串口。

另外，在 AM335x SDK 安装目录/opt/ti-processor-sdk-linux-am335x-evm-03.00.00.04/bin 下有 minicom 的设置脚本 setup-minicom.sh，执行后将对 minicom 进行设置，如图 12.88 所示。

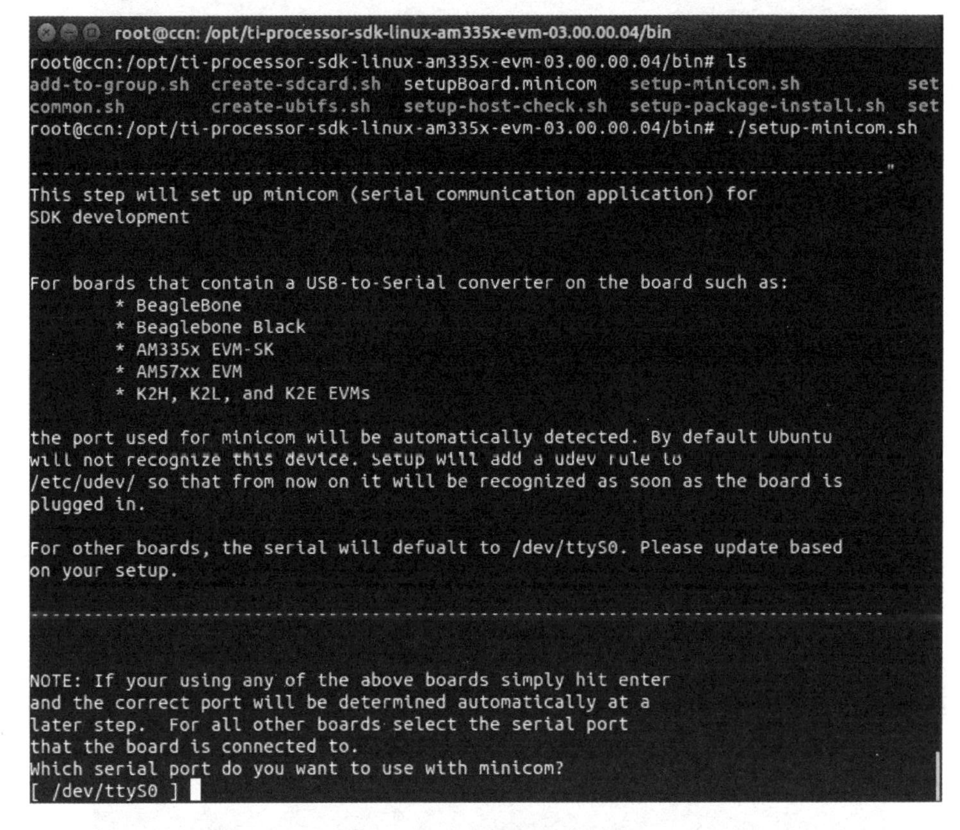

图 12.88　setup-minicom.sh

输入 USB 转串口设备/dev/ttyUSB1，继续执行 minicom 的脚本设置。脚本执行完成后，输入 minicom 命令，如图 12.89 所示，可启动 Minicom 串口调试终端。

启动 Minicom 后按回车键（AM335x Starter Kit 目标板的 Linux 系统已经启动完成），Minicom 终端将作为 AM335x Starter Kit 的系统控制台而显示对应的信息，如图 12.90 所示。

root 用户可直接登录到 AM335x Starter Kit 目标板的 Linux 系统。

使用网线连接目标板和 PC，并分别为其以太网 IP 地址进行设置、分配及测试，如图 12.91 和图 12.92 所示。

至此，嵌入式 Linux 平台开发所需的基础调试工具：串口终端 Minicom 和以太网通信设置测试完成。

```
root@ccn:/opt/ti-processor-sdk-linux-am335x-evm-03.00.00.04/bin# ls
add-to-group.sh  create-sdcard.sh  setupBoard.minicom   setup-minicom.sh
common.sh        create-ubifs.sh   setup-host-check.sh  setup-package-install.sh
root@ccn:/opt/ti-processor-sdk-linux-am335x-evm-03.00.00.04/bin# ./setup-minicom.

--------------------------------------------------------------------------------
This step will set up minicom (serial communication application) for
SDK development


For boards that contain a USB-to-Serial converter on the board such as:
        * BeagleBone
        * Beaglebone Black
        * AM335x EVM-SK
        * AM57xx EVM
        * K2H, K2L, and K2E EVMs

the port used for minicom will be automatically detected. By default Ubuntu
will not recognize this device. Setup will add a udev rule to
/etc/udev/ so that from now on it will be recognized as soon as the board is
plugged in.

For other boards, the serial will defualt to /dev/ttyS0. Please update based
on your setup.

--------------------------------------------------------------------------------


NOTE: If your using any of the above boards simply hit enter
and the correct port will be determined automatically at a
later step.  For all other boards select the serial port
that the board is connected to.
Which serial port do you want to use with minicom?
[ /dev/ttyS0 ] /dev/ttyUSB1

Copied existing /root/.minirc.dfl to /root/.minirc.dfl.old

Configuration saved to /root/.minirc.dfl. You can change it further from inside
minicom, see the Software Development Guide for more information.
--------------------------------------------------------------------------------
root@ccn:/opt/ti-processor-sdk-linux-am335x-evm-03.00.00.04/bin# minicom
```

图 12.89　输入 minicom 命令

```
root@ccn: /opt/ti-processor-sdk-linux-am335x-evm-03.00.00.04/bin

Welcome to minicom 2.7

OPTIONS: I18n
Compiled on Jan  1 2014, 17:13:19.
Port /dev/ttyUSB1, 01:11:36

Press CTRL-A Z for help on special keys

Arago Project http://arago-project.org am335x-evm /dev/ttyO0

Arago 2015.02 am335x-evm /dev/ttyO0

am335x-evm login:

CTRL-A Z for help | 115200 8N1 | NOR | Minicom 2.7 | VT102 | Offline | ttyUSB1
```

图 12.90　Minicom 控制台

```
root@ccn: /opt/ti-processor-sdk-linux-am335x-evm-03.00.00.04/bin
Port /dev/ttyUSB1, 01:11:36

Press CTRL-A Z for help on special keys

Arago Project http://arago-project.org am335x-evm /dev/ttyO0

Arago 2015.02 am335x-evm /dev/ttyO0

am335x-evm login: root
root@am335x-evm:~# ifconfig eth0 192.168.1.1
root@am335x-evm:~# ifconfig eth0
eth0      Link encap:Ethernet  HWaddr C4:ED:BA:88:1C:C7
          inet addr:192.168.1.1  Bcast:192.168.1.255  Mask:255.255.255.0
          UP BROADCAST RUNNING MULTICAST  MTU:1500  Metric:1
          RX packets:0 errors:0 dropped:0 overruns:0 frame:0
          TX packets:2 errors:0 dropped:0 overruns:0 carrier:0
          collisions:0 txqueuelen:1000
          RX bytes:0 (0.0 B)  TX bytes:684 (684.0 B)
          Interrupt:56

root@am335x-evm:~# ping 192.168.1.2
PING 192.168.1.2 (192.168.1.2): 56 data bytes
64 bytes from 192.168.1.2: seq=0 ttl=64 time=1.293 ms
64 bytes from 192.168.1.2: seq=1 ttl=64 time=0.642 ms
64 bytes from 192.168.1.2: seq=2 ttl=64 time=0.559 ms

CTRL-A Z for help | 115200 8N1 | NOR | Minicom 2.7 | VT102 | Offline | ttyUSB1
```

图 12.91 AM335x 目标板网络测试

```
root@ccn: /home/ccn
root@ccn:/home/ccn# ifconfig eth0 192.168.1.2
root@ccn:/home/ccn# ifconfig eth0
eth0      Link encap:Ethernet  HWaddr 98:4b:e1:a7:ed:07
          inet addr:192.168.1.2  Bcast:192.168.1.255  Mask:255.255.255.0
          inet6 addr: fe80::9a4b:e1ff:fea7:ed07/64 Scope:Link
          UP BROADCAST RUNNING MULTICAST  MTU:1500  Metric:1
          RX packets:49 errors:0 dropped:0 overruns:0 frame:0
          TX packets:302 errors:0 dropped:0 overruns:0 carrier:0
          collisions:0 txqueuelen:1000
          RX bytes:6564 (6.5 KB)  TX bytes:47929 (47.9 KB)

root@ccn:/home/ccn# ping 192.168.1.1
PING 192.168.1.1 (192.168.1.1) 56(84) bytes of data.
64 bytes from 192.168.1.1: icmp_seq=1 ttl=64 time=0.541 ms
64 bytes from 192.168.1.1: icmp_seq=2 ttl=64 time=0.409 ms
64 bytes from 192.168.1.1: icmp_seq=3 ttl=64 time=0.380 ms
64 bytes from 192.168.1.1: icmp_seq=4 ttl=64 time=0.482 ms
64 bytes from 192.168.1.1: icmp_seq=5 ttl=64 time=0.359 ms
64 bytes from 192.168.1.1: icmp_seq=6 ttl=64 time=0.379 ms
64 bytes from 192.168.1.1: icmp_seq=7 ttl=64 time=0.403 ms
^C
--- 192.168.1.1 ping statistics ---
7 packets transmitted, 7 received, 0% packet loss, time 6000ms
rtt min/avg/max/mdev = 0.359/0.421/0.541/0.066 ms
root@ccn:/home/ccn#
```

图 12.92 PC 宿主机网络设置

12.5.2　TFTP网络文件下载

在AM335x SDK安装目录/opt/ti-processor-sdk-linux-am335x-evm-03.00.00.04/bin，同样有TFTP的设置脚本setup-tftp.sh。之前运行setup.sh执行后会调用setup-package-install.sh以安装TFTP、Minicom等其他所需的工具软件，且会调用setup-tftp.sh执行配置；此时当然也可以重新单独执行。

/etc/xinetd.d/tftp为TFTP的配置文件，默认配置/tftpboot为TFTP服务器。上传、下载目录，如图12.93所示。

```
root@ccn: /opt/ti-processor-sdk-linux-am335x-evm-03.00.00.04/bin
e-install.sh
root@ccn:/opt/ti-processor-sdk-linux-am335x-evm-03.00.00.04/bin# vi setup-package-install.sh
root@ccn:/opt/ti-processor-sdk-linux-am335x-evm-03.00.00.04/bin# vi setup-tftp.sh
root@ccn:/opt/ti-processor-sdk-linux-am335x-evm-03.00.00.04/bin# ls /etc/xinetd.d/t
tftp  time
root@ccn:/opt/ti-processor-sdk-linux-am335x-evm-03.00.00.04/bin# ls /etc/xinetd.d/tftp
/etc/xinetd.d/tftp
root@ccn:/opt/ti-processor-sdk-linux-am335x-evm-03.00.00.04/bin# vi /etc/xinetd.d/tftp
root@ccn:/opt/ti-processor-sdk-linux-am335x-evm-03.00.00.04/bin# cat /etc/xinetd.d/tftp

service tftp
{
protocol = udp
port = 69
socket_type = dgram
wait = yes
user = nobody
server = /usr/sbin/in.tftpd
server_args = /tftpboot
disable = no
}

root@ccn:/opt/ti-processor-sdk-linux-am335x-evm-03.00.00.04/bin#
```

图12.93　TFTP配置文件

AM335x Starter Kit开发板的Linux系统已经包含了TFTP工具及命令支持，所以当宿主机和目标板的网络配置好后就可以使用TFTP命令将PC宿主机的应用程序等文件下载到目标板上执行等。如图12.94所示为宿主机IP地址及/tftpboot目录。

如图12.95所示为AM335x Starter Kit目标板网络配置和TFTP命令下载宿主机文件的操作。注：图12.95为在宿主机Ubuntu下，打开Minicom登录到目标板的Linux控制台。

AM335x Linux SDK使用的是Busybox，Busybox中TFTP命令的用法如下：

```
tftp [option] … host [port]
```

如果要下载或上传文件，则一定要用这些option：

-g 表示下载文件（get）；

-p 表示上传文件（put）；

-l 表示本地文件名（local file）；

-r 表示远程主机的文件名（remote file）。

```
root@ccn: /tftpboot
          collisions:0 txqueuelen:1000
          RX bytes:5459 (5.4 KB)  TX bytes:51205 (51.2 KB)

root@ccn:/tftpboot# ifconfig eth0
eth0      Link encap:Ethernet  HWaddr 98:4b:e1:a7:ed:07
          inet addr:192.168.1.2  Bcast:192.168.1.255  Mask:255.255.255.0
          inet6 addr: fe80::9a4b:e1ff:fea7:ed07/64 Scope:Link
          UP BROADCAST RUNNING MULTICAST  MTU:1500  Metric:1
          RX packets:80 errors:0 dropped:0 overruns:0 frame:0
          TX packets:152 errors:0 dropped:0 overruns:0 carrier:0
          collisions:0 txqueuelen:1000
          RX bytes:5459 (5.4 KB)  TX bytes:51205 (51.2 KB)

root@ccn:/tftpboot# ls -l
total 3812
-rw-r--r-- 1 root root   36841 10月  8  2016 am335x-boneblack.dtb
-rw-r--r-- 1 root root   34993 10月  8  2016 am335x-bone.dtb
-rw-r--r-- 1 root root   35249 10月  8  2016 am335x-bonegreen.dtb
-rw-r--r-- 1 root root   41713 10月  8  2016 am335x-evm.dtb
-rw-r--r-- 1 root root   40435 10月  8  2016 am335x-evmsk.dtb
-rw-r--r-- 1 root root   37508 10月  8  2016 am335x-icev2.dtb
-rw-r--r-- 1 root root 3664216 10月  8  2016 zImage-am335x-evm.bin
root@ccn:/tftpboot# 
```

图 12.94　PC 宿主机网络 IP 地址和 tftpboot 目录

```
root@ccn: /home/ccn

OPTIONS: I18n
Compiled on Jan  1 2014, 17:13:19.
Port /dev/ttyUSB1, 17:12:32

Press CTRL-A Z for help on special keys


root@am335x-evm:/tmp# ifconfig eth0
eth0      Link encap:Ethernet  HWaddr C4:ED:BA:88:1C:C7
          inet addr:192.168.1.1  Bcast:192.168.1.255  Mask:255.255.2550
          UP BROADCAST RUNNING MULTICAST  MTU:1500  Metric:1
          RX packets:83 errors:0 dropped:0 overruns:0 frame:0
          TX packets:79 errors:0 dropped:0 overruns:0 carrier:0
          collisions:0 txqueuelen:1000
          RX bytes:42074 (41.0 KiB)  TX bytes:5117 (4.9 KiB)
          Interrupt:56

root@am335x-evm:/tmp# tftp 192.168.1.2 -g -r am335x-evmsk.dtb
root@am335x-evm:/tmp# ls
am335x-boneblack.dtb  lighttpd
am335x-evmsk.dtb      qtembedded-0
root@am335x-evm:/tmp# 
```

图 12.95　AM335x Starter Kit 目标板执行 TFTP 命令下载文件

例如,图 12.95 是要从远程主机(或称安装开发环境的宿主机)192.168.1.2 上下载 am335x-evmsk.dtb,即对应输入的命令如下:

```
tftp 192.168.1.2 -g -r am335x-evmsk.dtb
```

12.5.3 Hello 测试程序

我们为目标板编译、运行最简单的 Hello 程序,以测试、熟悉本章前面介绍的嵌入式 Linux 开发环境和应用开发方法。

1. Hello 源程序

程序清单如下:

```
/* hello.c - 在终端上打印 Hello,Linux for AM335x */
#include <stdio.h>
int main(void){
    printf("Hello,Linux for AM335x\n");
    return 0;
}
```


2. 编　译

可以直接输入编译命令,也可以写 Makefile 脚本进行编译。命令如下:

```
#/opt/ti-processor-sdk-linux-am335x-evm-03.00.00.04/linux-devkit/
sysroots/x86_64-arago-linux/usr/bin/arm-linux-gnueabihf-gcc -o hello hello.c
```

如果已经将编译工具添加到系统的环境变量,则可以省略上述命令中的编译工具目录。脚本清单如下:

```
CROSS=/opt/ti-processor-sdk-linux-am335x-evm-03.00.00.04/linux-devkit/
sysroots/x86_64-arago-linux/usr/bin/arm-linux-gnueabihf-
all: hello
hello:
    $(CROSS)gcc -o hello hello.c
clean:
    rm -rf hello *.o
```

第一行是指定交叉编译器的安装路径,编译目标为 Hello,当执行 make 时将执行"/opt/ti-processor-sdk-linux-am335x-evm-03.00.00.04/linux-devkit/sysroots/x86_64-arago-linux/usr/bin/arm-linux-gnueabi-gcc -o hello hello.c"命令,当执行 make clean 时将执行"rm -rf hello *.o"命令。

```
#make
```

生成 hello 可执行文件。

3. 下载到目标板运行

当生成可执行文件 Hello 后，可用上节提到的 TFTP 方法将其下载到目标板的/tmp 目录下，再执行：

```
#./hello
```

此时目标板的 Linux 系统将在 console 里打印：

```
Hello,Linux for AM335x
```

第13章

嵌入式 Linux 驱动开发

嵌入式 Linux 驱动开发与基于 PC 的 Linux 驱动开发基本上是一样的，应用的知识、技术都是相互适用的，只是具体的硬件平台不同，编译工具不同。PC 编译后直接运行在 PC 上，而嵌入式 Linux 驱动程序经过 ARM 交叉编译后运行在 ARM 处理器上。

13.1 设备树

13.1.1 Linux 内核对硬件的描述

与之前 S3C2410 Linux 有所不同的是，AM335x 及现有 Linux 内核不会再将硬件驱动、板级描述等都包含进内核镜像中，而是独立的 DTB(the Device Tree Blob，设备树)镜像，即采用设备树机制来描述各板级的硬件。

在以前的 Linux 内核版本中：

① 内核包含了对硬件的全部描述；

② Bootloader 会加载一个二进制的内核镜像(uImage 或 zImage 等)，并执行它；

③ Bootloader 会提供内存大小和地址、kernel command line 参数及一些额外信息等，成为 ATAGS，且地址通过 r2 寄存器传给内核；

④ Bootloader 会告诉内核加载哪一款 board，通过 r1 寄存器存放 machine type integer。

如今的内核版本使用了 Device Tree：

① 内核不再包含对硬件的描述，它以 *.dtb 二进制文件形式单独存储在另外的位置；

② Bootloader 需要加载两个二进制文件：内核镜像(uImage 或 zImage)和 DTB；

③ Bootloader 通过 r2 寄存器来传递 DTB 地址，通过修改 DTB 可以修改内存信息、kernel command line，以及潜在的其他信息；

④ 不再有 machine type。

ATAGS 与 DTB 的对比图如图 13.1 所示。

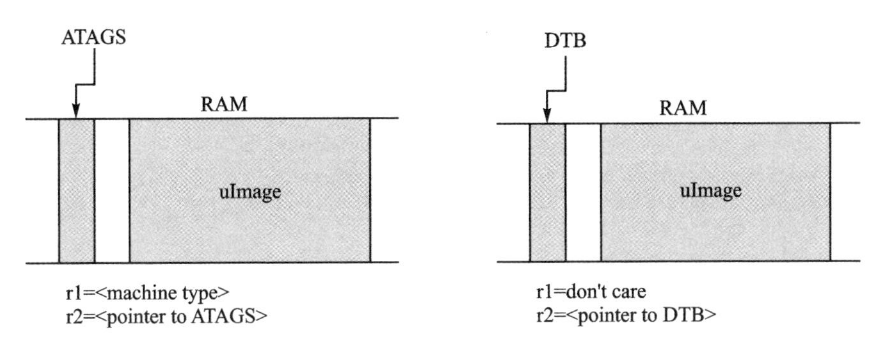

图 13.1 ATAGS 与 DTB 的对比图

13.1.2 设备树概述

设备树(Device Tree)是一种描述硬件的数据结构,起源于 OpenFirmware (OF)。Linux 之所以么会引入设备树,主要是在过去的 ARM Linux(ARM Linux V3.1 之前)中,ARM 架构的板极硬件目录 arch/arm/plat-xxx 和 arch/arm/mach-xxx 中充斥着大量的描述板级细节的代码,而这些板级细节对于内核来讲,并无用处,如板上的 platform 设备、resource、i2c_board_info、spi_board_info 以及各种硬件的 platform_data。在采用 Device Tree 后,许多硬件的细节可以直接通过 DTB 传递给 Linux,而不再需要在 Kernel 中包含大量的冗余代码。在内核文档及网络上都有关于设备树的详细介绍及使用:

- http://elinux.org/Device_Tree_Usage;
- http://blog.csdn.net/21cnbao/article/details/8457546;
- http://www.linuxidc.com/Linux/2016-12/137986.htm;
- http://www.linuxidc.com/Linux/2016-01/127337.htm;
- http://www.cnblogs.com/xiaojiang1025/p/6131381.html;
- http://www.wowotech.net/linux_kenrel/pin-control-subsystem.html。

下面为设备树的几个常用文件:

DTS(Device Tree Source)和 DTSI:设备树源文件,是一种 ASCII 文本格式的 Device Tree 描述。在 ARM Linux 中,一个.dts 文件通常对应一个 ARM 处理器,放置在内核的 arch/arm/boot/dts/目录下。而由于一个 SoC(即一个系列的 ARM 处理器)可能对应多个处理器型号(一个 SoC 也可能对应多个产品和电路板),所以使这些.dts 文件包含许多共同的部分。Linux 内核为了简化,把 SoC 共用的部分或者多个处理器共同的部分提炼为.dtsi(类似于 C 语言的头文件),其他的处理器对应的.dts 就包含(include)这个.dtsi,而.dtsi 又可以包含(include)其他的.dtsi。.dts 文件为板级定义,.dtsi 文件为 SoC 级定义。

.dts(或者其包含(include)的.dtsi)的基本元素为节点和属性,如图 13.2 所示。

```
/dts-v1/;

/ {
    node1 {
        a-string-property = "A string";
        a-string-list-property = "first string", "second string";
        // hex is implied in byte arrays. no '0x' prefix is required
        a-byte-data-property = [01 23 34 56];
        child-node1 {
            first-child-property;
            second-child-property = <1>;
            a-string-property = "Hello, world";
        };
        child-node2 {
        };
    };
    node2 {
        an-empty-property;
        a-cell-property = <1 2 3 4>; /* each number (cell) is a uint32 */
        child-node1 {
        };
    };
};
```

图 13.2 .dts 文件的基本数据格式

- 一个根节点:"/"。
- 根节点下面有一系列的子节点,如:node1 和 node2。
- 子节点下面有可能还会包含更多级的子节点,如:child-node1 和 child-node2。
- 各个节点都有一系列属性,这些属性可以为空,如:an-empty-property 可能为字符串,如:a-string-property 可能为字符串数组,如:a-string-list-property 可能为 cells(由 32 位无符号整数组成),如:a-cell-property,可能为十六进制数,如:a-byte-data-property 等。

DTB(Device Tree Blob):设备树 DTS 的二进制形式,下载到处理器 Flash 中,供处理器使用。

DTC(Device Tree Compiler):设备树编译工具,负责将 DTS 转换成 DTB。

使用中,应先根据硬件修改 DTS 文件,然后在编译的时候通过 DTC 工具将 DTS 文件转换成 DTB 文件,最后将 DTB 文件和 Bootloader、内核烧写到处理器的 Flash 存储器,或 TF 卡中。系统启动时,Bootloader(如 U-boot)在启动内核前将 DTB 文件读到内存中,在跳转到内核执行的同时将 DTB 起始地址传给内核。内核通过起始地址就可以根据 DTB 的结构解析整个设备树,并与内核中的驱动代码进行匹配。

13.1.3　AM335x Starter Kit 设备树分析

AM335x Starter Kit 开发板的设备树文件位于 Linux 内核源码根目录 arch\arm\boot\dts\下，am335x-evmsk.dtb 为最终下载到目标板的设备树二进制文件，而关于 am335x-evmsk.dtb 的编译、安装，第 12 章已经介绍过。am335x-evmsk.dts 为设备树的源文件，描述了 AM335x Starter Kit(即 TI AM335x EVM-SK)目标板的内存、电源及各个外设驱动等板上硬件，下面将简要概述，详细内容请参考源文件。

```
/*
 * AM335x Starter Kit
 * http://www.ti.com/tool/tmdssk3358
 */
/dts-v1/;
#include "am33xx.dtsi"//包含 am33xx 系列处理器公用的部分，即 SoC 级定义
#include <dt-bindings/pwm/pwm.h>
#include <dt-bindings/interrupt-controller/irq.h>
/{
    model = "TI AM335x EVM-SK";  //板级名称，即 TI AM335x Starter Kit
    compatible = "ti,am335x-evmsk","ti,am33xx";
    //该 compatible 定义系统的名称，组织形式为："<manufacturer>,<model>""兼容其
    //他系统"……
    //Linux 内核透过 root 节点"/"的 compatible 属性即可判断它启动的是什么 machine
    //即此处定义系统为 TI 的 am335x-evmsk，兼容 TI 的 am33xx 系列
    cpus {  //描述 CPU 数量和类别，该板为单核，即只有 1 个 CPU
        cpu@0 { //CPU0
            cpu0-supply = <&vdd1_reg>;
            //指定了 CPU0 的供电电源：vdd1_reg 会在下面单独定义
    };
    };
    memory { //定义内存地址和大小
        device_type = "memory";
        reg = <0x80000000 0x10000000>;
        //256 MB，基地址 = 0x80000000，大小 = 0x10000000 = 256MB
    };
    vbat: fixedregulator@0 { //固定电压调节器 0：输入电源 vbat
    //参考板原理图，VBAT 为外部输入电源，给电源管理 IC(TPS65910)各个 DC-DC 或 LDO 供电
        compatible = "regulator-fixed";
        //对应 regulator-fixed 所指的驱动，和下述多个 fixedregulator 一样
        //将来会在驱动 device_idtable 里查找 compatible 项，为 regulator-fixed 进行
       //匹配，属于 TPS65910 驱动
```

```
    regulator-name = "vbat";
    regulator-min-microvolt = <5000000>;//固定电压值 5 V
    regulator-max-microvolt = <5000000>;
    regulator-boot-on; //默认上电 boot 就打开 ON
};
lis3_reg: fixedregulator@1 {
//固定电压调节器 1: lis3_reg,加速度计 LIS331DLH 供电电源
    compatible = "regulator-fixed";
    regulator-name = "lis3_reg";
    regulator-boot-on; //默认上电 boot 就打开 ON
};
wl12xx_vmmc: fixedregulator@2 { //固定电压调节器 2: wl12xx_vmmc
    pinctrl-names = "default";
    //定义 pinctrl-names 属性(pinctrl-names property names the state)
    //共有 default、sleep、idle、init
    pinctrl-0 = <&wl12xx_gpio>;
    //GPIO 引脚复用,表示状态,对应 pinctrl-names 中的 default
    compatible = "regulator-fixed";
    regulator-name = "vwl1271";//WLAN 芯片
    //Starter Kit 板使用 LBEE5ZSTNC-523 (WLAN + BLUETOOTH)
    regulator-min-microvolt = <1800000>;//固定电压 1.8 V
    regulator-max-microvolt = <1800000>;
    gpio = <&gpio1 29 0>;//GPIO1_29:AM335X_MCASP0_AHCLKR(COMWL_RST) = WLAN_EN
    startup-delay-us = <70000>;
    enable-active-high; //高电平有效
};
vtt_fixed: fixedregulator@3 {//固定电压调节器 3: vtt_fixed
    compatible = "regulator-fixed";
    regulator-name = "vtt";
    regulator-min-microvolt = <1500000>;//1.5V
    regulator-max-microvolt = <1500000>;
    gpio = <&gpio0 7 GPIO_ACTIVE_HIGH>;//GPIO0_7: DDR_VTT_EN = TPS51200 EN
    regulator-always-on;//一直打开 ON
    regulator-boot-on; //默认上电 boot 就打开 ON
    enable-active-high; //高电平有效
};
leds {//共四个 LED
    pinctrl-names = "default","sleep";
    pinctrl-0 = <&user_leds_default>;//对应 pinctrl-names 中的 default
    pinctrl-1 = <&user_leds_sleep>;对应 pinctrl-names 中的 sleep
    compatible = "gpio-leds";//对应 gpio-leds 所指的驱动
    led@1 {
```

```
            label = "evmsk:green:usr0";//用户自定义 LED
            gpios = <&gpio1 4 GPIO_ACTIVE_HIGH>;//AM335X_GPIO_LED4
            default - state = "off";//默认 default 状态为灭
        };
        led@2 {
            label = "evmsk:green:usr1";//用户自定义 LED
            gpios = <&gpio1 5 GPIO_ACTIVE_HIGH>;//AM335X_GPIO_LED3
            default - state = "off";//默认 default 状态为灭
        };
        led@3 {
            label = "evmsk:green:mmc0";//MMC0 指示灯
            gpios = <&gpio1 6 GPIO_ACTIVE_HIGH>;//AM335X_GPIO_LED2
            linux,default - trigger = "mmc0";
            default - state = "off";//默认 default 状态为灭
        };
        led@4 {
            label = "evmsk:green:heartbeat";//心跳指示灯
            gpios = <&gpio1 7 GPIO_ACTIVE_HIGH>;//AM335X_GPIO_LED1
            linux,default - trigger = "heartbeat";
            default - state = "off";//默认 default 状态为灭
        };
    };
    gpio_buttons: gpio_buttons@0 {//共四个按键
        compatible = "gpio - keys";//对应 gpio - keys 所指的驱动
        //cell 在 Device Tree 表示 32 bit 的信息单位
        //# address - cells 用来描述子节点中 reg 属性的地址表中用来描述首地址的
        //cell 的数量
        //# size - cells,用来描述子节点 reg 属性的地址表中用来描述地址长度的 cell
        //的数量
        # address - cells = <1>;//用 1 个 u32 来描述地址域
        # size - cells = <0>;//地址长度为 0 个 cell
        switch@1 {
            label = "button0";
            linux,code = <0x100>;//1 个 address - cell,首地址为 0x100
            gpios = <&gpio2 3 GPIO_ACTIVE_HIGH>;//GPIO2_3:GPIO_KEY2,高电平有效
        };
        switch@2 {
            label = "button1";
            linux,code = <0x101>;//首地址为 0x101
            gpios = <&gpio2 2 GPIO_ACTIVE_HIGH>;//GPIO2_2:GPIO_KEY3,高电平有效
        };
        switch@3 {
```

```
            label = "button2";
            linux,code = <0x102>;//首地址为 0x102
            gpios = <&gpio0 30 GPIO_ACTIVE_HIGH>;//GPIO0_30:GPIO_KEY1,高电平有效
            gpio-key,wakeup;//待机唤醒按键
        };
        switch@4 {
            label = "button3";
            linux,code = <0x103>;//首地址为 0x103
            gpios = <&gpio2 5 GPIO_ACTIVE_HIGH>;//GPIO2_5:GPIO_KEY4,高电平有效
        };
    };
    backlight {//LCD 背光电源 PWM 控制输出
        compatible = "pwm-backlight";
        pwms = <&ecap2 0 50000 PWM_POLARITY_INVERTED>;//ECAP2_IN_PWM2_OUT
        //PWM_POLARITY_INVERTED:极性反转输出
        brightness-levels = <0 58 61 66 75 90 125 170 255>;//0~255,共 9 个 level
        default-brightness-level = <8>;//默认 level
    };
    sound {//声卡
        compatible = "simple-audio-card";
        simple-audio-card,name = "AM335x-EVMSK";
        simple-audio-card,widgets =
            "Headphone","Headphone Jack";//耳麦,耳机插孔
        simple-audio-card,routing =
            "Headphone Jack","HPLOUT",//耳机插孔左声道
            "Headphone Jack","HPROUT";//耳机插孔右声道
        simple-audio-card,format = "dsp_b";//音频格式:dsp_b
        simple-audio-card,bitclock-master = <&sound_master>;//主位时钟
        simple-audio-card,frame-master = <&sound_master>;//主帧
        simple-audio-card,bitclock-inversion;//位时钟反转
        simple-audio-card,cpu {//声卡 CPU
            sound-dai = <&mcasp1>;
        };
        sound_master: simple-audio-card,codec {//声卡 codec 解码器
            sound-dai = <&tlv320aic3106>;//声卡解码器芯片型号 tlv320aic3106
            system-clock-frequency = <24000000>;//系统时钟频率 24 MHz
        };
    };
    panel {//LCD
        compatible = "ti,tilcdc,panel";//匹配内核驱动的关键词
        pinctrl-names = "default","sleep";
        pinctrl-0 = <&lcd_pins_default>;
```

```
        pinctrl-1 = <&lcd_pins_sleep>;
        status = "okay";
        panel-info {//LCD 硬件配置信息
            ac-bias = <255>;//偏置电压,该数代表使用多少个电压参考点来驱动 LCD
            ac-bias-intrpt = <0>;
            dma-burst-sz = <16>;//DMA 传输请求时,
            //对于突发传输大小(burst size)的设置,即一次传送几个数据宽度(宽度由
            //LCD 数据线确定)
            bpp = <32>;//颜色分辨率,
            //如 24BPP=16M,即使用 24 位的数据来表示一个像素的颜色,每种原色使用 8
            //位(8:8:8)
            fdd = <0x80>;//FIFO DMA Request Delay
            sync-edge = <0>;//LCD_HSYNC 脉冲和 LCD_VSYNC 脉冲下降沿有效
            sync-ctrl = <1>;//允许配置同步脉冲上升沿还是下降沿有效,即配置
                            //sync-edge 有效
            raster-order = <0>;//行扫顺序
            fifo-th = <0>;//FIFO 阈值
        };
        display-timings {//LCD 时序
            480x272 {//LCD 分辨率
                hactive = <480>;//水平(行)方向有效像素点数
                vactive = <272>;//垂直(场)方向有效像素点数
                hback-porch = <43>; //Horizontal Back Porch Value(from 0~255)
                                    //水平方向后肩消影时间
                hfront-porch = <8>;//Horizontal Front Porch Value(from 0~255)
                                   //水平方向前肩消影时间
                hsync-len = <4>;//水平同步的长度
                vback-porch = <12>;//Vertical Back Porch Value(from 0~255)
                                   //垂直方向后肩消影时间
                vfront-porch = <4>;//Vertical Front Porch Value(from 0~255)
                                   //垂直方向前肩消影时间
                vsync-len = <10>; //垂直同步的长度
                clock-frequency = <9000000>;//像素时钟频率
                hsync-active = <0>;//水平方向同步的有效像素点
                vsync-active = <0>;//垂直方向同步的有效像素点
            };
        };
    };
};//根节点"/"结束
&am33xx_pinmux {//引脚复用配置信息
    pinctrl-names = "default";
    pinctrl-0 = <&gpio_keys_s0 &clkout2_pin &ddr3_vtt_toggle>;
```

```
ddr3_vtt_toggle: ddr3_vtt_toggle {
//芯片引脚复用功能的定义使用 pinctrl-single,pins 这个驱动
//格式: pinctrl-single,pins = <offset,function>,offset 为地址偏移量,
//function 为当前处理哪种模式
//内核解析该属性后根据 offset 和 function 配置对应的寄存器,为 GPIO 引脚配置为指
//定的复用功能
    pinctrl-single,pins = <
        0x164 (PIN_OUTPUT | MUX_MODE7) /* ecap0_in_pwm0_out.gpio0_7 */
        //#define   PIN_OUTPUT      (PULL_DISABLE)
        //定义输出时,禁止 GPIO 口内部上下拉功能,即开漏输出
        //#define   PULL_DISABLE   (1 << 3)
        //#define   MUX_MODE7     7
        //参考 am3359.pdf 关于引脚复用知: MODE0 = eCAP0_in_PWM0_out;MODE7 =
        //gpio0_7
        //Control Module 寄存器基地址为 0x44E1_0000,pin mux 的起始为 800h(conf_
        //gpmc_ad0)
        //964h 为 conf_ecap0_in_pwm0_out,所以这里的 offset = 0x164
        //这里该引脚设置为 gpio 输出,原理图中为 DDR_VTT_EN:VTT_DDR 电源使能
    >;
};
lcd_pins_default: lcd_pins_default {
//默认状态下: LCD 相关的多功能复用引脚的功能设置为 LCD
    pinctrl-single,pins = <
        0x20 (PIN_OUTPUT | MUX_MODE1)/* gpmc_ad8.lcd_data23 */
        //MODE0 = gpmc_ad8; MODE1 = lcd_data23
        ......
        0x3c (PIN_OUTPUT | MUX_MODE1)/* gpmc_ad15.lcd_data16 */
        //MODE0 = gpmc_ad15; MODE1 = lcd_data16
        0xa0 (PIN_OUTPUT | MUX_MODE0)/* lcd_data0.lcd_data0 */
        //MODE0 = lcd_data0
        ......
        0xdc (PIN_OUTPUT | MUX_MODE0)/* lcd_data15.lcd_data15 */
        //MODE0 = lcd_data15
        0xe0 (PIN_OUTPUT | MUX_MODE0)/* lcd_vsync.lcd_vsync */
        0xe4 (PIN_OUTPUT | MUX_MODE0)/* lcd_hsync.lcd_hsync */
        0xe8 (PIN_OUTPUT | MUX_MODE0)/* lcd_pclk.lcd_pclk */
        0xec (PIN_OUTPUT | MUX_MODE0)/* lcd_ac_bias_en.lcd_ac_bias_en */
    >;
};
lcd_pins_sleep: lcd_pins_sleep {
//待机休眠下: LCD 相关的多功能复用引脚的功能设置为 gpio 输入
    pinctrl-single,pins = <
```

```
            0x20 (PIN_INPUT_PULLDOWN | MUX_MODE7)/* gpmc_ad8.lcd_data23 */
            //MODE0 = gpmc_ad8; MODE1 = lcd_data23; MODE7 = gpio0_22
            ......
            0x3c (PIN_INPUT_PULLDOWN | MUX_MODE7)/* gpmc_ad15.lcd_data16 */
            //MODE0 = gpmc_ad8; MODE1 = lcd_data16; MODE7 = gpio1_15
            0xa0 (PULL_DISABLE | MUX_MODE7)/* lcd_data0.lcd_data0 */
            //MODE7 = gpio2_6
            ......
            0xdc (PULL_DISABLE | MUX_MODE7)/* lcd_data15.lcd_data15 */
            //MODE7 = gpio0_11
            0xe0 (PIN_INPUT_PULLDOWN | MUX_MODE7)/* lcd_vsync.lcd_vsync */
            0xe4 (PIN_INPUT_PULLDOWN | MUX_MODE7)/* lcd_hsync.lcd_hsync */
            0xe8 (PIN_INPUT_PULLDOWN | MUX_MODE7)/* lcd_pclk.lcd_pclk */
            0xec (PIN_INPUT_PULLDOWN | MUX_MODE7)/* lcd_ac_bias_en.lcd_ac_bias_en */
        >;
};
user_leds_default: user_leds_default {//默认 LED 引脚配置成 GPIO 口输出,且上拉
        pinctrl-single,pins = <
            0x10 (PIN_OUTPUT_PULLDOWN | MUX_MODE7)/* gpmc_ad4.gpio1_4 */
            0x14 (PIN_OUTPUT_PULLDOWN | MUX_MODE7)/* gpmc_ad5.gpio1_5 */
            0x18 (PIN_OUTPUT_PULLDOWN | MUX_MODE7)/* gpmc_ad6.gpio1_6 */
            0x1c (PIN_OUTPUT_PULLDOWN | MUX_MODE7)/* gpmc_ad7.gpio1_7 */
        >;
};
user_leds_sleep: user_leds_sleep {
//待机休眠时 LED 引脚配置成 GPIO 口输入并下拉为低电平
        pinctrl-single,pins = <
            0x10 (PIN_INPUT_PULLDOWN | MUX_MODE7)
            0x14 (PIN_INPUT_PULLDOWN | MUX_MODE7)
            0x18 (PIN_INPUT_PULLDOWN | MUX_MODE7)
            0x1c (PIN_INPUT_PULLDOWN | MUX_MODE7)
        >;
};
gpio_keys_s0: gpio_keys_s0 {//按键引脚配置成 GPIO 口输入
        pinctrl-single,pins = <
            0x94 (PIN_INPUT_PULLDOWN | MUX_MODE7)/* gpmc_oen_ren.gpio2_3 */
            0x90 (PIN_INPUT_PULLDOWN | MUX_MODE7)/* gpmc_advn_ale.gpio2_2 */
            0x70 (PIN_INPUT_PULLDOWN | MUX_MODE7)/* gpmc_wait0.gpio0_30 */
            0x9c (PIN_INPUT_PULLDOWN | MUX_MODE7)/* gpmc_ben0_cle.gpio2_5 */
        >;
};
i2c0_pins: pinmux_i2c0_pins {//I2C0
```

```
    pinctrl-single,pins = <
        0x188 (PIN_INPUT_PULLUP | MUX_MODE0)/* i2c0_sda.i2c0_sda */
        0x18c (PIN_INPUT_PULLUP | MUX_MODE0)/* i2c0_scl.i2c0_scl */
    >;
};
uart0_pins: pinmux_uart0_pins {//UART0
    pinctrl-single,pins = <
        0x170 (PIN_INPUT_PULLUP | MUX_MODE0)/* uart0_rxd.uart0_rxd */
        0x174 (PIN_OUTPUT_PULLDOWN | MUX_MODE0) /* uart0_txd.uart0_txd */
    >;
};
clkout2_pin: pinmux_clkout2_pin {
//Starter Kit 原理图 AM335X_XDMA_EVENT_INTR1 没有使用
    pinctrl-single,pins = <
        0x1b4 (PIN_OUTPUT_PULLDOWN | MUX_MODE3)/* xdma_event_intr1.clkout2 */
    >;
};
ecap2_pins: backlight_pins{
//LCD 背光驱动电源调节引脚 LCD_BACKLIGHTEN 设置成 PWM OUT
    pinctrl-single,pins = <
        0x19c 0x4 /* mcasp0_ahclkr.ecap2_in_pwm2_out MODE4 */
        //MODE4 = ecap2_in_pwm2_out
    >;
};
cpsw_default: cpsw_default {//默认状态下：双千兆以太网接口配置
    pinctrl-single,pins = <
        /* Slave 1 */
        0x114 (PIN_OUTPUT_PULLDOWN | MUX_MODE2)/* mii1_txen.rgmii1_tctl */
        ……
        0x140 (PIN_INPUT_PULLDOWN | MUX_MODE2)/* mii1_rxd0.rgmii1_rd0 */
        /* Slave 2 */
        0x40 (PIN_OUTPUT_PULLDOWN | MUX_MODE2)/* gpmc_a0.rgmii2_tctl */
        ……
        0x6c (PIN_INPUT_PULLDOWN | MUX_MODE2)/* gpmc_a11.rgmii2_rd0 */
    >;
};
cpsw_sleep: cpsw_sleep {//待机休眠状态下：双千兆以太网接口配置成 GPIO 输入
    pinctrl-single,pins = <
        /* Slave 1 reset value */
        0x114 (PIN_INPUT_PULLDOWN | MUX_MODE7)
        ……
        0x140 (PIN_INPUT_PULLDOWN | MUX_MODE7)
```

```
            /* Slave 2 reset value */
            0x40 (PIN_INPUT_PULLDOWN | MUX_MODE7)
            ……
            0x6c (PIN_INPUT_PULLDOWN | MUX_MODE7)
        >;
    };
    davinci_mdio_default: davinci_mdio_default {
    //默认情况下：GPIO 配置成千兆以太网的 MDIO 接口信号
        pinctrl-single,pins = <
        /* MDIO */
            0x148 (PIN_INPUT_PULLUP | SLEWCTRL_FAST | MUX_MODE0)    /* mdio_data.
mdio_data */
            0x14c (PIN_OUTPUT_PULLUP | MUX_MODE0)/* mdio_clk.mdio_clk */
        >;
    };
    davinci_mdio_sleep: davinci_mdio_sleep {
    //待机休眠：千兆以太网 MDIO 接口配置成 GPIO 输入
        pinctrl-single,pins = <
        /* MDIO reset value */
            0x148 (PIN_INPUT_PULLDOWN | MUX_MODE7)
            0x14c (PIN_INPUT_PULLDOWN | MUX_MODE7)
        >;
    };
    mmc1_pins: pinmux_mmc1_pins {//MicroSD 写保护检测输入引脚配置
        pinctrl-single,pins = <
            0x160 (PIN_INPUT | MUX_MODE7) /* spi0_cs1.gpio0_6 */
            //AM335X_SPI0_CS1 信号配置成 GPIO 输入
        >;
    };
    mcasp1_pins: mcasp1_pins {//mcasp1
        pinctrl-single,pins = <
            0x10c (PIN_INPUT_PULLDOWN | MUX_MODE4) /* mii1_crs.mcasp1_aclkx */
            0x110 (PIN_INPUT_PULLDOWN | MUX_MODE4) /* mii1_rxerr.mcasp1_fsx */
            0x108 (PIN_OUTPUT_PULLDOWN | MUX_MODE4) /* mii1_col.mcasp1_axr2 */
            0x144 (PIN_INPUT_PULLDOWN | MUX_MODE4) /* rmii1_ref_clk.mcasp1_axr3 */
        >;
    };
    mcasp1_pins_sleep: mcasp1_pins_sleep {
        pinctrl-single,pins = <
            0x10c (PIN_INPUT_PULLDOWN | MUX_MODE7)
            0x110 (PIN_INPUT_PULLDOWN | MUX_MODE7)
            0x108 (PIN_INPUT_PULLDOWN | MUX_MODE7)
```

```
            0x144 (PIN_INPUT_PULLDOWN | MUX_MODE7)
        >;
    };
    mmc2_pins: pinmux_mmc2_pins {//WLAN 接口配置
        pinctrl-single,pins = <
            0x74 (PIN_INPUT_PULLUP | MUX_MODE7) /* gpmc_wpn.gpio0_31 */
            0x80 (PIN_INPUT_PULLUP | MUX_MODE2) /* gpmc_csn1.mmc1_clk */
            0x84 (PIN_INPUT_PULLUP | MUX_MODE2) /* gpmc_csn2.mmc1_cmd */
            0x00 (PIN_INPUT_PULLUP | MUX_MODE1) /* gpmc_ad0.mmc1_dat0 */
            0x04 (PIN_INPUT_PULLUP | MUX_MODE1) /* gpmc_ad1.mmc1_dat1 */
            0x08 (PIN_INPUT_PULLUP | MUX_MODE1) /* gpmc_ad2.mmc1_dat2 */
            0x0c (PIN_INPUT_PULLUP | MUX_MODE1) /* gpmc_ad3.mmc1_dat3 */
        >;
    };
    wl12xx_gpio: pinmux_wl12xx_gpio {//WLAN_EN
        pinctrl-single,pins = <
            0x7c (PIN_OUTPUT_PULLUP | MUX_MODE7) /* gpmc_csn0.gpio1_29 */
        >;
    };
};//&am33xx_pinmux 结束
&uart0 {//UART0
    pinctrl-names = "default";
    pinctrl-0 = <&uart0_pins>;//UART0 引脚复用配置(指向之前的 uart0_pins)
    status = "okay";
};
&i2c0 {
    pinctrl-names = "default";
    pinctrl-0 = <&i2c0_pins>;
    status = "okay";
    clock-frequency = <400000>;//IIC 时钟 400 kHz
    tps: tps@2d {//电源管理芯片 TPS65910A3
        reg = <0x2d>;//TPS65910A3 IIC 接口器件地址
    };
    lis331dlh: lis331dlh@18 {//Accelerometer 加速度计 LIS331DLH
        compatible = "st,lis331dlh","st,lis3lv02d";
        reg = <0x18>;//LIS331DLH  IIC 接口器件地址
        Vdd-supply = <&lis3_reg>;
        Vdd_IO-supply = <&lis3_reg>;
        st,click-single-x;
        st,click-single-y;
        st,click-single-z;
        st,click-thresh-x = <10>;
```

```
        st,click-thresh-y = <10>;
        st,click-thresh-z = <10>;
        st,irq1-click;
        st,irq2-click;
        st,wakeup-x-lo;
        st,wakeup-x-hi;
        st,wakeup-y-lo;
        st,wakeup-y-hi;
        st,wakeup-z-lo;
        st,wakeup-z-hi;
        st,min-limit-x = <120>;
        st,min-limit-y = <120>;
        st,min-limit-z = <140>;
        st,max-limit-x = <550>;
        st,max-limit-y = <550>;
        st,max-limit-z = <750>;
    };
    tlv320aic3106: tlv320aic3106@1b {//音频编解码器 TLV320AIC3106
        #sound-dai-cells = <0>;
        compatible = "ti,tlv320aic3106";//音频驱动
        reg = <0x1b>;//TLV320AIC3106 IIC 接口器件地址
        status = "okay";
        /* Regulators */
        AVDD-supply = <&vaux2_reg>; //V3_3AUD(由 V3_3D 通过 FB11 磁珠所得)
        IOVDD-supply = <&vaux2_reg>;//V3_3D
        DRVDD-supply = <&vaux2_reg>;//V3_3AUD
        DVDD-supply = <&vbat>;//注:实际原理图中 DVDD 为 V1_8D 提供
    };
};
&usb {
    status = "okay";
};
&usb_ctrl_mod {
    status = "okay";
};
&usb0_phy {
    status = "okay";
};
&usb1_phy {
    status = "okay";
};
&usb0 {
```

```
    status = "okay";
};
&usb1 {
    status = "okay";
    dr_mode = "host";
};
&cppi41dma  {
    status = "okay";
};
&epwmss2 {//PWM2 LCD 背光调节输出
    status = "okay";
    ecap2: ecap@48304100 {//0x4830_4100 = PWMSS eCAP2 Registers
        status = "okay";
        pinctrl-names = "default";
        pinctrl-0 = <&ecap2_pins>;//指向 LCD_BACKLIGHTEN 引脚配置
    };
};
&wkup_m3_ipc {
    ti,needs-vtt-toggle;
    ti,vtt-gpio-pin = <7>;
    ti,scale-data-fw = "am335x-evm-scale-data.bin";
};
#include "tps65910.dtsi"
&tps {//电源管理芯片 TPS65910A3
    vcc1-supply = <&vbat>;//vcc1~vcc7 的电源输入都由 vbat 提供
    ......
    vcc7-supply = <&vbat>;
    vccio-supply = <&vbat>;
    regulators {
        vrtc_reg: regulator@0 {//VCC7
            regulator-always-on;
        };
        vio_reg: regulator@1 {//VCCIO
            regulator-always-on;
        };
        vdd1_reg: regulator@2 {//VCC1
        /* VDD_MPU voltage limits 0.95~1.26V with ±4% tolerance */
            regulator-name = "vdd_mpu";
            //MPU 提供电源 VDD_MPU,可根据频率功率等动态调整
            regulator-min-microvolt = <912500>;//最小电压值
            regulator-max-microvolt = <1351500>;//最大电压值
            regulator-boot-on;
```

```
            regulator-always-on;
        };
        vdd2_reg: regulator@3 {//VCC2
            /* VDD_CORE voltage limits 0.95~1.1 V with ±4% tolerance */
            regulator-name = "vdd_core";
            //CORE 提供电源 VDD_CORE,可根据频率功率等动态调整
            regulator-min-microvolt = <912500>;//最小电压值
            regulator-max-microvolt = <1150000>;//最大电压值
            regulator-boot-on;
            regulator-always-on;
        };
        vdd3_reg: regulator@4 {
            regulator-always-on;
        };
        vdig1_reg: regulator@5 {//VCC6
            regulator-always-on;
        };
        vdig2_reg: regulator@6 {//VCC6
            regulator-always-on;
        };
        vpll_reg: regulator@7 {//VCC5
            regulator-always-on;
        };
        vdac_reg: regulator@8 {//VCC5
            regulator-always-on;
        };
        vaux1_reg: regulator@9 {//VCC4
            regulator-always-on;
        };
        vaux2_reg: regulator@10 {//VCC4
            regulator-always-on;
        };
        vaux33_reg: regulator@11 {//VCC3
            regulator-always-on;
        };
        vmmc_reg: regulator@12 {//VCC3
            regulator-min-microvolt = <1800000>;//最小电压值
            regulator-max-microvolt = <3300000>;//最大电压值
            regulator-always-on;
        };
    };
};
```

```
&mac {//以太网 MAC
    pinctrl-names = "default","sleep";
    pinctrl-0 = <&cpsw_default>;//指向双千兆以太网接口配置
    pinctrl-1 = <&cpsw_sleep>;
    dual_emac = <1>;
    status = "okay";
};
&davinci_mdio {//以太网 MDIO 接口
    pinctrl-names = "default","sleep";
    pinctrl-0 = <&davinci_mdio_default>;//指向 MDIO 接口配置
    pinctrl-1 = <&davinci_mdio_sleep>;
    status = "okay";
};
&cpsw_emac0 {
    phy_id = <&davinci_mdio>,<0>;
    phy-mode = "rgmii-txid";
    dual_emac_res_vlan = <1>;
};
&cpsw_emac1 {
    phy_id = <&davinci_mdio>,<1>;
    phy-mode = "rgmii-txid";
    dual_emac_res_vlan = <2>;
};
&mmc1 {//MMC0:MicroSD
    status = "okay";
    vmmc-supply = <&vmmc_reg>;
    bus-width = <4>;
    pinctrl-names = "default";
    pinctrl-0 = <&mmc1_pins>;
    cd-gpios = <&gpio0 6 GPIO_ACTIVE_LOW>;//GPIO0_6 = AM335X_SPI0_CS1 = MMC0 CD
};
&sham {
    status = "okay";
};
&aes {
    status = "okay";
};
&gpio0 {
    ti,no-reset-on-init;
};
&mmc2 {//MMC1
    status = "okay";
```

```
    vmmc - supply = <&wl12xx_vmmc>;
    ti,non - removable;
    bus - width = <4>;
    cap - power - off - card;
    pinctrl - names = "default";
    pinctrl - 0 = <&mmc2_pins>;
    #address - cells = <1>;
    #size - cells = <0>;
    wlcore: wlcore@2 {
        compatible = "ti,wl1271";
        reg = <2>;
        interrupt - parent = <&gpio0>;
        interrupts = <31 IRQ_TYPE_LEVEL_HIGH>; /* gpio0_31 */
        ref - clock - frequency = <38400000>;
    };
};
&mcasp1 {//IIS
    #sound - dai - cells = <0>;
    pinctrl - names = "default","sleep";
    pinctrl - 0 = <&mcasp1_pins>;
    pinctrl - 1 = <&mcasp1_pins_sleep>;
    status = "okay";
    op - mode = <0>;            /* MCASP_IIS_MODE */
    tdm - slots = <2>;
    /* 4 serializers */
    serial - dir = <  /* 0: INACTIVE,1: TX,2: RX */
        0 0 1 2
    >;
    tx - num - evt = <32>;
    rx - num - evt = <32>;
};
&tscadc {//触摸屏 ADC
    status = "okay";
    tsc {
        ti,wires = <4>;
        ti,x - plate - resistance = <200>;
        ti,coordinate - readouts = <5>;
        ti,wire - config = <0x00 0x11 0x22 0x33>;
    };
};
&lcdc {
    status = "okay";
```

```
};
&sgx {
    status = "okay";
};
```

更详细的内容请读者参考：http://blog.csdn.net/girlkoo/article/details/41382663。

TI 的 GPIO 控制与三星芯片不同，三星的芯片引脚复用功能是放在 GPIO 寄存器中，而 TI 的芯片则有专门的控制模块叫 Control Module，该模块可以控制所有 GPIO 引脚功能的复用；此外，与三星芯片的另外一个不同点是，TI 芯片的描述分为技术参考手册和数据手册，技术参考手册非常详细地讲述同 family 的芯片功能及使用方法，数据手册则用来讲述同 family 中不同芯片特有的属性。因此，调试 TI 芯片时需要结合技术参考手册和数据手册，而配置 GPIO 则需要阅读技术手册的 GPIO、Control Module 两章和数据手册中相关的部分。

13.2　LED 显示驱动

13.2.1　AM335x 的 LED 控制

AM335x Starter Kit 板上有五个 LED，除了 D5 为常亮的电源指示灯外，其余的 D1～D4 四个 LED 都是可以对其进行亮灭控制的。

用 USB 线连接好 PC 和 AM335x Starter Kit 板，然后给 AM335x Starter Kit 板上电并启动 Linux 系统后，在 PC 上启动终端软件，即目标板的终端控制台。

cd/sys/class/leds 进入 led 子系统下的设备目录，ls 可以查看到包含四个设备，即四个 LED。当我们进入任何一个子设备目录后，都可以使用 echo 1 > brightness 点亮某个 LED 灯，如图 13.3 所示为点亮和熄灭板子上的 D1。

13.2.2　Linux 内核中的 leds 子系统概述

上一小节采用 echo 命令点亮 LED，与 S3C2410 中对 LED 的操作有很大的不同。S3C2410 中的 LED 驱动是字符型设备驱动，是在/dev/led 下生成文件，相对独立简单且易于理解。而 AM335x 则直接采用 Linux 系统自带的平台设备驱动 led 子系统，在/sys/class/leds/目录下。

Linux 内核中的平台设备 leds 子系统，主要是对 led 事件进行了分装和优化，是跨平台的 led 驱动，即不管是使用三星的平台，还是 TI 的平台，只要知道如何在 BSP 中添加平台数据，并且知道如何在应用程序中使用这个驱动，那么，就不用因为新的平台而再次编写 led 驱动。而对于应用层来说，由于不同平台都用 Linux 的 led 子系统，所以应用程序也不用做任何的改变，就可以在新的平台上运行，可移植性好。

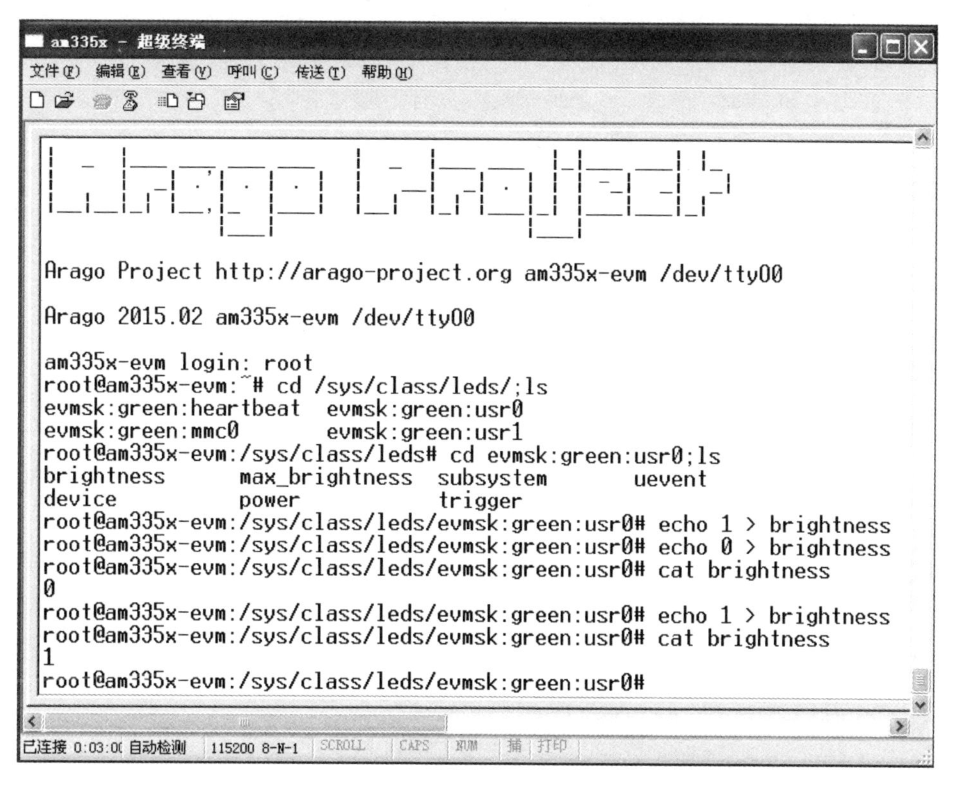

图 13.3 LED控制命令

内核中的leds子系统是将led抽象成platform_device,并有leds_class。这样,在/sys/class/leds/目录下面就可以利用sysfs文件系统来实现LED的操作。

13.2.3 leds子系统驱动代码分析

Linux的leds子系统的源码路径在/drivers/leds下(头文件为include/linux/leds.h),跨平台led驱动文件是/driver/leds/leds-gpio.c,子系统的核心文件是led-class.c和led-core.c。此外,还有led-triggers.c,分为timer、ide-disk、heartbeat、backlight、gpio、default-on等算法。

1. leds.h及主要的数据结构

在leds.h中定义了驱动中用到的数据结构、宏定义及声明的各接口函数。其中主要的数据结构如下:

```
enum led_brightness {
//led亮度,分为三级,灭、中间、最亮,但实际上led通常只有0(灭)、1(亮)
    LED_OFF = 0,
    LED_HALF = 127,
    LED_FULL = 255,
```

```
};

struct led_classdev {//led_classdev 代表 led 的实例
    const char    * name;//名字
    enum led_brightness brightness;//当前亮度值,也就是给 led 的值,写入 0 为灭,非 0 亮
    enum led_brightness max_brightness; //最大亮度值
    int   flags; //标志
#define LED_SUSPENDED              (1 << 0) //Lower 16 bits reflect status
#define LED_CORE_SUSPENDRESUME     (1 << 16) //Upper 16 bits reflect control information
#define LED_BLINK_ONESHOT          (1 << 17)
#define LED_BLINK_ONESHOT_STOP     (1 << 18)
#define LED_BLINK_INVERT           (1 << 19)
#define LED_SYSFS_DISABLE          (1 << 20)
#define SET_BRIGHTNESS_ASYNC       (1 << 21)
#define SET_BRIGHTNESS_SYNC        (1 << 22)
#define LED_DEV_CAP_FLASH          (1 << 23)
    //设置 led 的亮度,不可以睡眠,有必要的话可以使用工作队列
    void   ( * brightness_set)(struct led_classdev * led_cdev,enum led_brightness
brightness);
    //设置 led 的亮度(立即执行)
    int ( * brightness_set_sync)(struct led_classdev * led_cdev,enum led_brightness
brightness);
    enum led_brightness ( * brightness_get)(struct led_classdev * led_cdev); //获取亮度
    //闪烁时点亮和熄灭的时间设置
    int ( * blink_set)(struct led_classdev * led_cdev,unsigned long * delay_on,
unsigned long * delay_off);
    struct device   * dev;
    const struct attribute_group   * * groups;
    struct list_head   node;
    //LED Device list,即所有已经注册的 led_classdev 使用这个节点串联起来
    const char    * default_trigger; //默认触发器
    unsigned long   blink_delay_on,blink_delay_off; //闪烁的开关时间
    struct timer_list   blink_timer; //闪烁的定时器链表
    int blink_brightness; //闪烁的亮度
    void ( * flash_resume)(struct led_classdev * led_cdev);
    struct work_struct set_brightness_work;
    int delayed_set_value;
#ifdef CONFIG_LEDS_TRIGGERS //配置内核时使能触发器功能
    struct rw_semaphore trigger_lock;//trigger 锁,用于保护触发器数据
    struct led_trigger   * trigger;//触发器指针
    struct list_head   trig_list;
    //触发器使用的链表节点,用来连接同一触发器上的所有 led_classdev
```

```
    void   * trigger_data;//触发器使用的私有数据
    bool  activated;//true if activated - deactivate routine ues it sto do cleanup
#endif
    struct mutex   led_access; //Ensures consistent access to the LED Flash Class device
};

struct led_trigger {//触发器的结构体
    const char * name; //触发器名字
    void ( * activate)(struct led_classdev * led_cdev);
    //激活,led_classdev 和触发器建立连接时调用
    void ( * deactivate)(struct led_classdev * led_cdev);
    //取消激活,led_classdev 和触发器取消连接时调用
    rwlock_t  leddev_list_lock; //保护链表的锁
    struct list_head  led_cdevs; //链表头
    //连接下一个已注册触发器的链表节点 ,所有已注册的触发器都会被加入一个全局链表
    struct list_head  next_trig;
};

struct led_info {//平台设备相关的 led 数据结构
    const char * name;
    const char * default_trigger;
    int  flags;
};
struct led_platform_data {
    int  num_leds;
    struct led_info * leds;
};

/* For the leds-gpio driver */
struct gpio_led {//平台设备相关的 gpio led 数据结构
    const char * name;//led 的名字
    const char * default_trigger; //默认触发器的名字
    unsigned gpio;//使用的 gpio 编号
    unsigned active_low : 1; //如果为真则逻辑 1 代表低电平
    unsigned retain_state_suspended : 1;
    unsigned default_state : 2;
    //default_state should be one of LEDS_GPIO_DEFSTATE_(ON|OFF|KEEP)
    struct gpio_desc * gpiod;
};
#define LEDS_GPIO_DEFSTATE_OFF      0
#define LEDS_GPIO_DEFSTATE_ON       1
#define LEDS_GPIO_DEFSTATE_KEEP     2
```

```
struct gpio_led_platform_data {
    int num_leds;//led 的个数
    const struct gpio_led * leds;
#define GPIO_LED_NO_BLINK_LOW 0    /* No blink GPIO state low */
#define GPIO_LED_NO_BLINK_HIGH 1   /* No blink GPIO state high */
#define GPIO_LED_BLINK 2   /* Please,blink */
    int( * gpio_blink_set)(struct gpio_desc * desc,int state,unsigned long * delay_
on,unsigned long * delay_off);
    //硬件闪烁加速设置,可以为 NULL
};
```

2. led-core. c

led-core. c 主要声明了 leds 的链表及锁。


```
DECLARE_RWSEM(leds_list_lock);
EXPORT_SYMBOL_GPL(leds_list_lock);
LIST_HEAD(leds_list);
EXPORT_SYMBOL_GPL(leds_list);
```

另外还有其他的一些函数,这里就不一一列举了。

3. led-class. c

led-class. c 文件实现的功能总的来说是先建立一个 leds 类(led_classdev),然后在该类下建立一个设备节点,最后就在该设备节点下载建立几个属性文件(sysfs)。

现在假设有一个名为"evmsk:green:heartbeat"的 led_classdev 被注册了,那么会出现/sys/class/leds/evmsk:green:heartbeat 这个目录,这个目录下默认有 brightness 和 trigger 这两个属性文件,可以分别设置/读取 led 的亮度和触发器。如果和触发器 timer 建立了连接,还会有 delay_on 和 delay_off,这两个文件用于设置/读取熄灭和点亮的时间,单位是毫秒。当然还可以建立 max_brightness、subsystem、uevent、device、power 等其他属性文件。

```
static int __init leds_init(void){
//初始化,创建一个 leds 类 leds_class,即生成/sys/class/leds 类目录
    leds_class = class_create(THIS_MODULE,"leds");
        //将在/sys/classs/leds 目录下产生文件即产生 leds 类的文件名
        //第一个参数指定所属的模块,第二个参数指定设备的名字
    if (IS_ERR(leds_class))
        return PTR_ERR(leds_class);
    leds_class->pm = &leds_class_dev_pm_ops;
    leds_class->dev_groups = led_groups;
```

```
    return 0;
}
static void __exit leds_exit(void){//释放类 leds_class
    class_destroy(leds_class);
}
subsys_initcall(leds_init); //系统启动时就会被调用
module_exit(leds_exit);

void led_classdev_suspend(struct led_classdev * led_cdev){//挂起 led
    led_cdev->flags |= LED_SUSPENDED;//赋值 suspended 休眠挂起 led
    led_cdev->brightness_set(led_cdev,0);
    //将 led 的 flag 设为 LED_SUSPENDED,关闭 led
}
EXPORT_SYMBOL_GPL(led_classdev_suspend);
static int led_suspend(struct device * dev){//
    struct led_classdev * led_cdev = dev_get_drvdata(dev);
    if (led_cdev->flags & LED_CORE_SUSPENDRESUME)
        led_classdev_suspend(led_cdev);
    return 0;
}

void led_classdev_resume(struct led_classdev * led_cdev){//led 从挂起中恢复
    led_cdev->brightness_set(led_cdev,led_cdev->brightness);
    if (led_cdev->flash_resume)
        led_cdev->flash_resume(led_cdev);
    led_cdev->flags &= ~LED_SUSPENDED;//赋值 resume 恢复挂起
}
EXPORT_SYMBOL_GPL(led_classdev_resume);
static int led_resume(struct device * dev){
    struct led_classdev * led_cdev = dev_get_drvdata(dev);
    if (led_cdev->flags & LED_CORE_SUSPENDRESUME)
        led_classdev_resume(led_cdev);
    return 0;
}
```

brightness_show 和 brightness_store 分别负责显示和设置亮度,用户控件通过 /sys/class/leds/<device>/brightness 进行查看和设置亮度(如在上述目录时 cat brightness 和 echo 1 > brightness)就是和这两个函数交互的。

```
    static ssize_t brightness_show(struct device * dev,struct device_attribute * attr,
char * buf){//显示亮度
    struct led_classdev * led_cdev = dev_get_drvdata(dev);
```

```
        led_update_brightness(led_cdev); //no lock needed for this
        return sprintf(buf," % u\n",led_cdev - >brightness);//显示亮度值
    }
    static ssize_t brightness_store(struct device * dev,
            struct device_attribute * attr,const char * buf,size_t size){//设置亮度
        struct led_classdev * led_cdev = dev_get_drvdata(dev);
        unsigned long state;
        ssize_t ret;
        mutex_lock(&led_cdev - >led_access);
        if (led_sysfs_is_disabled(led_cdev)) {
            ret =  - EBUSY;
            goto unlock;
        }
        ret = kstrtoul(buf,10,&state);
        if (ret) goto unlock;
        if (state == LED_OFF) led_trigger_remove(led_cdev);
        led_set_brightness(led_cdev,state);//设置亮度值
        ret = size;
    unlock:
        mutex_unlock(&led_cdev - >led_access);
        return ret;
    }
    static DEVICE_ATTR_RW(brightness);//属性：亮度值,可读/写
    static ssize_t max_brightness_show(struct device * dev,struct device_attribute * at-
tr,char * buf){
        struct led_classdev * led_cdev = dev_get_drvdata(dev);
        return sprintf(buf," % u\n",led_cdev - >max_brightness);//显示最大亮度值
    }
    static DEVICE_ATTR_RO(max_brightness); //属性：最大亮度值,只读
    # ifdef CONFIG_LEDS_TRIGGERS
    static DEVICE_ATTR(trigger,0644,led_trigger_show,led_trigger_store);
    //生成 sysfs 属性文件 trigger
    //即/sys/class/leds/<device>/trigger,用于用户空间查看(cat trigger)和设置(echo
    //trigger)触发器
    //led_trigger_show 用于读取当前触发器的名字,led_trigger_store 用于指定触发器的名字,
    //它会寻找所有已注册的触发器,找到同名的并设置为当前 led 的触发器
    static struct attribute * led_trigger_attrs[] = {
        &dev_attr_trigger.attr,
        NULL,
    };
    static const struct attribute_group led_trigger_group = {
        .attrs = led_trigger_attrs,
```

```
};
#endif
static struct attribute * led_class_attrs[] = {
    &dev_attr_brightness.attr,
    &dev_attr_max_brightness.attr,
    NULL,
};
static const struct attribute_group led_group = {
    .attrs = led_class_attrs,
};
static const struct attribute_group * led_groups[] = {
    &led_group,
#ifdef CONFIG_LEDS_TRIGGERS
    &led_trigger_group,
#endif
    NULL,
};

static struct class * leds_class;//定义 leds 类
int led_classdev_register(struct device * parent,struct led_classdev * led_cdev){
//注册 struct led_classdev
//注册的 struct led_classdev 会被加入 leds_list 链表,这个链表定义在 driver/
//leds/led-core.c
    char name[64];
    int ret;
    ret = led_classdev_next_name(led_cdev->name,name,sizeof(name));
    if (ret < 0)   return ret;
    led_cdev->dev = device_create_with_groups(leds_class,parent,0,
        led_cdev,led_cdev->groups,"%s",name);
        //创建一个 struct device,它的父设备是 parent
        //drvdata 是 led_cdev,名字是 led_cdev->name,类别是 leds_class
    if (IS_ERR(led_cdev->dev))     return PTR_ERR(led_cdev->dev);
    if (ret)
        dev_warn(parent,"Led %s renamed to %s due to name collision",
            led_cdev->name,dev_name(led_cdev->dev));
#ifdef CONFIG_LEDS_TRIGGERS
    init_rwsem(&led_cdev->trigger_lock); //初始化 led_cdev 的触发器自旋锁
#endif
    mutex_init(&led_cdev->led_access);
    /* add to the list of leds */
```

```
    down_write(&leds_list_lock);
    list_add_tail(&led_cdev->node,&leds_list);
    //将新的 led 加入链表,全局链表是 leds_list
    up_write(&leds_list_lock);
    if (!led_cdev->max_brightness)
        led_cdev->max_brightness = LED_FULL;
    led_cdev->flags |= SET_BRIGHTNESS_ASYNC;
    led_update_brightness(led_cdev);
    //获取 led 当前的亮度更新 led_cdev 的 brightness 成员
    led_init_core(led_cdev);
#ifdef CONFIG_LEDS_TRIGGERS
    led_trigger_set_default(led_cdev);//为 led_cdev 设置默认的触发器
#endif
    dev_dbg(parent,"Registered led device: %s\n",led_cdev->name);
    return 0;
}
EXPORT_SYMBOL_GPL(led_classdev_register);
void led_classdev_unregister(struct led_classdev *led_cdev){//注销 struct led_classdev
#ifdef CONFIG_LEDS_TRIGGERS
    down_write(&led_cdev->trigger_lock);
    if (led_cdev->trigger)
        led_trigger_set(led_cdev,NULL);
    up_write(&led_cdev->trigger_lock);
#endif
    cancel_work_sync(&led_cdev->set_brightness_work);
    led_stop_software_blink(led_cdev);//Stop blinking
    led_set_brightness(led_cdev,LED_OFF);//关闭 LED 显示
    device_unregister(led_cdev->dev);
    down_write(&leds_list_lock);
    list_del(&led_cdev->node);
    up_write(&leds_list_lock);
    mutex_destroy(&led_cdev->led_access);
}
EXPORT_SYMBOL_GPL(led_classdev_unregister);
```

4. led-triggers.c

```
int led_trigger_register(struct led_trigger *trig){
//由该函数注册的 trigger 会被加入全局链表 trigger_list,链表头在/driver/leds/
//led-triggers.c 定义该函数会遍历所有已注册的 led_classdev,如果有哪个
//led_classdev 的默认触发器和自己同名,则调用 led_trigger_set 将自己设
//为那个 led 的触发器。led_classdev 注册的时候也会调用 led_trigger_set_default
```

```
//来遍历所有已注册的触发器，找到和 led_classdev.default_trigger 同名的触发
//器则将它设为自己的触发器
    struct led_classdev * led_cdev;
    struct led_trigger * _trig;
    rwlock_init(&trig->leddev_list_lock);//
    INIT_LIST_HEAD(&trig->led_cdevs);
    down_write(&triggers_list_lock);
    list_for_each_entry(_trig,&trigger_list,next_trig) {//确认触发器名不存在
        if (!strcmp(_trig->name,trig->name)) {
            up_write(&triggers_list_lock);
            return -EEXIST;
        }
    }
    list_add_tail(&trig->next_trig,&trigger_list);
    //Add to the list of led triggers
    up_write(&triggers_list_lock);//Register with any LEDs that have this as a default
trigger
    down_read(&leds_list_lock);
    list_for_each_entry(led_cdev,&leds_list,node) {
        down_write(&led_cdev->trigger_lock);
        if (!led_cdev->trigger && led_cdev->default_trigger &&
                !strcmp(led_cdev->default_trigger,trig->name))
            led_trigger_set(led_cdev,trig);
            //建立连接，连接时会调用触发器的 activate 方法
        up_write(&led_cdev->trigger_lock);
    }
    up_read(&leds_list_lock);
    return 0;
}
EXPORT_SYMBOL_GPL(led_trigger_register);
void led_trigger_unregister(struct led_trigger * trig){
//注销触发器，和注册触发器函数执行相反的工作
    struct led_classdev * led_cdev;
    if (list_empty_careful(&trig->next_trig))
        return;
    down_write(&triggers_list_lock);
    list_del_init(&trig->next_trig);//Remove from the list of led triggers
    up_write(&triggers_list_lock);
    //Remove anyone actively using this trigger
down_read(&leds_list_lock);
list_for_each_entry(led_cdev,&leds_list,node) {
    down_write(&led_cdev->trigger_lock);
```

```
        if (led_cdev->trigger == trig)
            led_trigger_set(led_cdev,NULL);
            //把所有和自己建立连接的 led 的 led_classdev.trigger 设为 NULL
        up_write(&led_cdev->trigger_lock);
    }
    up_read(&leds_list_lock);
}
EXPORT_SYMBOL_GPL(led_trigger_unregister);
void led_trigger_event(struct led_trigger * trig,enum led_brightness brightness){
//设置触发器上所有的 led 为某个亮度
    struct led_classdev * led_cdev;
    if (!trig)
        return;
    read_lock(&trig->leddev_list_lock);
    list_for_each_entry(led_cdev,&trig->led_cdevs,trig_list)
        led_set_brightness(led_cdev,brightness);
    read_unlock(&trig->leddev_list_lock);
}
EXPORT_SYMBOL_GPL(led_trigger_event);
void led_trigger_register_simple(const char * name,struct led_trigger ** tp){
//注册触发器的简单方法
//指定一个名字就可以注册一个触发器,注册的触发器通过 ** tp 返回,
//但是这样注册的触发器没有 active 和 deactivede
    struct led_trigger * trig;
    int err;
    trig = kzalloc(sizeof(struct led_trigger),GFP_KERNEL);
    if (trig) {
        trig->name = name;
        err = led_trigger_register(trig);
        if (err < 0) {
            kfree(trig);
            trig = NULL;
            pr_warn("LED trigger %s failed to register (%d)\n",name,err);
        }
    } else {
        pr_warn("LED trigger %s failed to register (no memory)\n",name);
    }
    * tp = trig;
}
EXPORT_SYMBOL_GPL(led_trigger_register_simple);
void led_trigger_unregister_simple(struct led_trigger * trig){
//"注册触发器的简单方法"相对应的注销函数
```

```
    if (trig)
        led_trigger_unregister(trig);
    kfree(trig);
}
EXPORT_SYMBOL_GPL(led_trigger_unregister_simple);
void led_trigger_set(struct led_classdev * led_cdev,struct led_trigger * trig){
//建立连接
//建立连接时会调用触发器的 activate 方法
    unsigned long flags;
    char * event = NULL;
    char * envp[2];
    const char * name;
    name = trig trig->name : "none";
    event = kasprintf(GFP_KERNEL,"TRIGGER = % s",name);
    if (led_cdev->trigger) {//Remove any existing trigger
        write_lock_irqsave(&led_cdev->trigger->leddev_list_lock,flags);
        list_del(&led_cdev->trig_list);
        write_unlock_irqrestore(&led_cdev->trigger->leddev_list_lock,flags);
        cancel_work_sync(&led_cdev->set_brightness_work);
        led_stop_software_blink(led_cdev);
        if (led_cdev->trigger->deactivate)
            led_cdev->trigger->deactivate(led_cdev);
        led_cdev->trigger = NULL;
        led_set_brightness(led_cdev,LED_OFF);
    }
    if (trig) {
        write_lock_irqsave(&trig->leddev_list_lock,flags);
        list_add_tail(&led_cdev->trig_list,&trig->led_cdevs);
        write_unlock_irqrestore(&trig->leddev_list_lock,flags);
        led_cdev->trigger = trig;
        if (trig->activate)
            trig->activate(led_cdev);
            }
    if (event) {
        envp[0] = event;
        envp[1] = NULL;
        kobject_uevent_env(&led_cdev->dev->kobj,KOBJ_CHANGE,envp);
        kfree(event);
    }
}
EXPORT_SYMBOL_GPL(led_trigger_set);
void led_trigger_remove(struct led_classdev * led_cdev){ //取消连接
```

```
//取消连接的时候会调用触发器的 deactivate 方法
    down_write(&led_cdev->trigger_lock);
    led_trigger_set(led_cdev,NULL);
    up_write(&led_cdev->trigger_lock);
}
EXPORT_SYMBOL_GPL(led_trigger_remove);
void led_trigger_set_default(struct led_classdev * led_cdev){
//在所有已注册的触发器中寻找 led_cdev 的默认触发器,并调用 led_trigger_set 建立连接
    struct led_trigger * trig;
    if (! led_cdev->default_trigger)
        return;
    down_read(&triggers_list_lock);
    down_write(&led_cdev->trigger_lock);
    list_for_each_entry(trig,&trigger_list,next_trig) {
        if (! strcmp(led_cdev->default_trigger,trig->name))
            led_trigger_set(led_cdev,trig);
    }
    up_write(&led_cdev->trigger_lock);
    up_read(&triggers_list_lock);
}
EXPORT_SYMBOL_GPL(led_trigger_set_default);
//它们之间的关系为 led←led 类设备←led trigger,也就是说 trigger 好比是控制 LED 类
//设备的算法,这个算法决定着 LED 什么时候亮什么时候暗,LED trigger 类设备可以是
//现实的硬件设备,比如 IDE 硬盘,也可以是系统心跳等事件
```

5. leds-gpio. c

在/drivers/leds/leds-gpio. c 下实现了 gpio-led 框架,这个 gpio-led 框架的作用是把传入的 GPIO 端口信息,注册成 led_classdev 等。在理解 drivers/leds/leds-gpio. c 之前,可以先看看 Makefile 和 Kconfig 两个编译配置的文件。

在/drivers/leds/Makefile 中:

```
obj-$(CONFIG_LEDS_GPIO)            += leds-gpio.o
```

在/drivers/leds/ Kconfig 中:

```
config LEDS_GPIO
    tristate "LED Support for GPIO connected LEDs"
    depends on LEDS_CLASS
    depends on GPIOLIB || COMPILE_TEST
    help
      This option enables support for the LEDs connected to GPIO
      outputs. To be useful the particular board must have LEDs
      and they must be connected to the GPIO lines.  The LEDs must be
```

```
defined as platform devices and/or OpenFirmware platform devices.
The code to use these bindings can be selected below.
```

因此在 make menuconfig 时可以找到对应的选项：

```
Device Drivers -> LED Support -> LED Support for GPIO connected LEDS
```

执行 make menuconfig 后，可以看到它默认是 * 号，意为已经编译进了内核。

再看 leds-gpio. c 里的一些代码：

```
static const struct of_device_id of_gpio_leds_match[] = {
    { .compatible = "gpio-leds",}, //这是和 dts 文件中 compatible 属性进行联系的
    {},
};
MODULE_DEVICE_TABLE(of,of_gpio_leds_match);
static struct platform_driver gpio_led_driver = {
    .probe = gpio_led_probe,
    .remove = gpio_led_remove,
    .shutdown = gpio_led_shutdown,
    .driver = {
        .name = "leds-gpio",
        .of_match_table = of_gpio_leds_match,  //
        .pm = &gpio_led_pm_ops,
    },
};
module_platform_driver(gpio_led_driver);
```

of_match_table 将设备 of_gpio_leds_match 与驱动 gpio_led_driver 联系起来。module_platform_driver 用来替代 module_init()和 module_exit()功能。

另外，leds-gpio. c 源文件中除了 gpio_led_probe()、gpio_led_remove()、gpio_led_shutdown()函数的实现外，还有 gpio_led_set()、gpio_led_suspend()、gpio_led_resume()、create_gpio_led()、delete_gpio_led()等。

6. ledtrig-default-on. c

在/drivers/leds/trigger/ledtrig-default-on. c 中实现了一个名为 default-on 的触发器。这个触发器只定义了 activate 成员函数。它的 activate 函数的定义如下：

```
static void defon_trig_activate(struct led_classdev * led_cdev){
    led_set_brightness_async(led_cdev,led_cdev->max_brightness);
}
static struct led_trigger defon_led_trigger = {
    .name = "default-on",
    .activate = defon_trig_activate,
};
```

```
static int __init defon_trig_init(void){
    return led_trigger_register(&defon_led_trigger);
}
static void __exit defon_trig_exit(void){
    led_trigger_unregister(&defon_led_trigger);
}
module_init(defon_trig_init);
module_exit(defon_trig_exit);
```

也就是说,点亮 LED 只能是最亮的亮度,无法调节。一旦 ledl_classdev 与之建立了连接,就一直处于最亮的状态,直到取消和触发器的连接。

7. ledtrig-heartbeat. c

在 /drivers/leds/ledtrig-heartbeat. c 中定义了一个名为 heartbeat 的心跳灯触发器,它可以控制所有与之建立连接的 LED 会不停地闪烁。这个触发器用来指示内核是否已经挂掉。如果与之建立连接的 LED 不再闪烁,则说明内核已经挂掉。这就是“心跳”的含义,和通过人的心脏是否跳动来判断人是否死亡的原理是类似的。

```
static void heartbeat_trig_activate(struct led_classdev * led_cdev){
    struct heartbeat_trig_data * heartbeat_data;
    int rc;
    heartbeat_data = kzalloc(sizeof( * heartbeat_data),GFP_KERNEL);
    if (!heartbeat_data)
        return;
    led_cdev->trigger_data = heartbeat_data;
    rc = device_create_file(led_cdev->dev,&dev_attr_invert);
    if (rc) {
        kfree(led_cdev->trigger_data);
        return;
    }
    setup_timer(&heartbeat_data->timer,
            led_heartbeat_function,(unsigned long) led_cdev);
    heartbeat_data->phase = 0;
    led_heartbeat_function(heartbeat_data->timer.data);
    led_cdev->activated = true;
}
static void heartbeat_trig_deactivate(struct led_classdev * led_cdev){
    struct heartbeat_trig_data * heartbeat_data = led_cdev->trigger_data;
    if (led_cdev->activated) {
        del_timer_sync(&heartbeat_data->timer);
        device_remove_file(led_cdev->dev,&dev_attr_invert);
        kfree(heartbeat_data);
```

```
        led_cdev->activated = false;
    }
}
static struct led_trigger heartbeat_led_trigger = {
    .name = "heartbeat", //这点很重要
    .activate = heartbeat_trig_activate,
    .deactivate = heartbeat_trig_deactivate,
};
static int __init heartbeat_trig_init(void){
    int rc = led_trigger_register(&heartbeat_led_trigger);
    if (! rc) {
        atomic_notifier_chain_register(&panic_notifier_list,
                                       &heartbeat_panic_nb);
        register_reboot_notifier(&heartbeat_reboot_nb);
    }
    return rc;
}
static void __exit heartbeat_trig_exit(void){
    unregister_reboot_notifier(&heartbeat_reboot_nb);
    atomic_notifier_chain_unregister(&panic_notifier_list,
                                     &heartbeat_panic_nb);
    led_trigger_unregister(&heartbeat_led_trigger);
}
module_init(heartbeat_trig_init);
module_exit(heartbeat_trig_exit);
```

8. ledtrig-ide-disk.c

在 /driver/leds/ledtrig-ide-disk.c 中定义了一个名为 ide-disk 的 IDE 硬盘指示灯触发器，与之建立连接的 LED 可以指示硬盘的忙碌状态。这个触发器并没有 active 接口，因此不会自动闪烁。当内核中的其他模块调用 ledtrig_ide_activity()函数时，硬盘指示灯就会亮闪一下。该函数为全局函数，内核空间都可以调用，每调用一次 LED 就闪烁一下，具体怎么用，完全依赖于 IDE 驱动。它可以有多个 led_classdev 和这个触发器建立连接。每次调用 ledtrig_ide_activity()，所有与之连接的 LED 都会闪烁一下。

使用 ledtrig_ide_activity()这个函数的模块应该包含<linux/leds.h>这个头文件。

/driver/leds/ledtrig-ide-disk.c 的主要代码如下：

```
#define BLINK_DELAY 30
DEFINE_LED_TRIGGER(ledtrig_ide);
static unsigned long ide_blink_delay = BLINK_DELAY;
```

```
void ledtrig_ide_activity(void){
    led_trigger_blink_oneshot(ledtrig_ide,&ide_blink_delay,&ide_blink_delay,0);
}
EXPORT_SYMBOL(ledtrig_ide_activity);
static int __init ledtrig_ide_init(void){
    led_trigger_register_simple("ide-disk",&ledtrig_ide);
    return 0;
}
static void __exit ledtrig_ide_exit(void){
    led_trigger_unregister_simple(ledtrig_ide);
}
module_init(ledtrig_ide_init);
module_exit(ledtrig_ide_exit);
```


9. ledtrig-timer. c

在 /driver/leds/ledtrig-timer. c 中定义了一个名为 timer 的闪烁定时触发器。当某个 led_classdev 与之连接后，这个触发器会在/sys/class/leds/<device>/下创建两个属性文件 delay_on/delay_off。用户往这两个文件中写入数据后，相应的 LED 会按照设置的高低电平的时间(单位毫秒)来闪烁。如果 led_classdev 注册了硬件闪烁的接口“led_cdev－>blink_set”，那么就是用硬件控制闪烁，否则就是用软件定时器来控制闪烁。

```
    static ssize_t led_delay_on_show(struct device * dev,struct device_attribute * attr,
char * buf){
    //读取闪烁的点亮时间，单位是毫秒
        struct led_classdev * led_cdev = dev_get_drvdata(dev);
        return sprintf(buf,"%lu\n",led_cdev->blink_delay_on);
    }
    static ssize_t led_delay_on_store(struct device * dev,
        struct device_attribute * attr,const char * buf,size_t size){
    //设置闪烁的点亮时间，单位是毫秒
        struct led_classdev * led_cdev = dev_get_drvdata(dev);
        unsigned long state;
        ssize_t ret = -EINVAL;
        ret = kstrtoul(buf,10,&state);
        if (ret)    return ret;
        led_blink_set(led_cdev,&state,&led_cdev->blink_delay_off);
        led_cdev->blink_delay_on = state;
        return size;
    }
```

```
    static ssize_t led_delay_off_show(struct device * dev,struct device_attribute * attr,
char * buf){
    //读取闪烁的熄灭时间,单位是毫秒
        struct led_classdev * led_cdev = dev_get_drvdata(dev);
        return sprintf(buf," % lu\n",led_cdev->blink_delay_off);
    }
    static ssize_t led_delay_off_store(struct device * dev,
        struct device_attribute * attr,const char * buf,size_t size){
    //设置闪烁的熄灭时间,单位是毫秒
        struct led_classdev * led_cdev = dev_get_drvdata(dev);
        unsigned long state;
        ssize_t ret = -EINVAL;
        ret = kstrtoul(buf,10,&state);
        if (ret)    return ret;
        led_blink_set(led_cdev,&led_cdev->blink_delay_on,&state);
        led_cdev->blink_delay_off = state;
        return size;
    }
    static DEVICE_ATTR(delay_on,0644,led_delay_on_show,led_delay_on_store);
    static DEVICE_ATTR(delay_off,0644,led_delay_off_show,led_delay_off_store);
    static void timer_trig_activate(struct led_classdev * led_cdev){
        int rc;
        led_cdev->trigger_data = NULL;
        rc = device_create_file(led_cdev->dev,&dev_attr_delay_on);
        if (rc)    return;
        rc = device_create_file(led_cdev->dev,&dev_attr_delay_off);
        if (rc)    goto err_out_delayon;
        led_blink_set(led_cdev,&led_cdev->blink_delay_on,
                &led_cdev->blink_delay_off);
        led_cdev->activated = true;
        return;
    err_out_delayon:
        device_remove_file(led_cdev->dev,&dev_attr_delay_on);
    }
    static void timer_trig_deactivate(struct led_classdev * led_cdev){
        if (led_cdev->activated) {
            device_remove_file(led_cdev->dev,&dev_attr_delay_on);
            device_remove_file(led_cdev->dev,&dev_attr_delay_off);
            led_cdev->activated = false;
```

```
        }
        led_set_brightness(led_cdev,LED_OFF); //Stop blinking
    }
    static struct led_trigger timer_led_trigger = {
        .name = "timer",
        .activate = timer_trig_activate,
        .deactivate = timer_trig_deactivate,
    };
    static int __init timer_trig_init(void){
        return led_trigger_register(&timer_led_trigger);//注册 led 触发器
    }
    static void __exit timer_trig_exit(void){
        led_trigger_unregister(&timer_led_trigger);
    }
    module_init(timer_trig_init);
    module_exit(timer_trig_exit);
```

13.2.4　leds 驱动与 DTS 中的联系

在 am335x-evmsk.dts 文件中有：

```
leds {
    pinctrl-names = "default","sleep";
    pinctrl-0 = <&user_leds_default>;
    pinctrl-1 = <&user_leds_sleep>;
    compatible = "gpio-leds";
    led@1 {
        label = "evmsk:green:usr0";
        gpios = <&gpio1 4 GPIO_ACTIVE_HIGH>;
        default-state = "off";
    };
    led@2 {
        label = "evmsk:green:usr1";
        gpios = <&gpio1 5 GPIO_ACTIVE_HIGH>;
        default-state = "off";
    };
    led@3 {
        label = "evmsk:green:mmc0";
        gpios = <&gpio1 6 GPIO_ACTIVE_HIGH>;
        linux,default-trigger = "mmc0";
```

```
        default - state = "off";
    };
    led@4 {
        label = "evmsk:green:heartbeat";
        gpios = <&gpio1 7 GPIO_ACTIVE_HIGH>;
        linux,default - trigger = "heartbeat";
        default - state = "off";
    };
};
```

其中,“compatible = "gpio-leds"”即 compatible 参数为 gpio-leds。

那么此时会在 drivers/leds/leds—gpio. c 文件中找到:

```
static const struct of_device_id of_gpio_leds_match[] = {
        { .compatible = "gpio - leds",},
        {},
    };
```

即从 of_gpio_leds_match 中的. compatible 项中匹配到 gpio-leds,这就是 dts 和驱动 leds-gpio. c 中的各个函数间的联系点。

与硬件相关的,比如使用的哪一个 GPIO 口引脚等配置都在 dts 中完成,而具体的功能函数及数据结构等都在 leds 驱动中完成,两者通过 compatible 及参数 gpio-leds 实现关联。

13.2.5　leds 驱动的测试

实际上,我们已经在 13.2.1 小节中讲述了通过 echo 命令实现对 LED 的亮灭控制,其也是 LED 驱动的一种测试方法,但是只能在控制台上手动输入命令执行,而不能在应用程序中调用,因此下面编写一个应用程序实现对 LED 的控制。

1. 测试程序——LED 的控制

下述测试代码实现的功能是让 usr0 和 usr1 两个 LED 灯进行亮灭闪烁。

```
#include <stdio.h>
#include <unistd.h>
#include <sys/types.h>
#include <sys/ipc.h>
#include <sys/ioctl.h>
#include <fcntl.h>
int main(void){
    int f_led0 = open("/sys/class/leds/evmsk:green:usr0/brightness",O_RDWR);
    int f_led1 = open("/sys/class/leds/evmsk:green:usr1/brightness",O_RDWR);
```

```
    unsigned char dat0,dat1;
    unsigned char i = 0;
    if (f_led0 < 0){
        printf("error in open /sys/class/leds/evmsk:green:usr0/brightness\n");
        return -1;
    }
    if (f_led1 < 0){
        printf("error in open /sys/class/leds/evmsk:green:usr1/brightness\n");
        return -1;
    }
    while(1){
        dat0 = '0';
        dat1 = '1';
        write(f_led0,&dat0,sizeof(dat0));//相当于 echo 0 > brightness
        write(f_led1,&dat1,sizeof(dat1));
        usleep(500000);
        dat0 = '1';
        dat1 = '0';
        write(f_led0,&dat0,sizeof(dat0));
        write(f_led1,&dat1,sizeof(dat1));
        usleep(500000);
    }
    close(f_led0);//实际上也没有执行
    close(f_led1);
}
```

2. Makefile

```
    CROSS = /opt/ti-processor-sdk-linux-am335x-evm-03.00.00.04/linux-devkit/
sysroots/x86_64-arago-linux/usr/bin/arm-linux-gnueabihf-
    all: leds_test

    leds_test:
        $(CROSS)gcc leds_test.c -o leds_test

    clean:
        rm -rf leds_test *.o
```

3. 测试过程

先用 make 编译生成目标文件 leds_test，再通过 TFTP 上传到目标板执行，如图 13.4 和图 13.5 所示。

```
root@ccn: /home/ccn/test
root@ccn:/home/ccn/test# vi leds_test.c
root@ccn:/home/ccn/test# vi Makefile
root@ccn:/home/ccn/test# make
/opt/ti-processor-sdk-linux-am335x-evm-03.00.00.04/linux-devkit/sysroots/x86_64
-arago-linux/usr/bin/arm-linux-gnueabihf-gcc leds_test.c -o leds_test
root@ccn:/home/ccn/test# ls
leds_test  leds_test.c  Makefile
root@ccn:/home/ccn/test# ifconfig eth0
eth0      Link encap:Ethernet  HWaddr 98:4b:e1:a7:ed:07
          inet addr:192.168.1.2  Bcast:192.168.1.255  Mask:255.255.255.0
          inet6 addr: fe80::9a4b:e1ff:fea7:ed07/64 Scope:Link
          UP BROADCAST RUNNING MULTICAST  MTU:1500  Metric:1
          RX packets:77 errors:0 dropped:0 overruns:0 frame:0
          TX packets:230 errors:0 dropped:0 overruns:0 carrier:0
          collisions:0 txqueuelen:1000
          RX bytes:5224 (5.2 KB)  TX bytes:58644 (58.6 KB)

root@ccn:/home/ccn/test# cp leds_test /tftpboot/;ls
leds_test  leds_test.c  Makefile
root@ccn:/home/ccn/test# ls /tftpboot/
am335x-boneblack.dtb  am335x-evm.dtb    hello
am335x-bone.dtb       am335x-evmsk.dtb  leds_test
am335x-bonegreen.dtb  am335x-icev2.dtb  zImage-am335x-evm.bin
root@ccn:/home/ccn/test#
```

图 13.4　PC 宿主机编译 LED 测试代码

```
root@ccn: /home/ccn
Stopping Bootlog daemon: bootlogd.

Arago Project http://arago-project.org am335x-evm /dev/ttyO0

Arago 2015.02 am335x-evm /dev/ttyO0

am335x-evm login: root
root@am335x-evm:~# ifconfig eth0 192.168.1.1
root@am335x-evm:~# cd /tmp
root@am335x-evm:/tmp# ls
lighttpd      qtembedded-0
root@am335x-evm:/tmp# tftp 192.168.1.2 -g -r leds_test
root@am335x-evm:/tmp# ./leds_test &
[1] 1546
-sh: ./leds_test: Permission denied
root@am335x-evm:/tmp# ls -l
-rw-r--r--    1 root     root         10612 Feb 11 10:00 leds_test
drwx------    3 root     root            60 Feb 11 09:59 lighttpd
drwx------    3 root     root            80 Feb 11 09:59 qtembedded-0
[1]+  Done(126)                  ./leds_test
root@am335x-evm:/tmp# chmod 777 leds_test
root@am335x-evm:/tmp# ls -l
-rwxrwxrwx    1 root     root         10612 Feb 11 10:00 leds_test
drwx------    3 root     root            60 Feb 11 09:59 lighttpd
drwx------    3 root     root            80 Feb 11 09:59 qtembedded-0
root@am335x-evm:/tmp# ./leds_test &
[1] 1550
root@am335x-evm:/tmp#
CTRL-A Z for help | 115200 8N1 | NOR | Minicom 2.7 | VT102 | Offline | ttyUSB1
```

图 13.5　目标板下载 LED 测试代码并执行

13.3 按键输入驱动

13.3.1 AM335x 的按键测试

现成的内核已经带有按键驱动，且按键（SW1～SW4）事件被映射到/dev/input/event1 设备节点上，因此可以使用 hexdump 工具读取按键事件。如图 13.6 所示为执行 hexdump 指令后，依次对按键 SW1、SW2、SW3、SW4 进行按下和释放后终端上的显示。

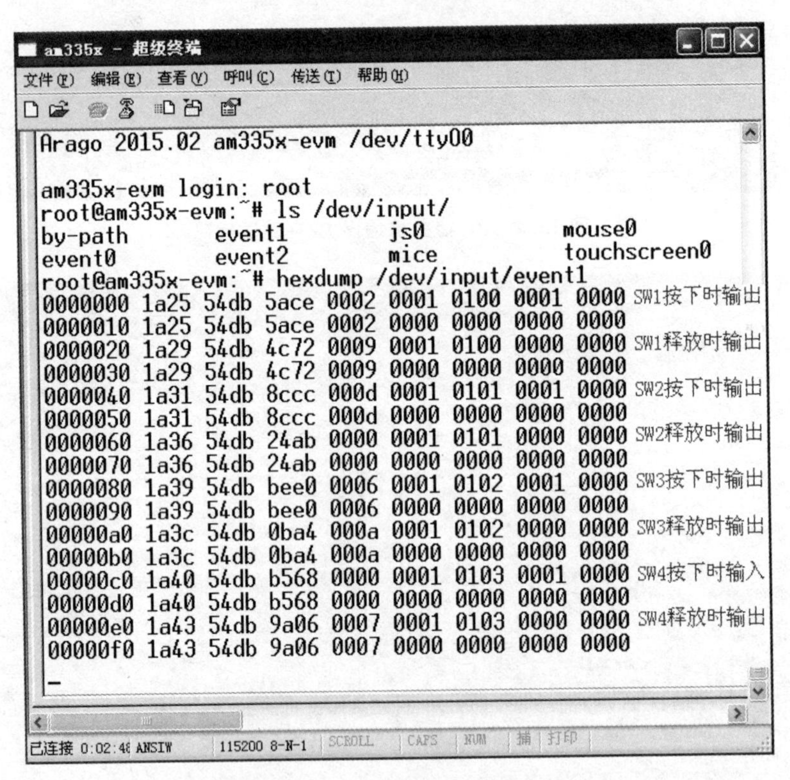

图 13.6 目标板按键测试

在 hexdump 中看到的数据为十六进制，数据格式需参考 Linux 的 input_event 事件的数据结构，还可以参考 documentation/input/input.txt 中的说明，且在 include/uapi/linux/input.h 文件中定义：

```
struct input_event {
    struct timeval time;
    __u16 type;
    __u16 code;
```

```
    __s32 value;
};
```

timeval 由 include/linux/time.h 文件定义：

```
struct timeval {
    __kernel_time_t tv_sec; /* seconds */
    __kernel_suseconds_t tv_usec; /* microseconds */
};
```

下面还有各个定义：

```
typedef __kernel_long_t __kernel_suseconds_t;
typedef __kernel_long_t __kernel_time_t;
typedef long __kernel_long_t;
```

hexdump 输出数据的各列定义：

第 1 列：行号。

第 2～5 列：输入事件时间戳，即结构体中的 time。

第 6 列：输入事件类型，即结构体中的 type，0 表示同步事件，1 表示键盘事件。

第 7 列：按键的键值，即结构体中的 code，也就是 dts 文件中的“linux，code”项，“linux，code ＝ ＜0x100＞”为 SW1，“linux，code ＝ ＜0x101＞”为 SW2，“linux，code ＝ ＜0x102＞”为 SW3，“linux，code ＝ ＜0x103＞”为 SW4。

第 8 列：按键的状态，即结构体中的 value，1 表示按下，0 表示松开。

另外，由图 2.26 按键输入电路可知，按键按下时为高电平(值为 1)，释放后为低电平(值为 0)。

可以使用 cat 命令查看确认 gpio_key 是否真的对应/dev/input/event1，命令如下：

```
#cat /proc/bus/input/devices
```

13.3.2 Linux 内核中的 input 子系统概述

与 leds 子系统一样，Linux 也将各种输入设备(如鼠标、键盘或按键和触摸屏等)做成统一的 input 子系统，/dev/input 目录下都是该子系统下的设备节点。

input 子系统已经在 Linux 中形成非常完整的结构和分层机制，网上有很多资料介绍，或者参考《Android 底层驱动分析和移植》一书，本书只做简要介绍。

输入设备都有一些共性，比如中断驱动和字符 I/O。基于分层的思想，Linux 内核将这些设备公有的部分提取出来，基于 cdev 提供接口，设计了输入子系统，所有使用输入子系统构建的设备都使用主设备号 13，同时输入子系统也支持自动创建设备文件，这些文件采用阻塞的 I/O 读/写方式，被创建在/dev/input/下。

内核中的输入子系统自底向上分为设备驱动层、输入核心层、事件处理层和用户空间。由于每种输入设备上报的事件都各不相同，所以为了应用层能够很好识别上报的事件，内核中也为应用层封装了标准的接口来描述一个事件，这些接口在/include/upai/linux/input 中。

设备驱动层是具体硬件相关的实现，也是驱动开发中主要完成的部分。

输入核心层主要提供一些 API 供设备驱动层调用，通过这些 API 设备驱动层上报的数据就可以传递到事件处理层。

事件处理层负责创建设备文件以及将上报的事件传递到用户空间。

有一篇关于输入子系统的文章笔者感觉还不错，读者也可以参考：http://www.cnblogs.com/xiaojiang1025/p/6414746.html。

13.3.3 输入子系统中按键驱动代码分析

Linux 内核中，整个输入子系统是相当庞大和复杂的，它的使用可以参考内核文档：Documentation/input/input-programming.txt。

下面我们主要以按键驱动为主线做简要的分析，从 am335x-evmsk.dts 关于 gpio_buttons 项的定义中可以找到：

```
compatible = "gpio - keys";
```

同时在 driver/input/keyboard/gpio-keys.c 中也能找到：

```
static const struct of_device_id gpio_keys_of_match[] = {
    { .compatible = "gpio - keys",},
    { },
};
```

driver/input/keyboard/gpio_key.c 就是 SW1～SW4 四个按键的统一的驱动程序，相关的头文件为 incldue/linux/ gpio_keys.h。

参考内核文档：Documentation\devicetree\bindings\input\gpio-keys.txt。

另外，这里需要特别强调的是：gpio-keys 是基于 input 架构实现的一个通用 GPIO 按键驱动。该驱动基于 platform_driver 架构，实现了驱动和设备分离，符合 Linux 设备驱动模型的思想。工程中的按键驱动一般都会基于 gpio-keys 来写，所以我们有必要对 gpio_keys 进行分析。也就是说，如果我们增加、减少，或修改，其实不需要修改 gpio_keys.c(该文件是和硬件无关的)及其他任何 input 架构的代码，只需要在 dts 文件中对 gpio_buttons 和 am33xx_pinmux 项进行修改就能满足要求。

```
incldue/linux/ gpio_keys.h
struct gpio_keys_button {
    unsigned int code;//输入事件代码 input event code (KEY_ * ,SW_ * )
    int gpio;//与 GPIO 的对应内容
    int active_low;
    //true indicates that button is considered depressed when gpio is low
```

```
    //如果设置为 1,则表明：当 gpio 是低电平时,按键被认为按下,即低电平有效
    const char * desc; //label that will be attached to button's gpio
    unsigned int type;//输入事件的类型(如 EV_KEY 按键事件,EV_ABS 绝对坐标事件)
    int wakeup;//配置按键作为待机唤醒源
    int debounce_interval;//按键去抖时间,单位 μs
    bool can_disable;
    //true indicates that userspace is allowed to disable button via sysfs
    int value; //EV_ABS 绝对坐标事件的坐标值
    unsigned int irq; //按键使能中断功能时的中断号
    struct gpio_desc * gpiod;//gpio 的描述符
};
struct gpio_keys_platform_data {
    struct gpio_keys_button * buttons;
    //指向描述按键的系列属性指针,应为一个数组的地址
    int nbuttons; //表示上述数组的数目
    unsigned int poll_interval;//polling interval in msecs - for polling driver only
    unsigned int rep:1;//使能自动重复功能
    int ( * enable)(struct device * dev);//platform hook for enabling the device
    void ( * disable)(struct device * dev);//platform hook for disabling the device
    const char * name;//input device name
};

driver/input/keyboard/gpio_key.c

//probe 函数,它会在 linux 内核中的 platform driver (of_device_id)
//和 dts 中的 platform device(.compatible)匹配上后被调用
static int gpio_keys_probe(struct platform_device * pdev){
//pdev 指向匹配成功的 platform device,通过它可以找到对应于 dts 文件中的设备节点
    struct device * dev = &pdev->dev;
    const struct gpio_keys_platform_data * pdata = dev_get_platdata(dev);
    struct gpio_keys_drvdata * ddata;//表示输入设备数据的结构
    struct input_dev * input;
    size_t size;
    int i,error;
    int wakeup = 0;
    if (!pdata) {
        pdata = gpio_keys_get_devtree_pdata(dev); //获取 dts 文件中设备节点及参数
        if (IS_ERR(pdata))  return PTR_ERR(pdata);
    }
    size = sizeof(struct gpio_keys_drvdata) +
           pdata->nbuttons * sizeof(struct gpio_button_data);
    ddata = devm_kzalloc(dev,size,GFP_KERNEL);
```

```
//kzalloc 对 kmalloc 的封装,会清 0 分配的空间
if (!ddata) {
    dev_err(dev,"failed to allocate state\n");
    return -ENOMEM;
}
input = devm_input_allocate_device(dev); //分配一个 input 设备
if (!input) {
    dev_err(dev,"failed to allocate input device\n");
    return -ENOMEM;
}
ddata->pdata = pdata;
ddata->input = input;//输入设备数据的构建
mutex_init(&ddata->disable_lock);
platform_set_drvdata(pdev,ddata); //保存平台数据
input_set_drvdata(input,ddata);
input->name = pdata->name ? : pdev->name; //设置 input 设备属性
input->phys = "gpio-keys/input0";
input->dev.parent = &pdev->dev;
input->open = gpio_keys_open;
input->close = gpio_keys_close;
input->id.bustype = BUS_HOST;
input->id.vendor = 0x0001;
input->id.product = 0x0001;
input->id.version = 0x0100;
//Enable auto repeat feature of Linux input subsystem
if (pdata->rep)   __set_bit(EV_REP,input->evbit);
for (i = 0; i < pdata->nbuttons; i++) { //各个按键的处理
    const struct gpio_keys_button * button = &pdata->buttons[i];
    struct gpio_button_data * bdata = &ddata->data[i];
    error = gpio_keys_setup_key(pdev,input,bdata,button);//gpio 按键的设置
    if (error)   return error;
    if (button->wakeup)   wakeup = 1;
}
error = sysfs_create_group(&pdev->dev.kobj,&gpio_keys_attr_group);
if (error) {
    dev_err(dev,"Unable to export keys/switches,error: %d\n",error);
    return error;
}
error = input_register_device(input); //注册输入设备
if (error) {
    dev_err(dev,"Unable to register input device,error: %d\n",error);
    goto err_remove_group;
```

```
        }
        device_init_wakeup(&pdev->dev,wakeup);
        return 0;
    err_remove_group:
        sysfs_remove_group(&pdev->dev.kobj,&gpio_keys_attr_group);
        return error;
    }
    //probe()函数最核心的功能是在一个循环中完成了各个按键的注册,
    //也就是根据 nbuttons 遍历 gpio_keys_platform_data 结构,对其中的每个按键进行事件的
    //注册处理
    //gpio_keys_setup_key()是实际的 GPIO 的按键初始化函数
    static int gpio_keys_setup_key(struct platform_device * pdev,struct input_dev * input,
                struct gpio_button_data * bdata,const struct gpio_keys_button * button){
        const char * desc = button->desc ? button->desc : "gpio_keys";
        struct device * dev = &pdev->dev;
        irq_handler_t isr; //中断处理程序
        unsigned long irqflags;
        int irq;
        int error;
        bdata->input = input;
        bdata->button = button;
        spin_lock_init(&bdata->lock);
        if (gpio_is_valid(button->gpio)) {
            error = devm_gpio_request_one(&pdev->dev,button->gpio,GPIOF_IN,desc);
            if (error < 0) {
                dev_err(dev,"Failed to request GPIO %d,error %d\n",button->gpio,
error);
                return error;
            }
            if (button->debounce_interval) {//按键去抖时间间隔
                error = gpio_set_debounce(button->gpio,button->debounce_interval
* 1000);
                /* use timer if gpiolib doesn't provide debounce */
                if (error < 0)
                    bdata->software_debounce = button->debounce_interval;
            }
            if (button->irq) {
                bdata->irq = button->irq;
            } else {
                irq = gpio_to_irq(button->gpio); //找到 GPIO 对应的中断
                if (irq < 0) {
                    error = irq;
```

```
                dev_err(dev,"Unable to get irq number for GPIO %d,error %d\n",
                    button->gpio,error);
                return error;
            }
            bdata->irq = irq;
        }
        INIT_DELAYED_WORK(&bdata->work,gpio_keys_gpio_work_func);
        //设置一个处理的队列
        isr = gpio_keys_gpio_isr;
        irqflags = IRQF_TRIGGER_RISING | IRQF_TRIGGER_FALLING;
    } else {
        if (!button->irq) {
            dev_err(dev,"No IRQ specified\n");
            return -EINVAL;
        }
        bdata->irq = button->irq;
        if (button->type && button->type != EV_KEY) {
            dev_err(dev,"Only EV_KEY allowed for IRQ buttons.\n");
            return -EINVAL;
        }
        bdata->release_delay = button->debounce_interval;
        setup_timer(&bdata->release_timer,gpio_keys_irq_timer,(unsigned long)
bdata);
        isr = gpio_keys_irq_isr;
        irqflags = 0;
    }
    input_set_capability(input,button->type ?: EV_KEY,button->code);
  /*
   * Install custom action to cancel release timer and
   * workqueue item.
   */
    error = devm_add_action(&pdev->dev,gpio_keys_quiesce_key,bdata);
    if (error) {
        dev_err(&pdev->dev,"failed to register quiesce action,error: %d\n",
error);
        return error;
    }
  /*
   * If platform has specified that the button can be disabled,
   * we don't want it to share the interrupt line.
   */
    if (!button->can_disable)
```

```
        irqflags |= IRQF_SHARED;
    error = devm_request_any_context_irq(&pdev->dev,bdata->irq,
                    isr,irqflags,desc,bdata);//注册中断
    if (error < 0) {
        dev_err(dev,"Unable to claim irq %d; error %d\n",bdata->irq,error);
        return error;
    }
    return 0;
}

static irqreturn_t gpio_keys_gpio_isr(int irq,void * dev_id){ //按键中断处理函数
//中断服务程序上半部分(就是去抖动检测),可对中断进行快速响应
    struct gpio_button_data * bdata = dev_id;
    BUG_ON(irq != bdata->irq);
    if (bdata->button->wakeup)
        pm_stay_awake(bdata->input->dev.parent);
    mod_delayed_work(system_wq,&bdata->work,
                msecs_to_jiffies(bdata->software_debounce)); //软件去抖动
    return IRQ_HANDLED;
}
static void gpio_keys_irq_timer(unsigned long _data){
//中断下半部分(按下):上报事件信息
    struct gpio_button_data * bdata = (struct gpio_button_data * )_data;
    struct input_dev * input = bdata->input;
    unsigned long flags;
    spin_lock_irqsave(&bdata->lock,flags);
    if (bdata->key_pressed) {//按键按下
        input_event(input,EV_KEY,bdata->button->code,0);//按键事件
        input_sync(input);
        bdata->key_pressed = false;//改成按键释放
    }
    spin_unlock_irqrestore(&bdata->lock,flags);
}
static irqreturn_t gpio_keys_irq_isr(int irq,void * dev_id){
//中断下半部分(释放):上报事件信息
    struct gpio_button_data * bdata = dev_id;
    const struct gpio_keys_button * button = bdata->button;
    struct input_dev * input = bdata->input;
    unsigned long flags;
    BUG_ON(irq != bdata->irq);
    spin_lock_irqsave(&bdata->lock,flags);
    if (!bdata->key_pressed) {//按键释放
```

```
            if (bdata->button->wakeup)
                pm_wakeup_event(bdata->input->dev.parent,0);
            input_event(input,EV_KEY,button->code,1);
            input_sync(input);
            if (!bdata->release_delay) {
                input_event(input,EV_KEY,button->code,0);//上报按键事件
                input_sync(input);
                goto out;
            }
            bdata->key_pressed = true;
        }
        if (bdata->release_delay)
            mod_timer(&bdata->release_timer,jiffies + msecs_to_jiffies(bdata->
release_delay));
    out:
        spin_unlock_irqrestore(&bdata->lock,flags);
        return IRQ_HANDLED;
    }

    MODULE_DEVICE_TABLE(of,gpio_keys_of_match);
    static struct platform_driver gpio_keys_device_driver = {
        .probe = gpio_keys_probe,
        .remove = gpio_keys_remove,
        .driver = {
            .name = "gpio-keys",
            .pm = &gpio_keys_pm_ops,
            .of_match_table = of_match_ptr(gpio_keys_of_match),
            //定义.compatible = "gpio-keys",以和 dts 中的匹配
        }
    };
    static int __init gpio_keys_init(void){
        return platform_driver_register(&gpio_keys_device_driver);//注册 platform 驱动
    }
    static void __exit gpio_keys_exit(void){
        platform_driver_unregister(&gpio_keys_device_driver);//释放 platform 驱动
    }
    late_initcall(gpio_keys_init); //初始化加载
    module_exit(gpio_keys_exit); //模块退出
```


13.3.4 按键驱动与 DTS 中的联系

在 am335x-evmsk.dts 文件中有：

```
gpio_buttons: gpio_buttons@0 {
    compatible = "gpio-keys";
    #address-cells = <1>;
    #size-cells = <0>;
    switch@1 {
        label = "button0";
        linux,code = <0x100>;
        gpios = <&gpio2 3 GPIO_ACTIVE_HIGH>; //GPIO_KEY2 = SW1
    };
    switch@2 {
        label = "button1";
        linux,code = <0x101>;
        gpios = <&gpio2 2 GPIO_ACTIVE_HIGH>; //GPIO_KEY3 = SW2
    };
    switch@3 {
        label = "button2";
        linux,code = <0x102>;
        gpios = <&gpio0 30 GPIO_ACTIVE_HIGH>;//GPIO_KEY1 = SW3
        gpio-key,wakeup; //配置可以待机唤醒
    };
    switch@4 {
        label = "button3";
        linux,code = <0x103>;
        gpios = <&gpio2 5 GPIO_ACTIVE_HIGH>;//GPIO_KEY4 = SW4
    };
};
```

其中,“compatible = "gpio-keys"”即compatible参数为gpio-keys。

那么此时会在driver/input/keyboard/gpio_key.c文件中找到:

```
static const struct of_device_id gpio_keys_of_match[] = {
    { .compatible = "gpio-keys",},
    { },
};
```

即从gpio_keys_of_match[]中的.compatible项中匹配到gpio-keys,这就是dts和驱动gpio-key.c中的各个函数间的联系点。

与硬件相关的,比如使用的哪一个GPIO口引脚等配置都在dts中完成,而具体的功能函数及数据结构等都在gpio_keys驱动中完成,两者通过compatible及参数gpio-keys实现关联。

13.3.5 按键驱动的测试

实际上,我们已经在13.3.1中讲述了通过hexdump工具实现对SW1、SW2、

SW3、SW4输入事件的触发测试，其也是按键驱动的一种测试方法，但是只能在控制台上手动输入命令执行，而不能在应用程序中调用，因此下面将编写一个应用程序实现对按键事件的捕获。

1. 测试程序——按键事件捕获

下述测试代码实现的功能是循环读取SW1、SW2、SW3、SW4四个按键的事件及值。

```
#include <stdio.h>
#include <unistd.h>
#include <sys/types.h>
#include <sys/ipc.h>
#include <sys/ioctl.h>
#include <fcntl.h>
int main(void){
    struct input_event key_evt;
    int ret;
    int f_key = open("/dev/input/event1",O_RDWR);
    if (f_key < 0){
        printf("error in open /dev/input/event1\n");
        return -1;
    }
    while(1){
        ret = read(fd_key,(unsigned char *)&key_evt,sizeof(struct input_event));
        //阻塞型读函数
        if(ret < 0) {
            printf("read key event failed : %d",ret);
        }else if( key_evt.code != 0x100 &&        //SW1
                key_evt.code != 0x101 &&         //SW2
                key_evt.code != 0x102 &&         //SW3
                key_evt.code != 0x103){          //SW4
                printf("unknown key code: %d",key_evt.code); //过滤掉无效按键
        }else{  //有效按键,读取按键值
                printf("get key event code: %x,value: %x,type: %x",
                            key_evt.code,key_evt.value,key_evt.type);
        }
    }
}
```

2. Makefile

```
CROSS = /opt/ti-processor-sdk-linux-am335x-evm-03.00.00.04/linux-devkit/sysroots/x86_64-arago-linux/usr/bin/arm-linux-gnueabihf-
```

```
all: keys_test
keys_test:
	$(CROSS)gcc keys_test.c -o keys_test
clean:
	rm -rf keys_test *.o
```

3. 测试过程

先用 make 编译生成目标文件 keys_test，再通过 TFTP 上传到目标板执行（参考 leds 测试实验的操作步骤），当按下某个按键时将会有打印信息。

13.4　PWM 的 LCD 背光调节驱动

13.4.1　AM335x 的背光调节测试

AM335x 集成多路 PWM 输出功能，官方驱动的说明及使用可参考：http://processors.wiki.ti.com/index.php/AM335x_PWM_Driver%27s_Guide#Controlling_backlight。

AM335x Starter Kit 板通过 ECAP2_IN_PWM2_OUT(LCD_BACKLIGHTEN)连接到背光驱动电路。

启动控制终端，通过设置不同的 brightness 值（0～8），以调节对应的背光亮度值，背光调节测试如图 13.7 所示。

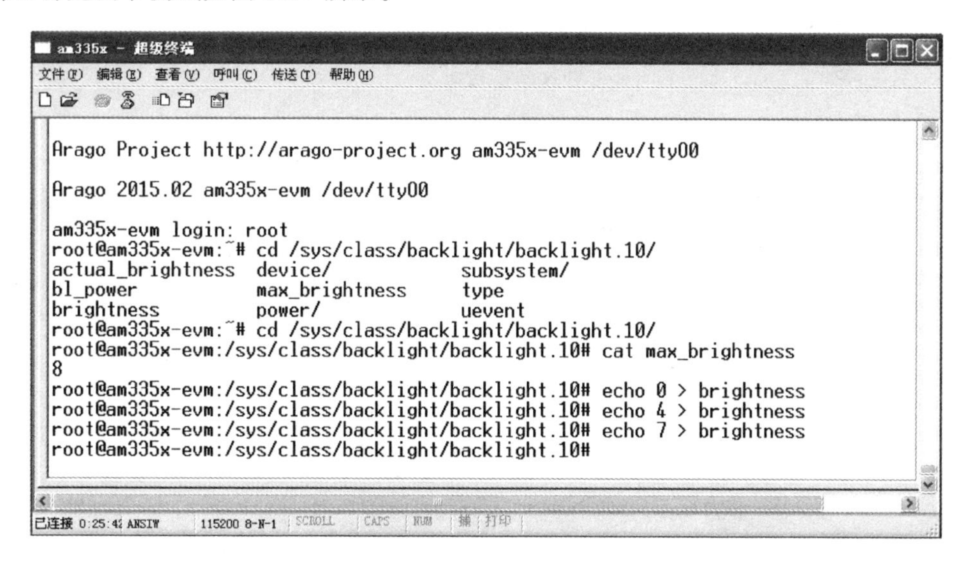

图 13.7　背光调节测试

13.4.2 Linux 内核中的 Backlight 背光子系统概述

关于背光子系统的详细介绍读者可参考下面几篇文章：

- http://blog.csdn.net/weiqing1981127/article/details/8511676；
- http://blog.csdn.net/weiqing1981127/article/details/8515847；
- http://blog.csdn.net/yuanlulu/article/details/7106821；
- http://blog.csdn.net/wh_19910525/article/details/19333455；
- https://wenku.baidu.com/view/62d08527f8c75fbfc67db293.html。

Linux 内核中的 Backlight 背光子系统位于/sys/class/backlight 目录下，提供给用户空间调节 LCD 背光的亮度，或者其他显示设备的背光亮度，甚至所有基于 PWM 输出控制的外设的接口。

也就是说，Backlight 子系统其实是基于 PWM 输出的一种驱动接口，不仅是 LCD 背光电源的调节，同样也可以对蜂鸣器的声音调节等。只要是基于 PWM 输出，通过调节 PWM 的周期或占空比实现调节的外设或电路，就可以设计成基于 Backlight 的背光子系统。

13.4.3 Backlight 背光子系统驱动代码分析

Backlight 背光子系统位于/driver/video/backlight 目录，其核心代码是/drivers/video/backlight.c，对应的头文件是/include/linux/backlight.h。

```
/include/linux/backlight.h
//几个重要的数据结构
struct backlight_ops { //背光操作函数结构体
    unsigned int options;
#define BL_CORE_SUSPENDRESUME (1 << 0)
    int (*update_status)(struct backlight_device *);
    //背光状态,用于通知驱动一些属性已经改变
    int (*get_brightness)(struct backlight_device *);//获取当前背光值
    int (*check_fb)(struct backlight_device *,struct fb_info *);
    //检查 framebuffer 设备
};

struct backlight_properties { //定义背光所有属性的结构体
    int brightness; //当前背光值,取值范围在 0~max_brightness
    int max_brightness; //最大背光值,为只读属性
    int power;
    int fb_blank;
    enum backlight_type type;
    unsigned int state;
```

```
#define BL_CORE_SUSPENDED (1 << 0) //backlight is suspended
#define BL_CORE_FBBLANK(1 << 1) //backlight is under an fb blank event
#define BL_CORE_DRIVER4(1 << 28) //reserved for driver specific use
#define BL_CORE_DRIVER3(1 << 29) //reserved for driver specific use
#define BL_CORE_DRIVER2(1 << 30) //reserved for driver specific use
#define BL_CORE_DRIVER1(1 << 31) //reserved for driver specific use
};

struct backlight_device { //背光子系统的设备结构
    struct backlight_properties props; //背光属性
    struct mutex update_lock; //状态更新时需要的互斥锁
    struct mutex ops_lock; //操作函数调用所需的互斥锁
    const struct backlight_ops * ops; //定义操作函数指针
    struct notifier_block fb_notif; //The framebuffer notifier block
    struct list_head entry; //list entry of all registered backlight devices
    struct device dev; //内嵌设备
    bool fb_bl_on[FB_MAX]; //Multiple framebuffers may share one backlight device
    int use_count;
};

struct generic_bl_info {
    const char * name; //名字字符指针,这个名字会出现在/sys/class/backlight/中
    int max_intensity; //最大亮度
    int default_intensity; //默认亮度
    int limit_mask; //亮度值的掩码,如 0xff
    void ( * set_bl_intensity)(int intensity); //设置亮度的函数
    void ( * kick_battery)(void); //设置亮度之后调用的函数,与电池相关,可以不定义
};

/drivers/video/backlight.c
//几个主要的函数
static void __exit backlight_class_exit(void){ //注销 backlight 类
    class_destroy(backlight_class); //销毁 backlight 类
}

static int __init backlight_class_init(void){ //backlight 类初始化
    backlight_class = class_create(THIS_MODULE,"backlight"); //注册 backlight 类
    if (IS_ERR(backlight_class)) {
        pr_warn("Unable to create backlight class; errno = %ld\n",
        PTR_ERR(backlight_class));
        return PTR_ERR(backlight_class);
    }
```

```
        backlight_class->dev_groups = bl_device_groups;
        backlight_class->pm = &backlight_class_dev_pm_ops;
        INIT_LIST_HEAD(&backlight_dev_list);
        mutex_init(&backlight_dev_list_mutex);
        BLOCKING_INIT_NOTIFIER_HEAD(&backlight_notifier);
        return 0;
    }

    /*
     * if this is compiled into the kernel,we need to ensure that the
     * class is registered before users of the class try to register lcd's
     */
    postcore_initcall(backlight_class_init);
    module_exit(backlight_class_exit);
```



```
//brightness是当前亮度,sysfs,用户层可通过cat或者echo命令来读/写该值,实现函数如下:
    static ssize_t brightness_show(struct device *dev,struct device_attribute *attr,
char *buf){
    //显示函数
        struct backlight_device *bd = to_backlight_device(dev);
        return sprintf(buf,"%d\n",bd->props.brightness); //输出当前亮度值
    }

    static ssize_t brightness_store(struct device *dev,
    struct device_attribute *attr,const char *buf,size_t count){
    //写函数
        int rc;
        struct backlight_device *bd = to_backlight_device(dev);
        unsigned long brightness;
        rc = kstrtoul(buf,0,&brightness);
        if (rc)
            return rc;
        rc = -ENXIO;
        mutex_lock(&bd->ops_lock);
        if (bd->ops) {
            if (brightness > bd->props.max_brightness)
                rc = -EINVAL;
            else {
                pr_debug("set brightness to %lu\n",brightness);
                bd->props.brightness = brightness; //保存当前亮度值
                backlight_update_status(bd);
                rc = count;
```

```
        }
    }
    mutex_unlock(&bd->ops_lock); //互斥
    backlight_generate_event(bd,BACKLIGHT_UPDATE_SYSFS);
    return rc;
}
static DEVICE_ATTR_RW(brightness); //sysfs,在/sys/devices/目录中增加 brightness 文件
                                   //include/linux/Device.h 中定义

//max_brightness 为最大背光值,sysfs,用户层可通过 cat 命令来读该值,实现函数如下:
static ssize_t max_brightness_show(struct device * dev,
struct device_attribute * attr,char * buf){
    struct backlight_device * bd = to_backlight_device(dev);
    return sprintf(buf,"%d\n",bd->props.max_brightness); //输出最大亮度值
}
static DEVICE_ATTR_RO(max_brightness);

EXPORT_SYMBOL(backlight_device_register); //注册背光设备
EXPORT_SYMBOL(backlight_device_unregister); //注销背光设备
```

另外与背光驱动直接相关的文件为/driver/video/backlight/pwm_bl.c,这里只列出了部分代码,详细内容请参考源代码。

```
struct pwm_bl_data {
    struct pwm_device * pwm;
    struct device * dev;
    unsigned int period;
    unsigned int lth_brightness;
    unsigned int * levels;
    bool enabled;
    struct regulator * power_supply;
    struct gpio_desc * enable_gpio;
    unsigned int scale;
    bool legacy;
    int ( * notify)(struct device * ,int brightness);
    void ( * notify_after)(struct device * ,int brightness);
    int ( * check_fb)(struct device * ,struct fb_info * );
    void ( * exit)(struct device * );
};
static void pwm_backlight_power_on(struct pwm_bl_data * pb,int brightness){
//背光电源打开
    int err;
    if (pb->enabled)
```

```
        return;
    err = regulator_enable(pb->power_supply);
    if (err < 0)
        dev_err(pb->dev,"failed to enable power supply\n");
    if (pb->enable_gpio)
        gpiod_set_value(pb->enable_gpio,1);
    pwm_enable(pb->pwm);
    pb->enabled = true;
}
static void pwm_backlight_power_off(struct pwm_bl_data * pb){//背光电源关断
    if (!pb->enabled)
        return;
    pwm_config(pb->pwm,0,pb->period);
    pwm_disable(pb->pwm);
    if (pb->enable_gpio)
        gpiod_set_value(pb->enable_gpio,0);
    regulator_disable(pb->power_supply);
    pb->enabled = false;
}
static int compute_duty_cycle(struct pwm_bl_data * pb,int brightness){//设置 PWM 占空比
    unsigned int lth = pb->lth_brightness;
    int duty_cycle;
    if (pb->levels)
        duty_cycle = pb->levels[brightness];//背光值对应的占空比值
    else
        duty_cycle = brightness;
    return (duty_cycle * (pb->period - lth) / pb->scale) + lth;
}
static int pwm_backlight_update_status(struct backlight_device * bl){ //更新背光亮度函数
    struct pwm_bl_data * pb = bl_get_data(bl);
    int brightness = bl->props.brightness;
    int duty_cycle;
    if (bl->props.power != FB_BLANK_UNBLANK ||bl->props.fb_blank != FB_BLANK_
UNBLANK || bl->props.state & BL_CORE_FBBLANK)
        brightness = 0;
    if (pb->notify)
        brightness = pb->notify(pb->dev,brightness);
    if (brightness > 0) { //背光值大于 0
        duty_cycle = compute_duty_cycle(pb,brightness); //设置背光 PWM 占空比
        pwm_config(pb->pwm,duty_cycle,pb->period);
```

```
        pwm_backlight_power_on(pb,brightness); //打开背光
    } else
        pwm_backlight_power_off(pb); //背光值为 0,关闭背光
    if (pb->notify_after)
        pb->notify_after(pb->dev,brightness);
    return 0;
}
static int pwm_backlight_check_fb(struct backlight_device * bl,struct fb_info * info)
{
    struct pwm_bl_data * pb = bl_get_data(bl);
    return !pb->check_fb || pb->check_fb(pb->dev,info);
}
static const struct backlight_ops pwm_backlight_ops = {
    .update_status = pwm_backlight_update_status, //更新背光亮度函数
    .check_fb = pwm_backlight_check_fb,
};

static struct of_device_id pwm_backlight_of_match[] = {
    { .compatible = "pwm-backlight" }, //和 DTS 相关联
    { }
};
MODULE_DEVICE_TABLE(of,pwm_backlight_of_match);

static int pwm_backlight_probe(struct platform_device * pdev){//探测函数
 ⋮
        //因代码很长,为节省篇幅故省略,读者可自行分析源代码
        //主要是注册 PWM 和 Backlight 设备,定义背光操作函数集合等
}
static int pwm_backlight_remove(struct platform_device * pdev){
    struct backlight_device * bl = platform_get_drvdata(pdev);
    struct pwm_bl_data * pb = bl_get_data(bl);
    backlight_device_unregister(bl);
    pwm_backlight_power_off(pb);
    if (pb->exit)
        pb->exit(&pdev->dev);
    if (pb->legacy)
        pwm_free(pb->pwm);
    return 0;
}
static void pwm_backlight_shutdown(struct platform_device * pdev){
    struct backlight_device * bl = platform_get_drvdata(pdev);
    struct pwm_bl_data * pb = bl_get_data(bl);
```

```
    pwm_backlight_power_off(pb);
}
static const struct dev_pm_ops pwm_backlight_pm_ops = {//电源管理相关的操作函数
    #ifdef CONFIG_PM_SLEEP //休眠相关的背光操作函数
    .suspend = pwm_backlight_suspend, //挂起
    .resume = pwm_backlight_resume, //恢复
    .poweroff = pwm_backlight_suspend,
    .restore = pwm_backlight_resume,
#endif
};
static struct platform_driver pwm_backlight_driver = {
    .driver = {
        .name = "pwm-backlight", //驱动名
        .pm = &pwm_backlight_pm_ops,
        .of_match_table = of_match_ptr(pwm_backlight_of_match),
        //关键点：pwm_backlight_of_match
    },
    .probe = pwm_backlight_probe, //探测函数
    .remove = pwm_backlight_remove,
    .shutdown = pwm_backlight_shutdown,
};
module_platform_driver(pwm_backlight_driver);
```

13.4.4　背光驱动与 DTS 的联系

在 am335x-evmsk.dts 文件中有：

```
backlight {
    compatible = "pwm-backlight";
    pwms = <&ecap2 0 50000 PWM_POLARITY_INVERTED>;
    //50 000,PWM 信号默认的周期
    //PWM_POLARITY_INVERTED 表示极性翻转
    //另有 PWM_POLARITY_NORMAL 表示极性正常
    brightness-levels = <0 58 61 66 75 90 125 170 255>; //分别对应 8 级的背光值
    default-brightness-level = <8>; //背光值共 8 级
};

&epwmss2 {
    status = "okay";
    ecap2: ecap@48304100 {
        status = "okay";
        pinctrl-names = "default";
```

```
        pinctrl-0 = <&ecap2_pins>;
    };
};

ecap2_pins: backlight_pins {
    pinctrl-single,pins = <
    0x19c 0x4 //mcasp0_ahclkr.ecap2_in_pwm2_out MODE4
    >;
};
```

其中,"compatible = "pwm-backlight""即 compatible 参数为 pwm-backlight。那么此时会在 driver/video/backlight/pwm_bl.c 文件中找到:

```
static struct of_device_id pwm_backlight_of_match[] = {
    { .compatible = "pwm-backlight" },
    { }
};
```

即从 pwm_backlight_of_match[]中的.compatible 项中匹配到 pwm_backlight,这就是 dts 和驱动 pwm_bl.c 中的各个函数间的联系点。

与硬件相关的,比如使用的哪一个 GPIO 口引脚等配置都在 dts 中完成,而具体的功能函数及数据结构等都在 pwm_bl.c 驱动中完成,两者通过 compatible 及参数 pwm-backlight 实现关联。

13.4.5　背光驱动的测试

实际上,我们已经通过图 13.7 中的内容讲述了通过 echo 命令实现对背光亮度的调节,也是背光驱动的一种测试方法,但是只能在控制台上手动输入命令执行,而不能在应用程序中调用,因此下面将编写一个应用程序实现对背光亮度的调节。

1. 测试程序——背光亮度调节

下述测试代码实现的功能让背光循环在 0~8 的背光值之间变化,时间间隔为 2 s。

```
#include <stdio.h>
#include <unistd.h>
#include <sys/types.h>
#include <sys/ipc.h>
#include <sys/ioctl.h>
#include <fcntl.h>
int main(void){
    int f_backlight = open("/sys/class/backlight/backlight.10/brightness",O_RDWR);
    unsigned char dat0;
    unsigned char i = 0;
```

```
        if (f_backlight < 0){
            printf("error in open /sys/class/backlight/backlight.10/brightness\n");
            return -1;
        }
        while(1){
            for (i = 0; i < 9; i++){
                switch(i){
                    case 0: dat0 = '0';
                            break;
                    case 1: dat0 = '1';
                            break;
                    case 2: dat0 = '2';
                            break;
                    case 3: dat0 = '3';
                            break;
                    case 4: dat0 = '4';
                            break;
                    case 5: dat0 = '5';
                            break;
                    case 6: dat0 = '6';
                            break;
                    case 7: dat0 = '7';
                            break;
                    default: dat0 = '8';
                            break;
                }
            write(f_backlight,&dat0,sizeof(dat0));//相当于 echo i > brightness
            sleep(2);
        }
        close(f_backlight);//实际上也没有执行
    }
```

2. Makefile

```
CROSS = /opt/ti-processor-sdk-linux-am335x-evm-03.00.00.04/linux-devkit/
sysroots/x86_64-arago-linux/usr/bin/arm-linux-gnueabihf-
all: backlight_test

backlight_test:
	$(CROSS)gcc backlight_test.c -o backlight_test

clean:
```

```
    rm -rf backlight_test *.o
```

3. 测试过程

先用 make 编译生成目标文件 backlight_test，再通过 TFTP 上传到目标板执行（参考 leds 测试实验的操作步骤）。

13.5 LCD 显示驱动及配置

Linux 下的 LCD 系统是非常完善的，通用的地方已经由驱动封装好，而如 LCD 分辨率、控制时序等都可以通过 DTS 配置完成显示，因此这里只需要讨论使用 DTS 方式配置内核完成 LCD 驱动。

关于 LCD 驱动参考如下：

- http://processors.wiki.ti.com/index.php/AM335x_LCD_Controller_Driver%27s_Guide;
- http://processors.wiki.ti.com/index.php/Linux_Core_LCD_Controller_User_Guide;
- http://processors.wiki.ti.com/index.php/Sitara_Linux_Training:_Linux_Board_Port#Adding_LCD_and_Touchscreen_Support。

LCD 时序配置的相关内容可以参考：AM335x LCDC Configuration In Linux.pdf 和 AM335x 关于 LCD 屏幕的配置.pdf。

13.5.1 LCD DTS 配置

am335x-evmsk.dts，关于 LCD 配置的相关代码如下：

```
panel {
    compatible = "ti,tilcdc,panel";//对应驱动 druvers/gpu/drm/tilcdc/tilcdc_panel.c
    pinctrl-names = "default","sleep";
    pinctrl-0 = <&lcd_pins_default>;
    pinctrl-1 = <&lcd_pins_sleep>;
    status = "okay";
    panel-info { //配置 LCD 的硬件信息,需参考 LCD 面板规格书
        ac-bias = <255>;
        ac-bias-intrpt = <0>;
        dma-burst-sz = <16>;
        bpp = <32>;
        fdd = <0x80>;
        sync-edge = <0>;
        sync-ctrl = <1>;
        raster-order = <0>;
```

```
            fifo-th = <0>;
        };
        display-timings { //LCD 刷屏的相关时序
            480x272 { //LCD 分辨率为 480 * 272
                hactive = <480>;
                vactive = <272>;
                hback-porch = <43>;
                hfront-porch = <8>;
                hsync-len = <4>;
                vback-porch = <12>;
                vfront-porch = <4>;
                vsync-len = <10>;
                clock-frequency = <9000000>;
                hsync-active = <0>;
                vsync-active = <0>;
            };
        };
    };
```

上述 display-timings 的各参数示意图如图 13.8 所示，具体的取值需要参考 LCD 的规格书。

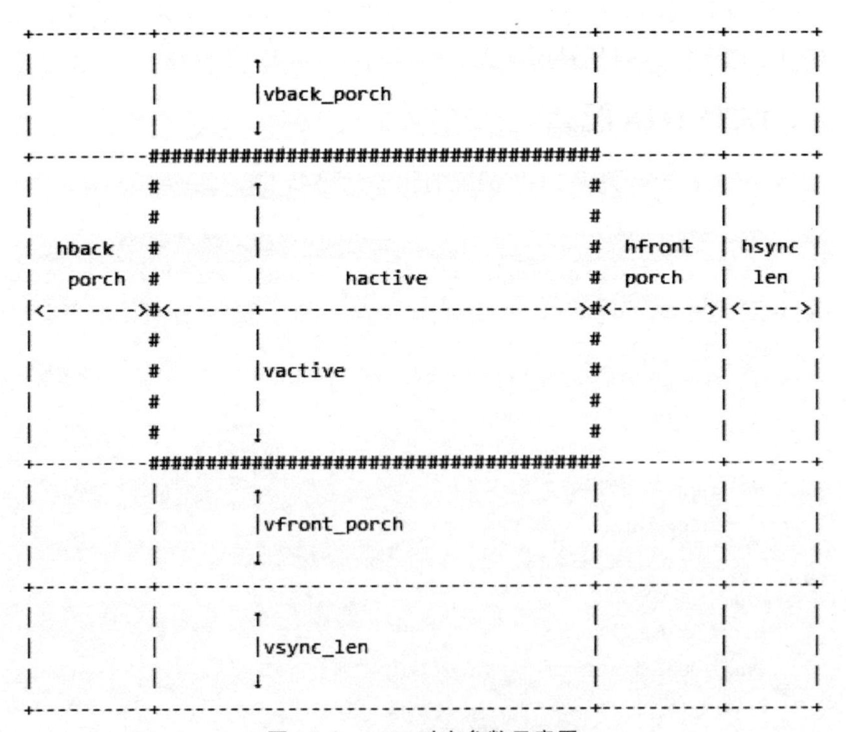

图 13.8　LCD 时序参数示意图

除时序外，还有 LCD 引脚配置。由于 AM335x 集成了 LCD 控制，该控制器与 LCD 的连接是通过 GPIO 引脚复用实现的，上述共提到 lcd_pins_default 和 lcd_pins_sleep 两种模式。

```
lcd_pins_default: lcd_pins_default {
    pinctrl-single,pins = <
        0x20 (PIN_OUTPUT | MUX_MODE1) /* gpmc_ad8.lcd_data23 */
        0x24 (PIN_OUTPUT | MUX_MODE1) /* gpmc_ad9.lcd_data22 */
        0x28 (PIN_OUTPUT | MUX_MODE1) /* gpmc_ad10.lcd_data21 */
        0x2c (PIN_OUTPUT | MUX_MODE1) /* gpmc_ad11.lcd_data20 */
        0x30 (PIN_OUTPUT | MUX_MODE1) /* gpmc_ad12.lcd_data19 */
        0x34 (PIN_OUTPUT | MUX_MODE1) /* gpmc_ad13.lcd_data18 */
        0x38 (PIN_OUTPUT | MUX_MODE1) /* gpmc_ad14.lcd_data17 */
        0x3c (PIN_OUTPUT | MUX_MODE1) /* gpmc_ad15.lcd_data16 */
        0xa0 (PIN_OUTPUT | MUX_MODE0) /* lcd_data0.lcd_data0 */
        0xa4 (PIN_OUTPUT | MUX_MODE0) /* lcd_data1.lcd_data1 */
        0xa8 (PIN_OUTPUT | MUX_MODE0) /* lcd_data2.lcd_data2 */
        0xac (PIN_OUTPUT | MUX_MODE0) /* lcd_data3.lcd_data3 */
        0xb0 (PIN_OUTPUT | MUX_MODE0) /* lcd_data4.lcd_data4 */
        0xb4 (PIN_OUTPUT | MUX_MODE0) /* lcd_data5.lcd_data5 */
        0xb8 (PIN_OUTPUT | MUX_MODE0) /* lcd_data6.lcd_data6 */
        0xbc (PIN_OUTPUT | MUX_MODE0) /* lcd_data7.lcd_data7 */
        0xc0 (PIN_OUTPUT | MUX_MODE0) /* lcd_data8.lcd_data8 */
        0xc4 (PIN_OUTPUT | MUX_MODE0) /* lcd_data9.lcd_data9 */
        0xc8 (PIN_OUTPUT | MUX_MODE0) /* lcd_data10.lcd_data10 */
        0xcc (PIN_OUTPUT | MUX_MODE0) /* lcd_data11.lcd_data11 */
        0xd0 (PIN_OUTPUT | MUX_MODE0) /* lcd_data12.lcd_data12 */
        0xd4 (PIN_OUTPUT | MUX_MODE0) /* lcd_data13.lcd_data13 */
        0xd8 (PIN_OUTPUT | MUX_MODE0) /* lcd_data14.lcd_data14 */
        0xdc (PIN_OUTPUT | MUX_MODE0) /* lcd_data15.lcd_data15 */
        0xe0 (PIN_OUTPUT | MUX_MODE0) /* lcd_vsync.lcd_vsync */
        0xe4 (PIN_OUTPUT | MUX_MODE0) /* lcd_hsync.lcd_hsync */
        0xe8 (PIN_OUTPUT | MUX_MODE0) /* lcd_pclk.lcd_pclk */
        0xec (PIN_OUTPUT | MUX_MODE0) /* lcd_ac_bias_en.lcd_ac_bias_en */
    >;
};
lcd_pins_sleep: lcd_pins_sleep {
    pinctrl-single,pins = <
        0x20 (PIN_INPUT_PULLDOWN | MUX_MODE7) /* gpmc_ad8.lcd_data23 */
        0x24 (PIN_INPUT_PULLDOWN | MUX_MODE7) /* gpmc_ad9.lcd_data22 */
        0x28 (PIN_INPUT_PULLDOWN | MUX_MODE7) /* gpmc_ad10.lcd_data21 */
        0x2c (PIN_INPUT_PULLDOWN | MUX_MODE7) /* gpmc_ad11.lcd_data20 */
```

```
        0x30 (PIN_INPUT_PULLDOWN | MUX_MODE7) /* gpmc_ad12.lcd_data19 */
        0x34 (PIN_INPUT_PULLDOWN | MUX_MODE7) /* gpmc_ad13.lcd_data18 */
        0x38 (PIN_INPUT_PULLDOWN | MUX_MODE7) /* gpmc_ad14.lcd_data17 */
        0x3c (PIN_INPUT_PULLDOWN | MUX_MODE7) /* gpmc_ad15.lcd_data16 */
        0xa0 (PULL_DISABLE | MUX_MODE7) /* lcd_data0.lcd_data0 */
        0xa4 (PULL_DISABLE | MUX_MODE7) /* lcd_data1.lcd_data1 */
        0xa8 (PULL_DISABLE | MUX_MODE7) /* lcd_data2.lcd_data2 */
        0xac (PULL_DISABLE | MUX_MODE7) /* lcd_data3.lcd_data3 */
        0xb0 (PULL_DISABLE | MUX_MODE7) /* lcd_data4.lcd_data4 */
        0xb4 (PULL_DISABLE | MUX_MODE7) /* lcd_data5.lcd_data5 */
        0xb8 (PULL_DISABLE | MUX_MODE7) /* lcd_data6.lcd_data6 */
        0xbc (PULL_DISABLE | MUX_MODE7) /* lcd_data7.lcd_data7 */
        0xc0 (PULL_DISABLE | MUX_MODE7) /* lcd_data8.lcd_data8 */
        0xc4 (PULL_DISABLE | MUX_MODE7) /* lcd_data9.lcd_data9 */
        0xc8 (PULL_DISABLE | MUX_MODE7) /* lcd_data10.lcd_data10 */
        0xcc (PULL_DISABLE | MUX_MODE7) /* lcd_data11.lcd_data11 */
        0xd0 (PULL_DISABLE | MUX_MODE7) /* lcd_data12.lcd_data12 */
        0xd4 (PULL_DISABLE | MUX_MODE7) /* lcd_data13.lcd_data13 */
        0xd8 (PULL_DISABLE | MUX_MODE7) /* lcd_data14.lcd_data14 */
        0xdc (PULL_DISABLE | MUX_MODE7) /* lcd_data15.lcd_data15 */
        0xe0 (PIN_INPUT_PULLDOWN | MUX_MODE7) /* lcd_vsync.lcd_vsync */
        0xe4 (PIN_INPUT_PULLDOWN | MUX_MODE7) /* lcd_hsync.lcd_hsync */
        0xe8 (PIN_INPUT_PULLDOWN | MUX_MODE7) /* lcd_pclk.lcd_pclk */
        0xec (PIN_INPUT_PULLDOWN | MUX_MODE7) /* lcd_ac_bias_en.lcd_ac_bias_en */
    >;
};
```

13.5.2 LCD 测试程序

Starter Kit 开发板上电启动即可以点亮 LCD 显示，这里不再做测试了，感兴趣的读者可以参考下述文档自行测试：http://processors.wiki.ti.com/index.php/AM335x_LCD_Controller_Driver%27s_Guide。

13.6 ADC 及触摸屏驱动

AM335x TS controller 支持如下四种模式：

- general-purpose ADC channels，即 8 路通用的 ADC；
- 4-wire TSC with 4 general-purpose ADC channels，即 4 路作为 4 线触摸屏连接，4 路作为通用 ADC；
- 5-wire TSC with 3 general-purpose ADC channels，即 5 路作为 5 线触摸屏连

接，3 路作为通用 ADC；

➢ 8-wire TSC，即 8 路作为 8 线触摸屏连接。

关于 4 线触摸屏的连接方式如图 13.9 所示，AM335x Starter Kit 板采用的正是 4 线模拟触摸屏。

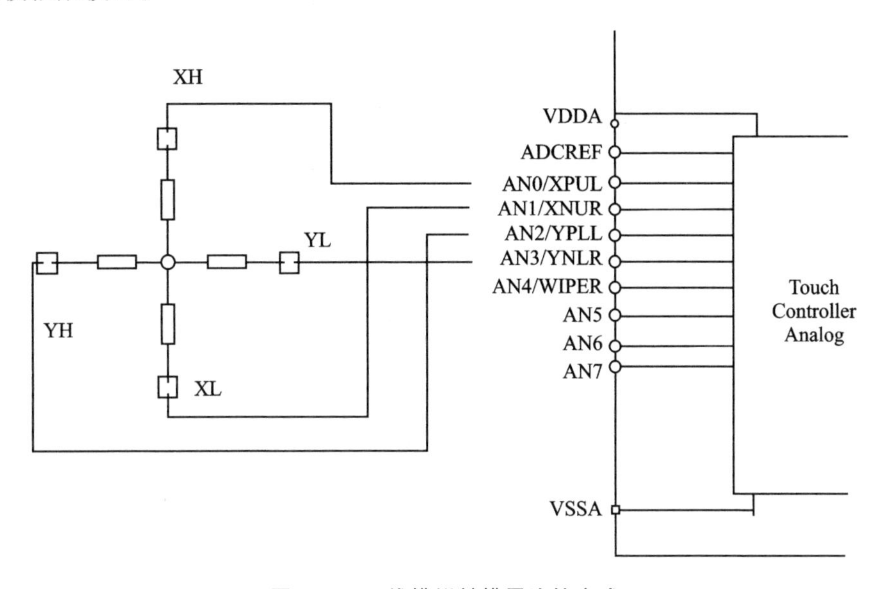

图 13.9　4 线模拟触摸屏连接方式

更详细的内容请读者参考以下资料：

➢ http://www.ti.com/lit/an/slyt209a/slyt209a.pdf；

➢ http://www.deyisupport.com/question_answer/w/faq/469.am335x-linux.aspx；

➢ http://processors.wiki.ti.com/index.php/AM335x_ADC_Driver-s_Guide；

➢ http://processors.wiki.ti.com/index.php/TI-Android-JB-4.2.2-DevKit-4.1.1_PortingGuide#Touchscreen。

13.6.1　AM335x 的触摸屏测试

现成的内核已经带有触摸屏驱动，且和按键一样，都属于 Linux 的输入子系统，被映射到/dev/input/event0 和/dev/input/touchscreen0 设备节点上，因此也可以使用 hexdump 工具读取触摸屏被按下的事件及坐标值。如图 13.10 所示为执行 hexdump 指令后，按中触摸屏后终端上的显示。

在 hexdump 中看到的数据为十六进制，数据格式需参考 Linux 的 input_event 事件的数据结构，还可以参考 documentation/input/input.txt 中的说明，且在 include/uapi/linux/input.h 文件中定义：

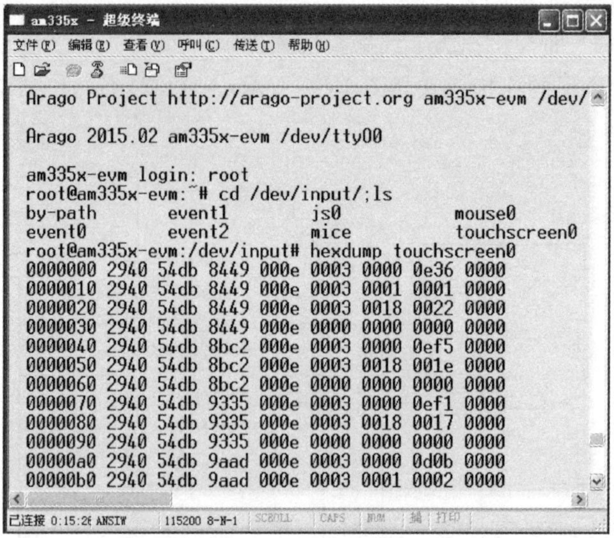

```
am335x - 超级终端
文件(F) 编辑(E) 查看(V) 呼叫(C) 传送(T) 帮助(H)
Arago Project http://arago-project.org am335x-evm /dev/

Arago 2015.02 am335x-evm /dev/ttyO0

am335x-evm login: root
root@am335x-evm:~# cd /dev/input/;ls
by-path        event1         js0            mouse0
event0         event2         mice           touchscreen0
root@am335x-evm:/dev/input# hexdump touchscreen0
0000000 2940 54db 8449 000e 0003 0000 0e36 0000
0000010 2940 54db 8449 000e 0003 0001 0001 0000
0000020 2940 54db 8449 000e 0003 0018 0022 0000
0000030 2940 54db 8449 000e 0000 0000 0000 0000
0000040 2940 54db 8bc2 000e 0003 0000 0ef5 0000
0000050 2940 54db 8bc2 000e 0003 0018 001e 0000
0000060 2940 54db 8bc2 000e 0000 0000 0000 0000
0000070 2940 54db 9335 000e 0003 0000 0ef1 0000
0000080 2940 54db 9335 000e 0003 0018 0017 0000
0000090 2940 54db 9335 000e 0000 0000 0000 0000
00000a0 2940 54db 9aad 000e 0003 0000 0d0b 0000
00000b0 2940 54db 9aad 000e 0003 0001 0002 0000
已连接 0:15:26 ANSIW   115200 8-N-1   SCROLL   CAPS   NUM   捕   打印
```

图 13.10　触摸屏测试

```
struct input_event {
    struct timeval time;
    __u16 type;
    __u16 code;
    __s32 value;
};
```

timeval 由 include/linux/time.h 文件定义：

```
struct timeval {
    __kernel_time_t tv_sec; /* seconds */
    __kernel_suseconds_t tv_usec; /* microseconds */
};
```

下述还有各个定义：

```
typedef __kernel_long_t __kernel_suseconds_t;
typedef __kernel_long_t __kernel_time_t;
typedef long __kernel_long_t;
```

hexdump 输出数据的各列定义：

第 1 列：行号。

第 2～5 列：输入事件时间戳，即结构体中的 time。

第 6 列：输入事件类型，即结构体中的 type，0 表示同步事件，3 表示绝对坐标事件。

第 7 列：坐标轴的值（X 或 Y），即结构体中的 code。

第 8 列：具体 A/D 值。

也可以使用 cat 命令查看确认 ti-tsc 是否真的对应/dev/input/event0，命令如下：

```
#cat /proc/bus/input/devices
```

13.6.2　触摸屏驱动代码分析

1. drivers\mfd\ti_am335x_tscadc.c

```
//只列出了主干函数，被调用的其他子功能函数请参考源码
static int ti_tscadc_probe(struct platform_device * pdev){
    struct ti_tscadc_dev * tscadc;
    struct resource * res;
    struct clk * clk;
    struct device_node * node = pdev->dev.of_node;
    struct mfd_cell * cell;
    struct property * prop;
    const __be32 * cur;
    u32 val;
    int err,ctrl;
    int clock_rate;
    int tsc_wires = 0,adc_channels = 0,total_channels;
    int readouts = 0;
    if (!pdev->dev.of_node) {//判断设备树参数是否有效
        dev_err(&pdev->dev,"Could not find valid DT data.\n");
        return -EINVAL;
    }
    node = of_get_child_by_name(pdev->dev.of_node,"tsc");
    of_property_read_u32(node,"ti,wires",&tsc_wires);//读取设备树中 tsc 的参数
    of_property_read_u32(node,"ti,coordiante-readouts",&readouts);
    //注：应为 coordinate
    node = of_get_child_by_name(pdev->dev.of_node,"adc");//读取设备树中 adc 的参数
    of_property_for_each_u32(node,"ti,adc-channels",prop,cur,val) {
        adc_channels++;
        if (val > 7) {
            dev_err(&pdev->dev," PIN numbers are 0..7 (not %d)\n",val);
            return -EINVAL;
        }
    }
    total_channels = tsc_wires + adc_channels;//AM335x 共 8 路
    if (total_channels > 8) {
        dev_err(&pdev->dev,"Number of i/p channels more than 8\n");
```

```
        return - EINVAL;
    }
    if (total_channels == 0) {
        dev_err(&pdev->dev,"Need atleast one channel.\n");
        return - EINVAL;
    }
    if (readouts * 2 + 2 + adc_channels > 16) {
        dev_err(&pdev->dev,"Too many step configurations requested\n");
        return - EINVAL;
    }
    tscadc = devm_kzalloc(&pdev->dev,
            sizeof(struct ti_tscadc_dev),GFP_KERNEL);//为 tscadc 设备分配内存
    if (!tscadc) {
        dev_err(&pdev->dev,"failed to allocate memory.\n");
        return - ENOMEM;
    }
    tscadc->dev = &pdev->dev;
    err = platform_get_irq(pdev,0);//获取 IRQ 中断 ID
    if (err < 0) {
        dev_err(&pdev->dev,"no irq ID is specified.\n");
        goto ret;
    } else{
        tscadc->irq = err;
    }
    res = platform_get_resource(pdev,IORESOURCE_MEM,0);//获取 resource 资源
    tscadc->tscadc_base = devm_ioremap_resource(&pdev->dev,res);//I/O 重映射
    if (IS_ERR(tscadc->tscadc_base))
    return PTR_ERR(tscadc->tscadc_base);
        tscadc->regmap_tscadc = devm_regmap_init_mmio(&pdev->dev,
                                tscadc->tscadc_base,&tscadc_regmap_config);
    if (IS_ERR(tscadc->regmap_tscadc)) {
        dev_err(&pdev->dev,"regmap init failed\n");
        err = PTR_ERR(tscadc->regmap_tscadc);
        goto ret;
    }
    spin_lock_init(&tscadc->reg_lock);
    init_waitqueue_head(&tscadc->reg_se_wait);
    pm_runtime_enable(&pdev->dev);
    pm_runtime_get_sync(&pdev->dev);
/*
 * The TSC_ADC_Subsystem has 2 clock domains
 * OCP_CLK and ADC_CLK.
```

```
 * The ADC clock is expected to run at target of 3MHz,
 * and expected to capture 12 - bit data at a rate of 200 KSPS.
 * The TSC_ADC_SS controller design assumes the OCP clock is
 * at least 6x faster than the ADC clock.
 */
    clk = clk_get(&pdev->dev,"adc_tsc_fck");
    if (IS_ERR(clk)) {
        dev_err(&pdev->dev,"failed to get TSC fck\n");
        err = PTR_ERR(clk);
        goto err_disable_clk;
    }
    clock_rate = clk_get_rate(clk);
    clk_put(clk);
    tscadc->clk_div = clock_rate / ADC_CLK;
/* TSCADC_CLKDIV needs to be configured to the value minus 1 */
    tscadc->clk_div--;
    tscadc_writel(tscadc,REG_CLKDIV,tscadc->clk_div);//配置时钟分频
/* Set the control register bits */
    ctrl = CNTRLREG_STEPCONFIGWRT | CNTRLREG_STEPID;
    tscadc_writel(tscadc,REG_CTRL,ctrl);
/* Set register bits for Idle Config Mode */
    if (tsc_wires > 0) {
        tscadc->tsc_wires = tsc_wires;
        if (tsc_wires == 5)
            ctrl |= CNTRLREG_5WIRE | CNTRLREG_TSCENB;
        else
            ctrl |= CNTRLREG_4WIRE | CNTRLREG_TSCENB;//4 线触摸屏
        tscadc_idle_config(tscadc);
    }
/* Enable the TSC module enable bit */
    ctrl |= CNTRLREG_TSCSSENB;
    tscadc_writel(tscadc,REG_CTRL,ctrl);
    tscadc->used_cells = 0;
    tscadc->tsc_cell = -1;
    tscadc->adc_cell = -1;
/* TSC Cell */
    if (tsc_wires > 0) {//触摸屏控制器通道
        tscadc->tsc_cell = tscadc->used_cells;
        cell = &tscadc->cells[tscadc->used_cells++];
        cell->name = "TI-am335x-tsc";
        cell->of_compatible = "ti,am3359-tsc";
        cell->platform_data = &tscadc;
```

```
        cell->pdata_size = sizeof(tscadc);
    }
/* ADC Cell */
    if (adc_channels > 0) {//通用 ADC 通道
        tscadc->adc_cell = tscadc->used_cells;
        cell = &tscadc->cells[tscadc->used_cells++];
        cell->name = "TI-am335x-adc";
        cell->of_compatible = "ti,am3359-adc";
        cell->platform_data = &tscadc;
        cell->pdata_size = sizeof(tscadc);
    }
    err = mfd_add_devices(&pdev->dev,pdev->id,tscadc->cells,
                                        tscadc->used_cells,NULL,0,NULL);
    if (err < 0)
        goto err_disable_clk;
    device_init_wakeup(&pdev->dev,true);
    platform_set_drvdata(pdev,tscadc);
    return 0;
err_disable_clk:
    pm_runtime_put_sync(&pdev->dev);
    pm_runtime_disable(&pdev->dev);
ret:
    return err;
}
static int ti_tscadc_remove(struct platform_device *pdev){
    struct ti_tscadc_dev *tscadc = platform_get_drvdata(pdev);
    tscadc_writel(tscadc,REG_SE,0x00);//STEPENABLE
    pm_runtime_put_sync(&pdev->dev);
    pm_runtime_disable(&pdev->dev);
    mfd_remove_devices(tscadc->dev);
    return 0;
}

static const struct of_device_id ti_tscadc_dt_ids[] = {
    { .compatible = "ti,am3359-tscadc",},
    { }
};
MODULE_DEVICE_TABLE(of,ti_tscadc_dt_ids);
static struct platform_driver ti_tscadc_driver = {
    .driver = {
        .name = "ti_am3359-tscadc",
        .pm = TSCADC_PM_OPS,
```

```
        .of_match_table = ti_tscadc_dt_ids,
    },
    .probe = ti_tscadc_probe,
    .remove = ti_tscadc_remove,
};
module_platform_driver(ti_tscadc_driver);
MODULE_DESCRIPTION("TI touchscreen / ADC MFD controller driver");
```

2. drivers\input\touchscreen\ti_am335x_tsc.c

```
//只列出了主干函数,被调用的其他子功能函数请参考源码
static int titsc_probe(struct platform_device * pdev){
    struct titsc * ts_dev;
    struct input_dev * input_dev;
    struct ti_tscadc_dev * tscadc_dev = ti_tscadc_dev_get(pdev);//获取平台设备
    int err;
    ts_dev = kzalloc(sizeof(structtitsc),GFP_KERNEL); //触摸屏设备分配内存空间
    input_dev = input_allocate_device();//为新的输入设备分配内存空间
    if (!ts_dev || !input_dev) {
        dev_err(&pdev->dev,"failed to allocate memory.\n");
        err = -ENOMEM;
        goto err_free_mem;
    }
    tscadc_dev->tsc = ts_dev;//tsc设备
    ts_dev->mfd_tscadc = tscadc_dev;//mfd多功能设备
    ts_dev->input = input_dev;//输入设备
    ts_dev->irq = tscadc_dev->irq;//触摸屏中断
    err = titsc_parse_dt(pdev,ts_dev);//解析触摸屏设备树中的相关参数
    if (err) {
        dev_err(&pdev->dev,"Could not find valid DT data.\n");
        goto err_free_mem;
    }
    err = request_irq(ts_dev->irq,titsc_irq,//申请中断,titsc_irq()为中断处理函数
            IRQF_SHARED,pdev->dev.driver->name,ts_dev);
    if (err) {
        dev_err(&pdev->dev,"failed to allocate irq.\n");
        goto err_free_mem;
    }
    if (device_may_wakeup(tscadc_dev->dev)) {//休眠时的中断唤醒功能
        err = dev_pm_set_wake_irq(tscadc_dev->dev,ts_dev->irq);
        if (err)
            dev_err(&pdev->dev,"irq wake enable failed.\n");
```

```
        }
    titsc_writel(ts_dev,REG_IRQSTATUS,IRQENB_MASK);//中断的相关配置
    titsc_writel(ts_dev,REG_IRQENABLE,IRQENB_FIFO0THRES);
    titsc_writel(ts_dev,REG_IRQENABLE,IRQENB_EOS);
    err = titsc_config_wires(ts_dev);
    //配置 wires,如 4 线电阻屏配置对应 XP、XN、YP、YN 及控制顺序等
    if (err) {
        dev_err(&pdev->dev,"wrong i/p wire configuration\n");
        goto err_free_irq;
    }
    titsc_step_config(ts_dev);//时序控制
    titsc_writel(ts_dev,REG_FIFO0THR,ts_dev->coordinate_readouts * 2 + 2 - 1);
    input_dev->name = "ti-tsc";
    input_dev->dev.parent = &pdev->dev;
    input_dev->evbit[0] = BIT_MASK(EV_KEY) | BIT_MASK(EV_ABS);
    input_dev->keybit[BIT_WORD(BTN_TOUCH)] = BIT_MASK(BTN_TOUCH);
    input_set_abs_params(input_dev,ABS_X,0,MAX_12BIT,0,0);//输入设备参数设置
    input_set_abs_params(input_dev,ABS_Y,0,MAX_12BIT,0,0);
    input_set_abs_params(input_dev,ABS_PRESSURE,0,MAX_12BIT,0,0);
    err = input_register_device(input_dev);//注册输入子系统
    if (err)
        goto err_free_irq;
        platform_set_drvdata(pdev,ts_dev);
        return 0;
err_free_irq://释放中断
        dev_pm_clear_wake_irq(tscadc_dev->dev);
        free_irq(ts_dev->irq,ts_dev);
err_free_mem://释放内存
        input_free_device(input_dev);
        kfree(ts_dev);
        return err;
}

static int titsc_remove(struct platform_device * pdev){
    struct titsc * ts_dev = platform_get_drvdata(pdev);
    u32 steps;
    dev_pm_clear_wake_irq(ts_dev->mfd_tscadc->dev);
    free_irq(ts_dev->irq,ts_dev); //释放中断
    /* total steps followed by the enable mask */
    steps = 2 * ts_dev->coordinate_readouts + 2;
    steps = (1 << steps) - 1;
    am335x_tsc_se_clr(ts_dev->mfd_tscadc,steps);
```

```
    input_unregister_device(ts_dev->input);
    kfree(ts_dev);
    return 0;
}

static const struct of_device_id ti_tsc_dt_ids[] = {
    { .compatible = "ti,am3359-tsc",},//和 DTS 相关联
    { }
};
MODULE_DEVICE_TABLE(of,ti_tsc_dt_ids);

static struct platform_driver ti_tsc_driver = {
    .probe = titsc_probe, //探测函数
    .remove = titsc_remove, //移除函数
    .driver = {
        .name = "TI-am335x-tsc",
        .pm = TITSC_PM_OPS,
        .of_match_table = ti_tsc_dt_ids,
    },
};
module_platform_driver(ti_tsc_driver);
MODULE_DESCRIPTION("TI touchscreen controller driver");
```

3. drivers/iio/adc/ti_am335x_adc.c

```
//只列出了主干函数,被调用的其他子功能函数请参考源码
static int tiadc_probe(struct platform_device *pdev){
    struct iio_dev *indio_dev;
    struct tiadc_device *adc_dev;
    struct device_node *node = pdev->dev.of_node;
    int err;
    if (! node) {//判断设备树参数是否有效
        dev_err(&pdev->dev,"Could not find valid DT data.\n");
        return -EINVAL;
    }
    indio_dev = devm_iio_device_alloc(&pdev->dev,
            sizeof(struct tiadc_device));//为 ADC 设备申请成为 iio 子系统
    if (indio_dev == NULL) {
        dev_err(&pdev->dev,"failed to allocate iio device\n");
        return -ENOMEM;
    }
    adc_dev = iio_priv(indio_dev);
```

```
    adc_dev->mfd_tscadc = ti_tscadc_dev_get(pdev);
    tiadc_parse_dt(pdev,adc_dev);//分析 DTS 参数
    indio_dev->dev.parent = &pdev->dev;
    indio_dev->name = dev_name(&pdev->dev);
    indio_dev->modes = INDIO_DIRECT_MODE;
    indio_dev->info = &tiadc_info;
    tiadc_step_config(indio_dev);//配置 STEP
    tiadc_writel(adc_dev,REG_FIFO1THR,FIFO1_THRESHOLD);
    err = tiadc_channel_init(indio_dev,adc_dev->channels);
    if (err < 0)
        return err;
    err = tiadc_iio_buffered_hardware_setup(indio_dev,
                                            &tiadc_worker_h,
                                            &tiadc_irq_h,
                                            adc_dev->mfd_tscadc->irq,
                                            IRQF_SHARED,
                                            &tiadc_buffer_setup_ops);

    if (err)
        goto err_free_channels;

    err = iio_device_register(indio_dev);
    if (err)
        goto err_buffer_unregister;
    platform_set_drvdata(pdev,indio_dev);
    return 0;
err_buffer_unregister:
    tiadc_iio_buffered_hardware_remove(indio_dev);
err_free_channels:
    tiadc_channels_remove(indio_dev);
    return err;
}
static int tiadc_remove(struct platform_device * pdev){
    struct iio_dev * indio_dev = platform_get_drvdata(pdev);
    struct tiadc_device * adc_dev = iio_priv(indio_dev);
    u32 step_en;
    iio_device_unregister(indio_dev);
    tiadc_iio_buffered_hardware_remove(indio_dev);
    tiadc_channels_remove(indio_dev);
    step_en = get_adc_step_mask(adc_dev);
    am335x_tsc_se_clr(adc_dev->mfd_tscadc,step_en);
    return 0;
}
```

```
static const struct of_device_id ti_adc_dt_ids[] = {
    { .compatible = "ti,am3359 - adc",},
    { }
};
MODULE_DEVICE_TABLE(of,ti_adc_dt_ids);
static struct platform_driver tiadc_driver = {
    .driver = {
        .name = "TI - am335x - adc",
        .pm = TIADC_PM_OPS,
        .of_match_table = ti_adc_dt_ids,
    },
    .probe = tiadc_probe,
    .remove = tiadc_remove,
};
module_platform_driver(tiadc_driver);
MODULE_DESCRIPTION("TI ADC controller driver");
```


13.6.3 触摸屏驱动与 DTS 的联系

在 am335x-evmsk. dts 文件中有：

```
……
#include "am33xx.dtsi"
……
&tscadc {
    status = "okay";
    tsc {
        ti,wires = <4>;
        ti,x - plate - resistance = <200>;
        ti,coordinate - readouts = <5>;
        ti,wire - config = <0x00 0x11 0x22 0x33>;
    };
};
……
```

在 am33xx. dtsi 文件中有：

```
tscadc: tscadc@44e0d000 {
    compatible = "ti,am3359 - tscadc";
    reg = <0x44e0d000 0x1000>;
    interrupt - parent = <&intc>;
    interrupts = <16>;
    ti,hwmods = "adc_tsc";
```

```
        status = "disabled";
        tsc {
            compatible = "ti,am3359 - tsc"; //
        };
        am335x_adc: adc {
            #io - channel - cells = <1>;
            compatible = "ti,am3359 - adc"; //
        };
    };
```

其中,“compatible = "ti,am3359-tscadc"”,即 compatible 参数为 ti,am3359-tscadc。

那么此时会在 drivers/mfd/ti_am335x_tscadc.c 文件中找到:


```
static const struct of_device_id ti_tscadc_dt_ids[] = {
    { .compatible = "ti,am3359 - tscadc",},
    { }
};
```

即从 ti_tscadc_dt_ids[]中的.compatible 项中匹配到 ti,am3359-tscadc,这就是 dts 和驱动 drivers/mfd/ti_am335x_tscadc.c 中的各个函数间的联系点。

另外,同理“compatible = "ti,am3359-tsc"”,即 compatible 参数为 ti,am3359-tsc。

那么此时会在 drivers/input/touchscreen/it_am335x_tsc.c 文件中找到:

```
static const struct of_device_id ti_tsc_dt_ids[] = {
    { .compatible = "ti,am3359 - tsc",},
    { }
};
```

同理“compatible = "ti,am3359-adc"”,即 compatible 参数为 ti,am3359-adc。

此时会在 drivers/iio/adc/ti_am335x_adc.c 文件中找到:

```
static const struct of_device_id ti_adc_dt_ids[] = {
    { .compatible = "ti,am3359 - adc",},
    { }
};
```

13.6.4 触摸屏驱动的测试

现成的内核已经带有触摸屏驱动,被映射到/dev/input/event0 和/dev/input/touchscreen0 设备节点上,因此可以使用 hexdump 工具读取触摸屏被按下的事件及坐标值。

但有些读者的驱动可能会存在一点小 BUG，就是执行 hexdump 指令后一直有事件被触发，而表现在 LCD 屏上时触摸屏光标会一直在闪烁。因此可以检查驱动代码中是否存在如下情况：

（1）ti_am335x_tscadc.c

```
static int ti_tscadc_probe(struct platform_device * pdev){
    ……
    of_property_read_u32(node,"ti,coordiante - readouts",&readouts);
//应修改为 of_property_read_u32(node,"ti,coordinate - readouts",&readouts);
    ……
}
```

（2）ti_am335x_tsc.c

```
static int titsc_parse_dt(struct platform_device * pdev,struct titsc * ts_dev){
    ……
    err = of_property_read_u32(node,"ti,coordiante - readouts",
                                   &ts_dev - >coordinate_readouts);
//应修改为 err = of_property_read_u32(node,"ti,coordinate - readouts",
//                                   &ts_dev - >coordinate_readouts);
}
static irqreturn_t titsc_irq(int irq,void * dev){
    ……
            z = z1 - z2; //应修改为 z = z2 - z1
            z * = x;
            z * = ts_dev - >x_plate_resistance;
            z / = z2; //应修改为 z / = z1
    ……
}
```

1. 测试程序——触摸屏事件捕获

下述测试代码实现的功能是循环读取触摸屏事件及值。

```
#include <stdio.h>
#include <unistd.h>
#include <sys/types.h>
#include <sys/ipc.h>
#include <sys/ioctl.h>
#include <fcntl.h>
int main(void){
    struct input_event tsc_evt;
    int ret;
    int f_tsc = open("/dev/input/touchscreen0",O_RDWR);
```

```
    if (f_tsc < 0){
        printf("error in open /dev/input/touchscreen0\n");
        return -1;
    }
    while(1){
        ret = read(fd_tsc,(unsigned char * )&tsc_evt,sizeof(struct input_event));
//阻塞型读函数
        if(ret < 0) {
            printf("read tsc event failed : %d",ret);
        }else{  //有效触摸屏事件
            printf("get tsc event code: %x,value: %x,type: %x",
                    tsc_evt.code,tsc_evt.value,tsc_evt.type);
        }
    }
}
```


2. Makefile

```
CROSS = /opt/ti-processor-sdk-linux-am335x-evm-03.00.00.04/linux-devkit/
sysroots/x86_64-arago-linux/usr/bin/arm-linux-gnueabihf-
all: tsc_test
tsc_test:
    $(CROSS)gcc tsc_test.c -o tsc_test
clean:
    rm -rf tsc_test *.o
```

3. 测试过程

先用 make 编译生成目标文件 tsc_test，再通过 TFTP 上传到目标板执行(参考 leds 测试实验的操作步骤)，当按下某个按键时将会有打印信息。

第 14 章 网络编程

《UNIX 网络编程》是学习 Linux 网络编程技术的经典书籍，只需学习前面或其中几章就可基本满足一般的网络编程应用。

14.1 常用函数

下面总结记录了《UNIX 网络编程》一书中阿南觉得比较常用的一些概念、函数，更详细的内容请参考该书。

1. 套接口地址结构：sockaddr_xx

IPv4 套接口地址结构通常称为网际套接口地址结构：sockaddr_in。

```
struct in_addr{
    in_addr_t  s_addr; //32 - bit IPv4 address
                       //network byte ordered
};
struct sockaddr_in{
    uint8_t           sin_len;
    sa_family_t       sin_family; //AF_INET
    in_port_t         sin_port; //16 - bit TCP or UDP port number
                                //network byte ordered
    struct in_addr_t sin_addr; //32 - bit Ipv4 address
                               //network byte ordered
    char              sin_zero[8]; //unused
}
```

通用套接口地址结构：

```
struct  sockaddr{
    uint8_t           sa_len;
    sa_family_t       sa_family; //address family:AF_xxx value
    char              sa_data[14]; //protocol specific address
}
```

由于套接口函数被定义为采用指向通用套接口地址结构的指针，如 bind 函数原型所示：

```
int  bind(int,struct sockaddr * ,socklen_t);
```

所以套接口函数的任何调用都必须将指向特定于协议的套接口地址结构的指针类型转换成指向通用套接口地址结构的指针。例如：

```
struct  sockaddr_in  serv;                //IPv4 socket address structure
bind(sockfd,(struct sockaddr * )&serv,sizeof(serv));
```

另外，在unp.h中有如下定义：

```
#defined  SA  struct sockaddr
```

可以简略地写成：

```
Bind(sockfd,(SA * )&serv,sizeof(serv));
```

2. 值-结果参数

当把套接口地址结构传递给套接口函数时，总是通过指针来传递的。结构的长度也作为参数来传递，其传递方式取决于结构的传递方向：从进程到内核，还是从内核到进程。

从进程到内核传递套接口地址结构有3个函数：bind、connect和sendto。它们结构的长度以整数大小传递。

```
struct sockaddr_in serv;
connect(sockfd,(SA * )&serv,sizeof(serv));
```

从内核到进程传递套接口地址结构有4个函数：accept、recvfrom、getsockname和getpeername。它们结构的长度以整数大小的指针传递。

```
struct sockaddr_un cli;             //Unix domain
socklen_t len;
len = sizeof(cli);                  //len is a value
getpeername(unixfd,(SA * )&cli,&len);
```

为何将结构大小由整数改为指向整数的指针呢？这是因为：当函数被调用时，结构大小是一个值(value，此值告诉内核该结构的大小，使内核在写此结构时不至于越界)；当函数返回时，结构大小又是一个结果(result，它告诉进程内核在此结构中确切存储了多少信息)。这种参数类型叫做值-结果(value-result)参数。

3. 字节排序函数

内存中存储多个字节，有"小端"和"大端"两种字节序。我们把系统所用的字节序称为主机字节序(host byte order)，把网络协议中的字节序称为网络字节序(network byte order)，它们不一定相同，所以就有了主机字节序和网络字节序间的相互转换函数：

```
#include <netinet/in.h>
uint16_t htons(uint16_t host16bitvalue);
uint16_t htonl(uint32_t host32bitvalue);
```

返回：网络字节序值。

```
uint16_t ntohs(uint16_t net16bitvalue);
uint16_t ntohl(uint32_t net32bitvalue);
```

返回：主机字节序值。

4. 字节操纵函数

多字节字段的操作有两组函数。

第一组函数以字母 b(表示 byte)开头，起源于 4.2BSD，由支持套接口函数的系统提供，如下：

```
#include <strings.h>
void bzero(void * dest,size_t nbytes);
void bcopy(const void * src,void * dest,size_t nbytes);
void bcmp(const void * ptr1,const void * ptr2,size_t nbytes);
```

返回：0——相等，非 0——不相等。

第二组函数其名字以 mem(表示 memory)开头，起源于 ANSI C 标准，由支持 ANSI C 库的系统提供，如下：

```
#include <string.h>
void * memset(void * dest,int c,size_t len);
```

其与 bzero 不同的是：将目标指定数目的字节置为值 c。

```
void * memcpy(void * dest,const void * src,size_t nbytes);
```

其与 bcopy 相比交换了两个指针参数的顺序，memcpy 两个指针的顺序是按照 C 的赋值语句从左到右的顺序书写的：

```
dest = src;
int memcmp(const void * ptr1,const void * ptr2,size_t nbytes);
```

返回：0——相同，大于 0 或小于 0——不相同，具体是大于 0 还是小于 0 取决于第一个不等字节的大小。

另外，memXXX 函数都要求有一个长度参数，且它总是最后一个。

5. 地址转换函数——ASCII 字符串与网络字节序的二进制间转换地址

inet_aton、inet_addr 和 inet_ntoa 在点分十进制数串(例如，"202.168.112.96")与它的 32 位网络字节序二进制值间转换 IPv4 地址，如下：

```
#include <arpa/inet.h>
int inet_aton(const char * strptr,struct in_addr * addrptr);
//返回：1——串有效，0——串无效
in_addr_t inet_addr(const char * strptr);
//返回：若成功，返回 32 位二进制的网络字节序地址；若出错，则返回 INADDR_NONE(一般为一
//个 32 位均为 1 的值)，这意味着点分十进制数串 255.255.255.255(IPv4 的有限广播地址)
//不能由此函数处理，因为它的二进制值被用来指示函数的失败。目前一般由 inet_aton 替代它
char * inet_ntoa(struct in_addr inaddr);
//返回：指向点分十进制数串的指针
```

下面两个函数较新，对 IPv4 和 IPv6 地址都能处理。

```
#include <arpa/inet.h>
int inet_pton(int family,const char * strptr,void * addrptr);
//返回：1 成功，0 输入不是有效的表达格式，-1 出错
const char * inet_ntop(int family,const void * addrptr,char * strptr,size_t len);
//返回：指向结果的指针——成功，NULL——出错
```


参数 len 是目标的大小，以免函数溢出其调用者的缓冲区，如果太小，无法容纳表达格式结果（包括终止的空字符），则返回一个空指针，并置 errno 为 ENOSPC。

另外在头文件<netinet/in.h>中有如下定义：

```
#define   INET_ADDRSTRLEN    16
#define   INET6_ADDRSTRLEN   46
```

6. 字节流套接口函数

字节流套接口（如 TCP 套接口）上的 read 和 write 函数所表示的行为不同于通常的文件 I/O。字节流套接口上的读/写输入或输出的字节数可能比要求的数量少，但这不是错误状况，原因是内核中套接口的缓冲区可能已达到了极限。此时需要再次调用 read 或 write 函数，以输入或输出剩余的字节。

```
ssize_t readn(int filedes,void * buff,size_t nbytes);
ssize_t writen(int filedes,const void * );
ssize_t readline(int filedes,void * buff,size_t maxien);
```

下述为 writen 函数原码，往一个描述字写 n 字节数据。

```
ssize_t writen(int fd,const void * vptr,size_t n){
    size_t          nleft;
    ssize_t         nwritten;
    const char       * ptr;
    ptr = vptr;
    nleft = n;
    while (nleft > 0) {
```

```
        if ( (nwritten = write(fd,ptr,nleft)) <= 0){
            if (nwritten < 0 && errno == EINTR)
                nwritten = 0;           /* and call write() again */
            else
                return(-1);             /* error */
        }
        nleft -= nwritten;
        ptr += nwritten;
    }
    return(n);
}
```

recv 和 send 函数类似于标准的 read 和 write 函数，不过需要一个额外的参数，如下：

```
#include <sys/socket.h>
ssize_t recv(int sockfd,void * buff,size_t nbytes,int flags);
ssize_t send(int sockfd,const void * buff,size_t nbytes,int flags);
```

返回：读入或写出字节数——成功，－1——出错。

前 3 个参数等同于 read 和 write 的 3 个参数，flags 参数的值或为 0，或为下述一个或多个常值的逻辑或。

MSG_DONTROUTE (send)——绕过路由表查找。

MSG_DONTWAIT (recv 和 send)——仅本操作非阻塞。

MSG_OOB (recv 和 send)——发送或接收数据。

MSG_PEEK (recv)——窥看外来消息。

MSG_WAITALL (recv)——等待所有数据，它告知内核不要在尚未读入请求数目的字节之前让一个读操作返回。出现下述情况例外：① 捕获一个信号；② 连接被终止；③ 套接口发生一个错误。

7. 基本套接口函数

所有客户和服务器都从调用 socket 开始，它返回一个套接口描述字。客户随后调用 connect 函数来建立与服务器的连接，服务器则调用 bind、listen 和 accept。套接口通常使用标准的 close 函数关闭，不过有时也会用 shutdown 关闭。另外还有获得本地协议地址(getsockname)和远地协议地址(getsockname)等函数。

无论是客户还是服务器，为了执行网络 I/O，必须做的第一件事情就是调用 sock 函数，指定期望的通信协议类型来创建套接口描述字。

```
#include <sys/socket.h>
int socket(int family,int type,int protocol);
//返回：非负描述字(套接口描述字，简称套接字：sockfd)——成功，-1——出错
```

family 为下述常值,指明协议族(family):

AF_INET——IPv4 协议;

AF_INET6——IPv6 协议;

AF_LOCAL——Unix 域协议;

AF_ROUTE——路由套接口;

AF_KEY——密钥套接口。

type 为下述常值,指明套接口类型:

SOCK_STREAM——字节流套接口;

SOCK_DGRAM——数据报套接口;

SOCK_SEQPACKET——有序分组套接口;

SOCK_RAW——原始套接口。

Protocol 为下述常值,指明某个协议类型,或设为 0,以选择 family 和 type 组合的系统缺省值。

IPPROTO_TCP——TCP 传输协议;

IPPROTO_UDP——UDP 传输协议;

IPPROTO_SCTP——SCTP 传输协议。

客户端调用 connect 函数来建立与服务器的连接,如下:

```
#include <sys/socket.h>
int connect(int sockfd,const struct sockaddr * servaddr,socklen_t addrlen);
//返回:0——成功,-1——出错
```

sockfd 是由 socket 函数返回的套接口描述字;servaddr 为必须含有所连接服务器 IP 和端口号的套接口地址结构指针;addrlen 为 servaddr 结构的大小。

如果是 TCP 客户在调用 connect 函数将激发 TCP 的三路握手过程,而且仅在连接建立成功或出错时才返回。另外,如果 connect 失败则该套接口不可再用,必须 close 该套接口描述字,重新调用 socket。

bind 函数把一个本地协议地址赋予一个套接口:

```
#include <sys/socket.h>
int bind(int sockfd,const struct sockaddr * myaddr,socklen_t addrlen);
```

sockfd 是由 socket 函数返回的套接口描述字;myaddr 是一个指向特定于协议的地址结构指针;addrlen 是该地址结构的长度。

本地地址结构参数:

sin_family: 协议族。

sin_port: 服务器在启动时必须捆绑它们的端口,只有这样才能被大家(客户端)认识。

sin_addr.s_addr: 对于 TCP 客户,为在该套接口上发送的 IP 数据报指派的源

IP 地址。对于 TCP 服务器，则限定该套接口只接收那些目的地为这个 IP 地址的客户连接。

调用 bind 函数可以指定 IP 地址(本地地址)或端口，也可以不指定 IP 地址(通配地址)和端口。对于 IPv4，通配地址由常值 INADDR_ANY 来指定，它允许服务器在任意网络接口上接收客户连接(假定服务器主机有多个网络接口)。

listen 函数仅由 TCP 服务器调用，通常应该在调用 socket 和 bind 两个函数之后，并在调用 accept 函数之前调用，它做两件事情：当 socket 函数创建一个套接口时，它被假设为一个主动套接口，也就是说，它是一个将调用 connect 发起连接的客户套接口，指示内核应接收指向该套接口的连接请求；调用 listen 导致套接口从 CLOSED 状态转换到 LISTEN(被动)状态。

listen 函数的第二个参数规定了内核应该为相应套接口排队的最大连接个数。

```
#include <sys/socket.h>
int listen(int sockfd,int backlog);
```

返回：0——成功，－1——出错。

accept 函数由 TCP 服务器调用，用于从已完成连接队列的队头返回下一个已完成连接。如果已完成连接队列为空，那么进程被投入睡眠(假定套接口为缺省的阻塞方式)。

```
#include <sys/socket.h>
int accept(int sockfd,struct sockaddr * cliaddr,socklen_t * addrlen);
```

返回：非负描述字——成功，－1——出错。

参数 cliaddr 和 addrlen 用来返回已连接的对端进程(客户端)的协议地址。addrlen 是值-结果参数。如果 accept 成功，那么其返回值是由内核自动生成的一个全新描述字，代表与所返回客户的 TCP 连接。在讨论 accept 函数时，我们称它的第一个参数为监听套接口(listening socket)描述字(由 socket 创建，随后用作 bind 和 listen 的第一个参数的描述字)，称它的返回值为已连接套接口(connected socket)描述字。一个服务器通常仅仅创建一个监听套接口，它在该服务器的生命期内一直存在。内核为每个由服务器进程接收的客户连接创建了一个已连接套接口(也就是说对于它的 TCP 三路握手过程已经完成)。当服务器完成对于某个给定客户的服务时，相应的已连接套接口就被关闭。如果对返回客户协议地址不感兴趣，那么可以把 cliaddr 和 addrlen 均置为空指针。

本地协议地址(getsockname)和远地协议地址(getpeername)函数：

```
#include <sys/socket.h>
int getsockname(int sockfd,struct sockaddr * localaddr,socklen_t * addrlen);
int getpeername(int sockfd,struct sockaddr * peeraddr,socklen_t * addrlen);
```

两者均返回：0——成功，－1——出错。

wait 和 waitpid 等待子进程终止函数：

```
#include <sys/wait.h>
pid_t wait(int * statloc);
pid_t waitpid(pid_t pid,int * statloc,int options);
```

两者均返回：终止的进程 ID 为成功，-1 为出错。

statloc 指向子进程的终止状态，如果不关心终止状态，则可指定为空指针。

pid 参数允许指定想等待的进程 ID，值-1 表示等待第一个终止的子进程。

option 允许指定附加选项，最常用的选项是 WNOHANG，它告知内核在没有已终止子进程时不要阻塞。

下面为等待多个子进程终止的例子：

```
#include "unp.h"
void sig_chld(int signo){
    pid_t pid;
    int stat;
    while ((pid = waitpid(-1,&stat,WNOHANG)) > 0){
        printf("child %d terminated\n",pid);
    }
    return;
}
```

shutdown 函数与 close 函数相比有两点不同：① close 把描述字的引用计数值减 1，仅在该计数变为 0 时才关闭套接口。使用 shutdown 可以不管引用计数就激发 TCP 的正常连接终止。② close 终止数据传送的两个方向：读和写。而 shutdown 可以只关闭一半的 TCP 连接，即只终止读连接或写连接。

```
#include <sys/socket.h>
int shutdown(int sockfd,int howto);
```

返回：0——成功，-1——出错。

函数的行为依赖于 howto 参数的值：

SHUT_RD——关闭连接的读这一半，套接口中不再有数据可接收，而且套接口接收缓冲区中的现有数据都被丢弃。进程不能再对这样的套接口调用任何读函数。对一个 TCP 套接口这样调用 shutdown 函数后，由该套接口接收的来自对端的任何数据都被确认，然后悄然丢弃。

SHUT_WR——关闭连接的写这一半，当前留在套接口发送缓冲区中的数据将被发送，后跟 TCP 的正常连接终止序列。进程不能再对这样的套接口调用任务写函数。

SHUT_RDWR——连接的读这一半和写这一半都关闭。

8. I/O复用：select和poll函数

阻塞型I/O只能等待一个描述字就绪，即进程调用系统调用(如：accept，read等)后阻塞直到数据报到达且被拷贝到应用进程的缓冲区中或者发生错误才返回。而有了I/O复用，就可以调用select或poll函数，阻塞在这(select和poll)两个系统调用中的某一个之上，而不是阻塞在真正的I/O系统调用，这样就可以等待多个描述字就绪。

select函数：

```
#include <sys/select.h>
#include <sys/time.h>
int select(int maxfdp1,fd_set * readset,fd_set * writeset,fd_set * exceptset,const struct timeval * timeout);
```

返回：就绪描述字的正数目，0——超时，-1——出错。

timeout参数告知内核等待所指定描述字中的任何一个就绪可花多长时间，其结构如下：

```
struct timeval{
    long tv_sec;    //seconds
    long tv_usec;   //microseconds
}
```

timeout的取值有三种可能：① 空指针，永远等待下去，直到有一个描述字准备就绪。② 结构中的定时器值为0，根本不等待，即检查描述字后立即返回。③ 结构中的定时器值大于0，等待一段固定时间。

参数readset、writeset和exceptset指定要让内核测试读、写和异常条件的描述字，如果对某一个条件不感兴趣，则可以将它设为空指针，表示不测试该类型(或读，或写，或异常)的任何描述字，如下述select函数只测试读就绪的描述字：

```
select(maxfdp1,&rset,NULL,NULL,NULL);
```

给这三个参数的每一个指定一个或多个描述字，值隐藏在名为fd_set的数据类型和以下四个宏中：

```
void FD_ZERO(fd_set * fdset);
void FD_SET(int fd,fd_set * fdset);
void FD_CLR(int fd,fd_set * fdset);
void FD_ISSET(int fd,fd_set * fdset);
```

maxfdp1参数指定待测试的描述字个数，描述字0,1,2…一直到maxfdp1 - 1均将被测试。

poll函数在功能上与select函数类似，不过在处理流设备时，它能够提供额外的信息。

```
#include <poll.h>
int poll(struct pollfd * fdarray,unsigned long nfds,int timeout);
```

返回：就绪描述字的正数目，0——超时，-1——出错

fdarray 参数指向 pollfd 结构数组第一个元素的指针，每个数组元素都是一个 pollfd 结构，用于指定测试某个给定描述字 fd 的条件。

```
struct pollfd{
    int fd; //descriptor to check
    short events; //events of interest on fd
    short revents;//events that occurred on fd
}
```

测试的条件由 events 成员指定，函数在相应的 revents 成员中返回该描述字的状态。下面为 events 和 revents 标志的常值：

POLLIN——普通或优先级带数据可读，POLLRDNORM 和 POLLRDBAND 逻辑或；

POLLRDNORM——普通数据可读；

POLLRDBAND——优先级带数据可读；

POLLPRI——高优先级数据可读；

POLLOUT——普通数据可写；

POLLWRNORM——普通数据可写，等同于 POLLOUT；

POLLWRBAND——优先级带数据可写；

POLLERR——发生错误；

POLLHUP——发生挂起；

POLLNVAL——描述字不是一个打开的文件。

nfds 参数指定结构数组中元素的个数。

timeout 参数指定 poll 函数返回前等待多长时间，取值如下：

INFTIM——永远等待；0——立即返回，不阻塞进程；大于 0——等待指定数目的毫秒数。

9. 信号处理

信号(signal)就是通知某个进程发生了某个事件，有时也称软件中断(software interrupt)。信号通常是异步发生的，也就是说进程预先不知道信号准确发生的时刻。它可以由一个进程发给另一进程(或自身)，也可以由内核发给某个进程。

SIGCHLD 信号就是由内核在任何一个进程终止时发给它的父进程的一个信号。

```
Sigfunc * signal(int signo,Sigfunc * func){
    struct sigaction act,oact;
```

```
    act.sa_handler = func;
    sigemptyset(&act.sa_mask);
    act.sa_flags = 0;
    if (signo == SIGALRM) {
#ifdef SA_INTERRUPT
        act.sa_flags |= SA_INTERRUPT; /* SunOS 4.x */
#endif
    }
    else{
#ifdef SA_RESTART
        act.sa_flags |= SA_RESTART;   /* SVR4,44BSD */
#endif
    }
    if (sigaction(signo,&act,&oact) < 0){
        return(SIG_ERR);
    }
    return(oact.sa_handler);
}
/* end signal */
Sigfunc *Signal(int signo,Sigfunc *func) {  /* for our signal() function */
    Sigfunc *sigfunc;
    if ( (sigfunc = signal(signo,func)) == SIG_ERR){
        err_sys("signal error");
    }
    return(sigfunc);
}
```

在服务器中当子进程关闭时，捕获 SIGCHLD，对其处理以防止僵死进程。信号处理函数如下：

```
void sig_chld(int signo){
    pid_t pid;
    int stat;
    while ( (pid = waitpid(-1,&stat,WNOHANG)) > 0){
    //当没有信号处理时返回-1退出循环
        printf("child %d terminated\n",pid);
    }
    return;
}
```

要在创建第一个子进程之前调用："Signal(SIGCHLD,sig_chld);"。

10. 调试命令

netstat -a——检查服务器监听套接口的状态，命令显示如下：

```
#netstat -a |grep 9877
Active Internet connections (servers and established)
Proto    Recv-Q    Send-Q    Local Addres    Foreign Address    State
tcp         0         0        *:9877            *:*             LISTEN
```

上例为只显示与端口号 9877 相关的状态，Local Addres 为本地地址结构，Foreign Addres 为远程地址结构。“*”表示一个为 0 的 IP 地址(INADDR_ANY，通配地址)，或为 0 的端口号。State 为网络正处于的状态，LISTEN 为正在监听，另外还有 ESTABLISHED 表示已确定连接的，TIME_WAIT 等待超时等，相关内容可参考 TCP 状态转换图。

14.2　服务器实例

```
#include <stdio.h>
#include <sys/socket.h>
#include <sys/wait.h>
#include <netinet/in.h>
#include <arpa/inet.h>
#include <netdb.h>
#include <signal.h>
#include <errno.h>
#define SERV_PORT 9877
#define MAXLINE 256
struct SERIAL_DATA{
    unsigned short Length;
    unsigned short wReserved;
    unsigned char *pBuffer;
};
struct DOWN_LOAD_PORT{
    unsigned short CommandNum;
    unsigned short MaxLength;
    struct SERIAL_DATA *psCommand;
    struct SERIAL_DATA sOldInput;
    struct SERIAL_DATA sOldHttpInput;
    struct SERIAL_DATA sCommandHead;
    struct SERIAL_DATA sDelimiter;
    struct SERIAL_DATA sAckOK;
    struct SERIAL_DATA sAckDetect;
    struct SERIAL_DATA sAckERR;
    unsigned char byParity;
    unsigned char bReserved;
```

```
    unsigned short wReserved;
};
const char * gpCommandHead = "\x02\x1b\x43";
const char * gpSerialCommand[] = {
    "\x08","\xcf\xad\xf0\xe7\xd9\xb6\xe0\xf8",      //0  update user program
    "\x08","\xfc\xad\xf0\xe7\xd9\xb6\xe0\x8f",      //1  update system program
    "\x04","INFO",                                  //2  look over user information
    "\x03","VER",                                   //3  look over system version
    "\x07","Control",                               //4  switch run times show
    "\x07","IRLearn",                               //5  IR study
    "\x07","GetTime",                               //6  read time
    "\x07","SetTime",                               //7  set time
    "\x03","Del",                                   //8  delete user program
    "\x0d","GSQRestartGSQ",                         //9  reboot user program
    "\x0f","GSQHardResetGSQ",                       //10 systerm hardware reset
    "\x06","Search",                                //11 find devices for 485NET
    "\x01","\xf0",                                  //12 terminal check command,
                                                         test connecting
    "\x08","\xfc\xda\x67\x85\x2d\x7f\x9e\x88",      //13 send IR code test
    "\x0c","SearchPlugin",                          //14 search plugin devices
    "\x05","Digit",                                 //15 Digit Signal
    "\x06","Analog",                                //16 Analog Signal
    "\x06","Serial",                                //17 Serial Signal
    "\x06","IRLong",                                //18 IR study of long
    "\x0b","SetBaudrate",                           //19 setup baudrate
    "\x0c","ReadBaudrate",                          //20 read baudrate
};
const char * gpDelimiter = "\x02\x0d\x0a";
const char * gpAckOK = "\x08\xe7\xa8\xb6\xceOK\x0d\x0a";
const char * gpAckERR = "\x01\x9f";
const char * gpAckDetect = "\x01\x0f";
static struct DOWN_LOAD_PORT * gpDownPort;
void * InitCommand(void){
    int i;
    struct SERIAL_DATA * pSerial;
    gpDownPort = (struct DOWN_LOAD_PORT * )malloc(sizeof(struct DOWN_LOAD_PORT));
    gpDownPort->CommandNum = sizeof(gpSerialCommand)/8;
    gpDownPort->byParity = 0;
    pSerial = gpDownPort->psCommand = (struct SERIAL_DATA * )malloc(sizeof(struct
SERIAL_DATA) * gpDownPort->CommandNum);
    gpDownPort->MaxLength = 0;
    for (i = 0; i < gpDownPort->CommandNum * 2; pSerial ++){
```

```
            pSerial->Length = (unsigned short)(**(gpSerialCommand+(i++)));
            pSerial->pBuffer = (unsigned char *)(*(gpSerialCommand+(i++)));
            if (gpDownPort->MaxLength < pSerial->Length){
                gpDownPort->MaxLength = pSerial->Length;
            }
        }
        gpDownPort->sCommandHead.Length = (unsigned short)(*gpCommandHead);
        gpDownPort->sCommandHead.pBuffer = (unsigned char *)gpCommandHead + 1;
        gpDownPort->sDelimiter.Length = (unsigned short)(*gpDelimiter);
        gpDownPort->sDelimiter.pBuffer = (unsigned char *)gpDelimiter + 1;
        gpDownPort->sAckOK.Length = (unsigned short)(*gpAckOK);
        gpDownPort->sAckOK.pBuffer = (unsigned char *)gpAckOK + 1;
        gpDownPort->sAckDetect.Length = (unsigned short)(*gpAckDetect);
        gpDownPort->sAckDetect.pBuffer = (unsigned char *)gpAckDetect + 1;
        gpDownPort->sAckERR.Length = (unsigned short)(*gpAckERR);
        gpDownPort->sAckERR.pBuffer = (unsigned char *)gpAckERR + 1;
        gpDownPort->MaxLength += gpDownPort->sCommandHead.Length + gpDownPort->
sDelimiter.Length;
        gpDownPort->sOldInput.Length = 0;
        gpDownPort->sOldInput.pBuffer = (unsigned char *)malloc(gpDownPort->Max-
Length + 4);
        gpDownPort->sOldHttpInput.Length = 0;
        gpDownPort->sOldHttpInput.pBuffer = (unsigned char *)malloc(256);
        memset(gpDownPort->sOldHttpInput.pBuffer,0,256);
        return gpDownPort;
    }
    int Process_Command(int nIDCommand,int sockfd){ //命令处理函数
        struct SERIAL_DATA *pSerial = gpDownPort->psCommand + nIDCommand;
        int k;
        if ( (k = writen(sockfd,gpDownPort->sAckOK.pBuffer,gpDownPort->sAckOK.
Length)) < 0){
            printf("write faill\n");
            return (-1);
        }
    printf("command %d: %s\n",nIDCommand,pSerial->pBuffer);//打印查找到的命令串
    }

    int Communication(int sockfd){ //与客户机通信处理程序
        struct SERIAL_DATA *pSerial;
        ssize_t nread,nbuf,len,i;
        char buf[MAXLINE];
        char *ptr;
```

```
    again:
        ptr = buf;
        nbuf = 0;
        while (1){
            if ( (nread = recv(sockfd,ptr,MAXLINE,0)) <= 0){ //等待接收客户机数据
                if (errno == EINTR){
                    continue;
                }
                else{
                    perror("Communication: read error");
                    return (-1);
                }
            }
            else{
                nbuf += nread;
                ptr += nread;
                if (nbuf <= gpDownPort->sCommandHead.Length){ //匹配命令头
                    if (memcmp(buf,gpDownPort->sCommandHead.pBuffer,nbuf) == 0){
                        continue;
                    }
                    else{ //命令头不匹配,应答出错
                        if(writen(sockfd,gpDownPort->sAckERR.pBuffer,gpDownPort->
sAckERR.Length)<0){
                            return (-1);
                        }
                        else{
                            goto again; //清缓冲区重新开始接收
                        }
                    }
                }
                else if (nbuf <= (gpDownPort->sCommandHead.Length + gpDownPort->
sDelimiter.Length)){
                    continue;
                }
                else {
                    if(memcmp(&buf[nbuf - gpDownPort->sDelimiter.Length],gpDownPort
->sDelimiter.pBuffer,gpDownPort->sDelimiter.Length) == 0){//匹配结束符
                        len = nbuf - gpDownPort->sCommandHead.Length - gpDownPort-
>sDelimiter.Length;
                        pSerial = gpDownPort->psCommand;
                        for (i = 0; i < gpDownPort->CommandNum; i++,pSerial++){
                        //查找命令
```

```
                        if (len == pSerial->Length){
                            if(memcmp(&buf[gpDownPort->sCommandHead.Length],pS-
erial->pBuffer,pSerial->Length) == 0){
                                Process_Command(i,sockfd); //命令处理
                                goto again;
                            }
                        }
                    }
                    printf("Bad Command\n");
                    if (writen(sockfd,gpDownPort->sAckERR.pBuffer,gpDownPort->
sAckERR.Length) < 0){
                        return (-1);
                    }
                    else{
                        goto again; //无效命令,重新开始接收
                    }
                }
            }
        }
    }
}
int main(int argc,char * argv[]){
    struct sockaddr_in cliaddr,servaddr; //客户和服务器地址结构
    int listenfd,connfd; //监听和连接描述字
    socklen_t clilen; //客户地址结构长度
    pid_t childpid; //子进程描述符
    if ( (listenfd = socket(AF_INET,SOCK_STREAM,IPPROTO_TCP)) < 0){ //创建套接口
        perror("call to socket");
        exit(1);
    }
    bzero(&servaddr,sizeof(servaddr)); //服务器地址结构清 0
    servaddr.sin_family = AF_INET; //网络协议:IPv4 协议
    servaddr.sin_addr.s_addr = INADDR_ANY; //通配地址
    servaddr.sin_port = htons(SERV_PORT); //网络端口
    if (bind(listenfd,(struct sockaddr *)&servaddr,sizeof(servaddr)) < 0){
        perror("call to bind");
        exit(1);
    }
    if (listen(listenfd,20) < 0){
        perror("call to listen");
        exit(1);
    }
```

```
    printf("Accepting connections ...\n");
    Signal(SIGCHLD,sig_chld); //创建子进程结束信号处理
    InitCommand(); //初始化与客户机通信协议及命令
    while (1){
        clilen = sizeof(cliaddr);
        if ( (connfd = accept(listenfd,(struct sockaddr *)&cliaddr,&clilen)) < 0){
            if (errno == EINTR){
                continue;
            }
            else{
                perror("call to accept");
                exit(1);
            }
        }
        if ( (childpid = fork()) == 0){ //child process
            close(listenfd); //close listening socket
            Communication(connfd); //处理与客户机间的通信
            exit(0);
        }
        close(connfd);
    }
}
```

另外，与信号相关及 writen 等函数的程序均在前面列出这里不再赘述。

14.3　客户端测试

可以在另一台 PC(或者本机的另一终端，指向 IP：127.0.0.1)的 Linux 下运行下述客户程序进行测试，上面的服务器程序可以先在 PC 的 Linux 下运行测试，也可以编译成在 ARM 板上运行，效果一样。

```
#include <unistd.h>
#include <stdio.h>
#include <sys/socket.h>
#include <netdb.h>
#include <errno.h>
#define MAXLINE 256
#define SERV_PORT 9877
#define max(a,b) ((a) > (b) ? (a) : (b))
const char * gpCommandHead = "\x02\x1b\x43";
const char * gpDelimiter = "\x02\x0d\x0a";
const char * gpAckOK = "\x08\xe7\xa8\xb6\xceOK\x0d\x0a";
```

```
const char * gpAckERR = "\x01\x9f";
void str_cli(FILE * fp,int sockfd){
    int maxfdp1,stdineof;
    fd_set rset;
    char sendline[MAXLINE],recvline[MAXLINE];
    int n,i;
    stdineof = 0;
    FD_ZERO(&rset);
    while (1){
        if (stdineof == 0){
            FD_SET(fileno(fp),&rset);
        }
        FD_SET(sockfd,&rset);
        maxfdp1 = max(fileno(fp),sockfd) + 1;
        select(maxfdp1,&rset,NULL,NULL,NULL);
        if (FD_ISSET(sockfd,&rset)){
            if ( (n = read(sockfd,recvline,MAXLINE)) <= 0){
                return;
            }
            if (memcmp(recvline,gpAckOK + 1,n) == 0){
                write(fileno(fp),"AckOK\n",6);
            }
            else if (memcmp(recvline,gpAckERR + 1,n) == 0){
                write(fileno(fp),"AckERR\n",7);
            }
        }
        if (FD_ISSET(fileno(fp),&rset)){
            if ( (n = read(fileno(fp),sendline,MAXLINE)) == 0) {
                stdineof = 1;
                shutdown(sockfd,SHUT_WR);
                FD_CLR(fileno(fp),&rset);
                continue;
            }
            writen(sockfd,gpCommandHead + 1,2);
            writen(sockfd,sendline,n - 1);
            writen(sockfd,gpDelimiter + 1,2);
        }
    }
}
int main(int argc,char * argv[]){
    int sockfd;
    struct sockaddr_in servaddr;
```

```
    sockfd = socket(AF_INET,SOCK_STREAM,0);
    bzero(&servaddr,sizeof(servaddr));
    servaddr.sin_family = AF_INET;
    servaddr.sin_port = htons(SERV_PORT);
    inet_pton(AF_INET,"192.168.1.1",&servaddr.sin_addr);
    connect(sockfd,(struct sockaddr * ) &servaddr,sizeof(servaddr));
    str_cli(stdin,sockfd);
    exit(0);
}
```

14.4　利用 I/O 复用替代多进程的并发服务器

如果连接服务器的客户端不是几个，而是几十个，此时再一一创建这么多进程，可想而知这需要多少资源。如果一个客户改变了服务器某个设备的状态，而在其他客户上的该设备状态也要更新，这怎么办？利用进程间通信吗？此时笔者选择了利用 I/O 复用来实现。

```
int SoftwareTPServer(int port){
    struct sockaddr_in cliaddr,servaddr;
    int listenfd,connfd,maxfd,nready,i,clientfd[MAX_CLIENT],maxi,msgqlen;
    socklen_t clilen;
    fd_set rset,allset;
    msg_SoftwareTP pmsg;
    InitCommand();
    maxi = -1;
    for (i = 0; i < MAX_CLIENT; i++){
        clientfd[i] = -1;
    }
    FD_ZERO(&allset);
    if ( (listenfd = socket(AF_INET,SOCK_STREAM,IPPROTO_TCP)) < 0){
        perror("call to socket");
        exit(1);
    }
    bzero(&servaddr,sizeof(servaddr));
    servaddr.sin_family = AF_INET;
    servaddr.sin_addr.s_addr = htonl(INADDR_ANY);
    servaddr.sin_port = htons(port);
    if (bind(listenfd,(struct sockaddr * )&servaddr,sizeof(servaddr)) < 0){
        perror("call to bind");
```

```
        exit(1);
    }
    if (listen(listenfd,20) < 0){
        perror("call to listen");
        exit(1);
    }
    printf("SoftwareTP Accepting connections ...\n");
    FD_SET(listenfd,&allset);
    FD_SET(g_Variant.qid_SoftwareTP,&allset);
    if (listenfd > g_Variant.qid_SoftwareTP){
        maxfd = listenfd;
    }
    else{
        maxfd = g_Variant.qid_SoftwareTP;
    }
    while (1){
        rset = allset;
        nready = select(maxfd + 1,&rset,NULL,NULL,NULL);
        if (FD_ISSET(listenfd,&rset)){
            clilen = sizeof(cliaddr);
            if ( (connfd = accept(listenfd,(struct sockaddr * )&cliaddr,&clilen)) < 0)
            {
                if (errno == EINTR){
                    goto drop;
                }
                else{
                    perror("call to accept");
                    exit(1);
                }
            }
            for (i = 0; i < MAX_CLIENT && clientfd[i] > 0; i++);
            if (i < MAX_CLIENT){
                clientfd[i] = connfd;
                FD_SET(connfd,&allset);
                if (connfd > maxfd){
                    maxfd = connfd;
                }
                if (i > maxi){
                    maxi = i;
```

```
                }
            }
            else{
                printf("too many clients");
                close(connfd);
            }
    drop:       if (--nready <= 0){
                continue;
            }
        }
        if (FD_ISSET(g_Variant.qid_SoftwareTP,&rset)){
            if ( (msgqlen = msgrcv(g_Variant.qid_SoftwareTP,&pmsg,256,type_out,
IPC_NOWAIT)) > 0)
            {
                for (i = 0; i <= maxi; i++){
                    if (clientfd[i] > 0){
                        printf("msgqlen:%d,SoftwareTP out:%d\n",msgqlen,type_
out);

                        writen(clientfd[i],pmsg.buf,pmsg.Length);
                    }
                }
            }
            if (--nready <= 0){
                continue;
            }
        }
        for (i = 0; i <= maxi; i++){
            if (clientfd[i] < 0){
                continue;
            }
            if (FD_ISSET(clientfd[i],&rset)){
                if (Communication_SoftwareTP(clientfd[i],10006) != 0){
                    printf("client[%d] had quit out.\n",i);
                    close(clientfd[i]);
                    FD_CLR(clientfd[i],&allset);
                    clientfd[i] = -1;
                }
            }
            if (--nready <= 0){
```

```
                    break;
                }
            }
            if (i == maxi){
                for (; i >= 0; i--){
                    if (clientfd[i] > 0){
                        break;
                    }
                }
                maxi = i;
            }
        }
    }
```

参考文献

[1] 杜春雷. ARM 体系结构与编程[M]. 北京：清华大学出版社，2003.
[2] Richard Stevens W. UNIX 环境高级编程[M]. 尤晋元，译. 北京：机械工业出版社，2000.
[3] Kurt Wall. GNU/Linux 编程指南[M]. 张辉，译. 北京：清华大学出版社，2002.
[4] Richard Stevens W. UNIX 网络编程[M]. 杨继张，译. 北京：清华大学出版社，2006.
[5] Rubini A. Linux 设备驱动程序[M]. 魏永明，译. 2 版. 北京：中国电力出版社，2002.